Deep Learning in Natural Language Processing

Li Deng · Yang Liu
Editors

Deep Learning in Natural Language Processing

Editors
Li Deng
AI Research at Citadel
Chicago, IL
USA

and

AI Research at Citadel
Seattle, WA
USA

Yang Liu
Tsinghua University
Beijing
China

ISBN 978-981-13-3848-9 ISBN 978-981-10-5209-5 (eBook)
https://doi.org/10.1007/978-981-10-5209-5

Printed on acid-free paper

This Springer imprint is published by the registered company Springer Nature Singapore Pte Ltd.
part of Springer Nature
The registered company address is: 152 Beach Road, #21-01/04 Gateway East, Singapore 189721, Singapore

Foreword

"Written by a group of the most active researchers in the field, led by Dr. Deng, an internationally respected expert in both NLP and deep learning, this book provides a comprehensive introduction to and up-to-date review of the state of art in applying deep learning to solve fundamental problems in NLP. Further, the book is highly timely, as demands for high-quality and up-to-date textbooks and research references have risen dramatically in response to the tremendous strides in deep learning applications to NLP. The book offers a unique reference guide for practitioners in various sectors, especially the Internet and AI start-ups, where NLP technologies are becoming an essential enabler and a core differentiator."

Hongjiang Zhang (Founder, Sourcecode Capital; former CEO of KingSoft)

"This book provides a comprehensive introduction to the latest advances in deep learning applied to NLP. Written by experienced and aspiring deep learning and NLP researchers, it covers a broad range of major NLP applications, including spoken language understanding, dialog systems, lexical analysis, parsing, knowledge graph, machine translation, question answering, sentiment analysis, and social computing.

The book is clearly structured and moves from major research trends, to the latest deep learning approaches, to their limitations and promising future work. Given its self-contained content, sophisticated algorithms, and detailed use cases, the book offers a valuable guide for all readers who are working on or learning about deep learning and NLP."

Haifeng Wang (Vice President and Head of Research, Baidu; former President of ACL)

"In 2011, at the dawn of deep learning in industry, I estimated that in most speech recognition applications, computers still made 5 to 10 times more errors than human subjects, and highlighted the importance of knowledge engineering in future directions. Within only a handful of years since, deep learning has nearly closed the gap in the accuracy of conversational speech recognition between human and computers. Edited and written by Dr. Li Deng—a pioneer in the recent speech recognition revolution using deep learning—and his colleagues, this book elegantly describes this part of the fascinating history of speech recognition as an important subfield of natural language processing (NLP). Further, the book expands this historical perspective from speech recognition to more general areas of NLP, offering a truly valuable guide for the future development of NLP.

Importantly, the book puts forward a thesis that the current deep learning trend is a revolution from the previous data-driven (shallow) machine learning era, although ostensibly deep learning appears to be merely exploiting more data, more computing power, and more complex models. Indeed, as the book correctly points out, the current state of the art of deep learning technology developed for NLP applications, despite being highly successful in solving individual NLP tasks, has not taken full advantage of rich world knowledge or human cognitive capabilities. Therefore, I fully embrace the view expressed by the book's editors and authors that more advanced deep learning that seamlessly integrates knowledge engineering will pave the way for the next revolution in NLP.

I highly recommend speech and NLP researchers, engineers, and students to read this outstanding and timely book, not only to learn about the state of the art in NLP and deep learning, but also to gain vital insights into what the future of the NLP field will hold."

Sadaoki Furui (President, Toyota Technological Institute at Chicago)

Preface

Natural language processing (NLP), which aims to enable computers to process human languages intelligently, is an important interdisciplinary field crossing artificial intelligence, computing science, cognitive science, information processing, and linguistics. Concerned with interactions between computers and human languages, NLP applications such as speech recognition, dialog systems, information retrieval, question answering, and machine translation have started to reshape the way people identify, obtain, and make use of information.

The development of NLP can be described in terms of three major waves: rationalism, empiricism, and deep learning. In the first wave, rationalist approaches advocated the design of handcrafted rules to incorporate knowledge into NLP systems based on the assumption that knowledge of language in the human mind is fixed in advance by generic inheritance. In the second wave, empirical approaches assume that rich sensory input and the observable language data in surface form are required and sufficient to enable the mind to learn the detailed structure of natural language. As a result, probabilistic models were developed to discover the regularities of languages from large corpora. In the third wave, deep learning exploits hierarchical models of nonlinear processing, inspired by biological neural systems to learn intrinsic representations from language data, in ways that aim to simulate human cognitive abilities.

The intersection of deep learning and natural language processing has resulted in striking successes in practical tasks. Speech recognition is the first industrial NLP application that deep learning has strongly impacted. With the availability of large-scale training data, deep neural networks achieved dramatically lower recognition errors than the traditional empirical approaches. Another prominent successful application of deep learning in NLP is machine translation. End-to-end neural machine translation that models the mapping between human languages using neural networks has proven to improve translation quality substantially. Therefore, neural machine translation has quickly become the new de facto technology in major commercial online translation services offered by large technology companies: Google, Microsoft, Facebook, Baidu, and more. Many other areas of NLP, including language understanding and dialog, lexical analysis and parsing, knowledge graph, information retrieval, question answering from text, social computing, language generation, and text sentiment analysis, have also seen much significant progress using deep learning, riding on the third wave of NLP. Nowadays, deep learning is a dominating method applied to practically all NLP tasks.

The main goal of this book is to provide a comprehensive survey on the recent advances in deep learning applied to NLP. The book presents state of the art of NLP-centric deep learning research, and focuses on the role of deep learning played in major NLP applications including spoken language understanding, dialog systems, lexical analysis, parsing, knowledge graph, machine translation, question answering, sentiment analysis, social computing, and natural language generation (from images). This book is suitable for readers with a technical background in computation, including graduate students, post-doctoral researchers, educators, and industrial researchers and anyone interested in getting up to speed with the latest techniques of deep learning associated with NLP.

The book is organized into eleven chapters as follows:

- Chapter 1: A Joint Introduction to Natural Language Processing and to Deep Learning (Li Deng and Yang Liu)
- Chapter 2: Deep Learning in Conversational Language Understanding (Gokhan Tur, Asli Celikyilmaz, Xiaodong He, Dilek Hakkani-Tür, and Li Deng)
- Chapter 3: Deep Learning in Spoken and Text-Based Dialog Systems (Asli Celikyilmaz, Li Deng, and Dilek Hakkani-Tür)
- Chapter 4: Deep Learning in Lexical Analysis and Parsing (Wanxiang Che and Yue Zhang)
- Chapter 5: Deep Learning in Knowledge Graph (Zhiyuan Liu and Xianpei Han)
- Chapter 6: Deep Learning in Machine Translation (Yang Liu and Jiajun Zhang)
- Chapter 7: Deep Learning in Question Answering (Kang Liu and Yansong Feng)
- Chapter 8: Deep Learning in Sentiment Analysis (Duyu Tang and Meishan Zhang)
- Chapter 9: Deep Learning in Social Computing (Xin Zhao and Chenliang Li)
- Chapter 10: Deep Learning in Natural Language Generation from Images (Xiaodong He and Li Deng)
- Chapter 11: Epilogue (Li Deng and Yang Liu)

Chapter 1 first reviews the basics of NLP as well as the main scope of NLP covered in the following chapters of the book, and then goes in some depth into the historical development of NLP summarized as three waves and future directions. Subsequently, in Chaps. 2–10, an in-depth survey on the recent advances in deep learning applied to NLP is organized into nine separate chapters, each covering a largely independent application area of NLP. The main body of each chapter is written by leading researchers and experts actively working in the respective field.

The origin of this book was the set of comprehensive tutorials given at the 15th China National Conference on Computational Linguistics (CCL 2016) held in October 2016 in Yantai, Shandong, China, where both of us, editors of this book, were active participants and were taking leading roles. We thank our Springer's senior editor, Dr. Celine Lanlan Chang, who kindly invited us to create this book and who has been providing much of timely assistance needed to complete this book. We are grateful also to Springer's Assistant Editor, Jane Li, for offering invaluable help through various stages of manuscript preparation.

We thank all authors of Chaps. 2–10 who devoted their valuable time carefully preparing the content of their chapters: Gokhan Tur, Asli Celikyilmaz, Dilek Hakkani-Tur, Wanxiang Che, Yue Zhang, Xianpei Han, Zhiyuan Liu, Jiajun Zhang, Kang Liu, Yansong Feng, Duyu Tang, Meishan Zhang, Xin Zhao, Chenliang Li, and Xiaodong He. The authors of Chaps. 4–9 are CCL 2016 tutorial speakers. They spent a considerable amount of time in updating their tutorial material with the latest advances in the field since October 2016.

Further, we thank numerous reviewers and readers, Sadaoki Furui, Andrew Ng, Fred Juang, Ken Church, Haifeng Wang, and Hongjiang Zhang, who not only gave us much needed encouragements but also offered many constructive comments which substantially improved earlier drafts of the book.

Finally, we give our appreciations to our organizations, Microsoft Research and Citadel (for Li Deng) and Tsinghua University (for Yang Liu), who provided excellent environments, supports, and encouragements that have been instrumental for us to complete this book. Yang Liu is also supported by National Natural Science Foundation of China (No.61522204, No.61432013, and No.61331013).

Seattle, USA
Beijing, China
October 2017

Li Deng
Yang Liu

Contents

Contributors

Asli Celikyilmaz Microsoft Research, Redmond, WA, USA

Wanxiang Che Harbin Institute of Technology, Harbin, China

Li Deng Citadel, Seattle & Chicago, USA

Yansong Feng Peking University, Beijing, China

Dilek Hakkani-Tür Google, Mountain View, CA, USA

Xianpei Han Institute of Software, Chinese Academy of Sciences, Beijing, China

Xiaodong He Microsoft Research, Redmond, WA, USA

Chenliang Li Wuhan University, Wuhan, China

Kang Liu Institute of Automation, Chinese Academy of Sciences, Beijing, China

Yang Liu Tsinghua University, Beijing, China

Zhiyuan Liu Tsinghua University, Beijing, China

Duyu Tang Microsoft Research Asia, Beijing, China

Gokhan Tur Google, Mountain View, CA, USA

Jiajun Zhang Institute of Automation, Chinese Academy of Sciences, Beijing, China

Meishan Zhang Heilongjiang University, Harbin, China

Yue Zhang Singapore University of Technology and Design, Singapore

Xin Zhao Renmin University of China, Beijing, China

Acronyms

AI	Artificial intelligence
AP	Averaged perceptron
ASR	Automatic speech recognition
ATN	Augmented transition network
BiLSTM	Bidirectional long short-term memory
BiRNN	Bidirectional recurrent neural network
BLEU	Bilingual evaluation understudy
BOW	Bag-of-words
CBOW	Continuous bag-of-words
CCA	Canonical correlation analysis
CCG	Combinatory categorial grammar
CDL	Collaborative deep learning
CFG	Context free grammar
CYK	Cocke–Younger–Kasami
CLU	Conversational language understanding
CNN	Convolutional neural network
CNNSM	Convolutional neural network based semantic model
cQA	Community question answering
CRF	Conditional random field
CTR	Collaborative topic regression
CVT	Compound value typed
DA	Denoising autoencoder
DBN	Deep belief network
DCN	Deep convex net
DNN	Deep neural network
DSSM	Deep structured semantic model
DST	Dialog state tracking
EL	Entity linking
EM	Expectation maximization
FSM	Finite state machine

GAN	Generative adversarial network
GRU	Gated recurrent unit
HMM	Hidden Markov model
IE	Information extraction
IRQA	Information retrieval-based question answering
IVR	Interactive voice response
KBQA	Knowledge-based question answering
KG	Knowledge graph
L-BFGS	Limited-memory Broyden–Fletcher–Goldfarb–Shanno
LSI	Latent semantic indexing
LSTM	Long short-term memory
MC	Machine comprehension
MCCNN	Multicolumn convolutional neural network
MDP	Markov decision process
MERT	Minimum error rate training
METEOR	Metric for evaluation of translation with explicit ordering
MIRA	Margin infused relaxed algorithm
ML	Machine learning
MLE	Maximum likelihood estimation
MLP	Multiple layer perceptron
MMI	Maximum mutual information
M-NMF	Modularized nonnegative matrix factorization
MRT	Minimum risk training
MST	Maximum spanning tree
MT	Machine translation
MV-RNN	Matrix-vector recursive neural network
NER	Named entity recognition
NFM	Neural factorization machine
NLG	Natural language generation
NMT	Neural machine translation
NRE	Neural relation extraction
OOV	Out-of-vocabulary
PA	Passive aggressive
PCA	Principal component analysis
PMI	Point-wise mutual information
POS	Part of speech
PV	Paragraph vector
QA	Question answering
RAE	Recursive autoencoder
RBM	Restricted Boltzmann machine
RDF	Resource description framework
RE	Relation extraction
RecNN	Recursive neural network
RL	Reinforcement learning
RNN	Recurrent neural network

ROUGE	Recall-oriented understudy for gisting evaluation
RUBER	Referenced metric and unreferenced metric blended evaluation routine
SDS	Spoken dialog system
SLU	Spoken language understanding
SMT	Statistical machine translation
SP	Semantic parsing
SRL	Semantic role labeling
SRNN	Segmental recurrent neural network
STAGG	Staged query graph generation
SVM	Support vector machine
UAS	Unlabeled attachment score
UGC	User-generated content
VIME	Variational information maximizing exploration
VPA	Virtual personal assistant

Chapter 1
A Joint Introduction to Natural Language Processing and to Deep Learning

Li Deng and Yang Liu

Abstract In this chapter, we set up the fundamental framework for the book. We first provide an introduction to the basics of natural language processing (NLP) as an integral part of artificial intelligence. We then survey the historical development of NLP, spanning over five decades, in terms of three waves. The first two waves arose as rationalism and empiricism, paving ways to the current deep learning wave. The key pillars underlying the deep learning revolution for NLP consist of (1) distributed representations of linguistic entities via embedding, (2) semantic generalization due to the embedding, (3) long-span deep sequence modeling of natural language, (4) hierarchical networks effective for representing linguistic levels from low to high, and (5) end-to-end deep learning methods to jointly solve many NLP tasks. After the survey, several key limitations of current deep learning technology for NLP are analyzed. This analysis leads to five research directions for future advances in NLP.

1.1 Natural Language Processing: The Basics

Natural language processing (NLP) investigates the use of computers to process or to understand human (i.e., natural) languages for the purpose of performing useful tasks. NLP is an interdisciplinary field that combines computational linguistics, computing science, cognitive science, and artificial intelligence. From a scientific perspective, NLP aims to model the cognitive mechanisms underlying the understanding and production of human languages. From an engineering perspective, NLP is concerned with how to develop novel practical applications to facilitate the interactions between computers and human languages. Typical applications in NLP include speech recognition, spoken language understanding, dialogue systems, lexical analysis, parsing, machine translation, knowledge graph, information retrieval, question answering,

L. Deng (✉)
Citadel, Seattle & Chicago, USA
e-mail: l.deng@ieee.org

Y. Liu
Tsinghua University, Beijing, China
e-mail: liuyang2011@tsinghua.edu.cn

L. Deng and Y. Liu (eds.), *Deep Learning in Natural Language Processing*, https://doi.org/10.1007/978-981-10-5209-5_1

sentiment analysis, social computing, natural language generation, and natural language summarization. These NLP application areas form the core content of this book.

Natural language is a system constructed specifically to convey meaning or semantics, and is by its fundamental nature a symbolic or discrete system. The surface or observable "physical" signal of natural language is called text, always in a symbolic form. The text "signal" has its counterpart—the speech signal; the latter can be regarded as the continuous correspondence of symbolic text, both entailing the same latent linguistic hierarchy of natural language. From NLP and signal processing perspectives, speech can be treated as "noisy" versions of text, imposing additional difficulties in its need of "de-noising" when performing the task of understanding the common underlying semantics. Chapters 2 and 3 as well as current Chap. 1 of this book cover the speech aspect of NLP in detail, while the remaining chapters start directly from text in discussing a wide variety of text-oriented tasks that exemplify the pervasive NLP applications enabled by machine learning techniques, notably deep learning.

The symbolic nature of natural language is in stark contrast to the continuous nature of language's neural substrate in the human brain. We will defer this discussion to Sect. 1.6 of this chapter when discussing future challenges of deep learning in NLP. A related contrast is how the symbols of natural language are encoded in several continuous-valued modalities, such as gesture (as in sign language), handwriting (as an image), and, of course, speech. On the one hand, the word as a symbol is used as a "signifier" to refer to a concept or a thing in real world as a "signified" object, necessarily a categorical entity. On the other hand, the continuous modalities that encode symbols of words constitute the external signals sensed by the human perceptual system and transmitted to the brain, which in turn operates in a continuous fashion. While of great theoretical interest, the subject of contrasting the symbolic nature of language versus its continuous rendering and encoding goes beyond the scope of this book.

In the next few sections, we outline and discuss, from a historical perspective, the development of general methodology used to study NLP as a rich interdisciplinary field. Much like several closely related sub- and super-fields such as conversational systems, speech recognition, and artificial intelligence, the development of NLP can be described in terms of three major waves (Deng 2017; Pereira 2017), each of which is elaborated in a separate section next.

1.2 The First Wave: Rationalism

NLP research in its first wave lasted for a long time, dating back to 1950s. In 1950, Alan Turing proposed the Turing test to evaluate a computer's ability to exhibit intelligent behavior indistinguishable from that of a human (Turing 1950). This test is based on natural language conversations between a human and a computer designed to generate human-like responses. In 1954, the Georgetown-IBM experiment demonstrated the first machine translation system capable of translating more than 60 Russian sentences into English.

The approaches, based on the belief that knowledge of language in the human mind is fixed in advance by generic inheritance, dominated most of NLP research between about 1960 and late 1980s. These approaches have been called rationalist ones (Church 2007). The dominance of rationalist approaches in NLP was mainly due to the widespread acceptance of arguments of Noam Chomsky for an innate language structure and his criticism of N-grams (Chomsky 1957). Postulating that key parts of language are hardwired in the brain at birth as a part of the human genetic inheritance, rationalist approaches endeavored to design hand-crafted rules to incorporate knowledge and reasoning mechanisms into intelligent NLP systems. Up until 1980s, most notably successful NLP systems, such as ELIZA for simulating a Rogerian psychotherapist and MARGIE for structuring real-world information into concept ontologies, were based on complex sets of handwritten rules.

This period coincided approximately with the early development of artificial intelligence, characterized by expert knowledge engineering, where domain experts devised computer programs according to the knowledge about the (very narrow) application domains they have (Nilsson 1982; Winston 1993). The experts designed these programs using symbolic logical rules based on careful representations and engineering of such knowledge. These knowledge-based artificial intelligence systems tend to be effective in solving narrow-domain problems by examining the "head" or most important parameters and reaching a solution about the appropriate action to take in each specific situation. These "head" parameters are identified in advance by human experts, leaving the "tail" parameters and cases untouched. Since they lack learning capability, they have difficulty in generalizing the solutions to new situations and domains. The typical approach during this period is exemplified by the expert system, a computer system that emulates the decision-making ability of a human expert. Such systems are designed to solve complex problems by reasoning about knowledge (Nilsson 1982). The first expert system was created in 1970s and then proliferated in 1980s. The main "algorithm" used was the inference rules in the form of "if-then-else" (Jackson 1998). The main strength of these first-generation artificial intelligence systems is its transparency and interpretability in their (limited) capability in performing logical reasoning. Like NLP systems such as ELIZA and MARGIE, the general expert systems in the early days used hand-crafted expert knowledge which was often effective in narrowly defined problems, although the reasoning could not handle uncertainty that is ubiquitous in practical applications.

In specific NLP application areas of dialogue systems and spoken language understanding, to be described in more detail in Chaps. 2 and 3 of this book, such rationalistic approaches were represented by the pervasive use of symbolic rules and templates (Seneff et al. 1991). The designs were centered on grammatical and ontological constructs, which, while interpretable and easy to debug and update, had experienced severe difficulties in practical deployment. When such systems worked, they often worked beautifully; but unfortunately this happened just not very often and the domains were necessarily limited.

Likewise, speech recognition research and system design, another long-standing NLP and artificial intelligence challenge, during this rationalist era were based heavily on the paradigm of expert knowledge engineering, as elegantly analyzed in (Church and Mercer 1993). During 1970s and early 1980s, the expert system approach to speech recognition was quite popular (Reddy 1976; Zue 1985). However, the lack of abilities to learn from data and to handle uncertainty in reasoning was acutely recognized by researchers, leading to the second wave of speech recognition, NLP, and artificial intelligence described next.

1.3 The Second Wave: Empiricism

The second wave of NLP was characterized by the exploitation of data corpora and of (shallow) machine learning, statistical or otherwise, to make use of such data (Manning and Schtze 1999). As much of the structure of and theory about natural language were discounted or discarded in favor of data-driven methods, the main approaches developed during this era have been called empirical or pragmatic ones (Church and Mercer 1993; Church 2014). With the increasing availability of machine-readable data and steady increase of computational power, empirical approaches have dominated NLP since around 1990. One of the major NLP conferences was even named "Empirical Methods in Natural Language Processing (EMNLP)" to reflect most directly the strongly positive sentiment of NLP researchers during that era toward empirical approaches.

In contrast to rationalist approaches, empirical approaches assume that the human mind only begins with general operations for association, pattern recognition, and generalization. Rich sensory input is required to enable the mind to learn the detailed structure of natural language. Prevalent in linguistics between 1920 and 1960, empiricism has been undergoing a resurgence since 1990. Early empirical approaches to NLP focused on developing generative models such as the hidden Markov model (HMM) (Baum and Petrie 1966), the IBM translation models (Brown et al. 1993), and the head-driven parsing models (Collins 1997) to discover the regularities of languages from large corpora. Since late 1990s, discriminative models have become the *de facto* approach in a variety of NLP tasks. Representative discriminative models and methods in NLP include the maximum entropy model (Ratnaparkhi 1997), supporting vector machines (Vapnik 1998), conditional random fields (Lafferty et al. 2001), maximum mutual information and minimum classification error (He et al. 2008), and perceptron (Collins 2002).

Again, this era of empiricism in NLP was paralleled with corresponding approaches in artificial intelligence as well as in speech recognition and computer vision. It came about after clear evidence that learning and perception capabilities are crucial for complex artificial intelligence systems but missing in the expert systems popular in the previous wave. For example, when DARPA opened its first Grand Challenge for autonomous driving, most vehicles then relied on the knowledge-based artificial intelligence paradigm. Much like speech recognition and NLP, the autonomous driving and computer vision researchers immediately realized the limitation of the knowledge-based paradigm due to the necessity for machine learning with uncertainty handling and generalization capabilities.

The empiricism in NLP and speech recognition in this second wave was based on data-intensive machine learning, which we now call "shallow" due to the general lack of abstractions constructed by many-layer or "deep" representations of data which would come in the third wave to be described in the next section. In machine learning, researchers do not need to concern with constructing precise and exact rules as required for the knowledge-based NLP and speech systems during the first wave. Rather, they focus on statistical models (Bishop 2006; Murphy 2012) or simple neural networks (Bishop 1995) as an underlying engine. They then automatically learn or "tune" the parameters of the engine using ample training data to make them handle uncertainty, and to attempt to generalize from one condition to another and from one domain to another. The key algorithms and methods for machine learning include EM (expectation-maximization), Bayesian networks, support vector machines, decision trees, and, for neural networks, backpropagation algorithm.

Generally speaking, the machine learning based NLP, speech, and other artificial intelligence systems perform much better than the earlier, knowledge-based counterparts. Successful examples include almost all artificial intelligence tasks in machine perception—speech recognition (Jelinek 1998), face recognition (Viola and Jones 2004), visual object recognition (Fei-Fei and Perona 2005), handwriting recognition (Plamondon and Srihari 2000), and machine translation (Och 2003).

More specifically, in a core NLP application area of machine translation, as to be described in detail in Chap. 6 of this book as well as in (Church and Mercer 1993), the field has switched rather abruptly around 1990 from rationalistic methods outlined in Sect. 1.2 to empirical, largely statistical methods. The availability of sentence-level alignments in the bilingual training data made it possible to acquire surface-level translation knowledge not by rules but from data directly, at the expense of discarding or discounting structured information in natural languages. The most representative work during this wave is that empowered by various versions of IBM translation models (Brown et al. 1993). Subsequent developments during this empiricist era of machine translation further significantly improved the quality of translation systems (Och and Ney 2002; Och 2003; Chiang 2007; He and Deng 2012), but not at the level of massive deployment in real world (which would come after the next, deep learning wave).

In the dialogue and spoken language understanding areas of NLP, this empiricist era was also marked prominently by data-driven machine learning approaches. These approaches were well suited to meet the requirement for quantitative evaluation and concrete deliverables. They focused on broader but shallow, surface-level coverage of text and domains instead of detailed analyses of highly restricted text and domains. The training data were used not to design rules for language understanding and response action from the dialogue systems but to learn parameters of (shallow) statistical or neural models automatically from data. Such learning helped reduce the cost of hand-crafted complex dialogue manager's design, and helped improve robustness against speech recognition errors in the overall spoken language understanding and dialogue systems; for a review, see He and Deng (2013). More specifically, for the dialogue policy component of dialogue systems, powerful reinforcement learning based on Markov decision processes had been introduced during this era; for a review, see Young et al. (2013). And for spoken language understanding, the dominant methods moved from rule- or template-based ones during the first wave to generative models like hidden Markov models (HMMs) (Wang et al. 2011) to discriminative models like conditional random fields (Tur and Deng 2011).

Similarly, in speech recognition, over close to 30 years from early 1980 s to around 2010, the field was dominated by the (shallow) machine learning paradigm using the statistical generative model based on the HMM integrated with Gaussian mixture models, along with various versions of its generalization (Baker et al. 2009a, b; Deng and O'Shaughnessy 2003; Rabiner and Juang 1993). Among many versions of the generalized HMMs were statistical and neural-network-based hidden dynamic models (Deng 1998; Bridle et al. 1998; Deng and Yu 2007). The former adopted EM and switching extended Kalman filter algorithms for learning model parameters (Ma and Deng 2004; Lee et al. 2004), and the latter used backpropagation (Picone et al. 1999). Both of them made extensive use of multiple latent layers of representations for the generative process of speech waveforms following the long-standing framework of analysis-by-synthesis in human speech perception. More significantly, inverting this "deep" generative process to its counterpart of an end-to-end discriminative process gave rise to the first industrial success of deep learning (Deng et al. 2010, 2013; Hinton et al. 2012), which formed a driving force of the third wave of speech recognition and NLP that will be elaborated next.

1.4 The Third Wave: Deep Learning

While the NLP systems, including speech recognition, language understanding, and machine translation, developed during the second wave performed a lot better and with higher robustness than those during the first wave, they were far from human-level performance and left much to desire. With a few exceptions, the (shallow) machine learning models for NLP often did not have the capacity sufficiently large to absorb the large amounts of training data. Further, the learning algorithms, methods, and infrastructures were not powerful enough. All this changed several years ago, giving rise to the third wave of NLP, propelled by the new paradigm of deep-structured machine learning or deep learning (Bengio 2009; Deng and Yu 2014; LeCun et al. 2015; Goodfellow et al. 2016).

In traditional machine learning, features are designed by humans and feature engineering is a bottleneck, requiring significant human expertise. Concurrently, the associated shallow models lack the representation power and hence the ability to form levels of decomposable abstractions that would automatically disentangle complex factors in shaping the observed language data. Deep learning breaks away the above difficulties by the use of deep, layered model structure, often in the form of neural networks, and the associated end-to-end learning algorithms. The advances in deep learning are one major driving force behind the current NLP and more general artificial intelligence inflection point and are responsible for the resurgence of neural networks with a wide range of practical, including business, applications (Parloff 2016).

More specifically, despite the success of (shallow) discriminative models in a number of important NLP tasks developed during the second wave, they suffered from the difficulty of covering all regularities in languages by designing features manually with domain expertise. Besides the incompleteness problem, such shallow models also face the sparsity problem as features usually only occur once in the training data, especially for highly sparse high-order features. Therefore, feature design has become one of the major obstacles in statistical NLP before deep learning comes to rescue. Deep learning brings hope for addressing the human feature engineering problem, with a view called "NLP from scratch" (Collobert et al. 2011), which was in early days of deep learning considered highly unconventional. Such deep learning approaches exploit the powerful neural networks that contain multiple hidden layers to solve general machine learning tasks dispensing with feature engineering. Unlike shallow neural networks and related machine learning models, deep neural networks are capable of learning representations from data using a cascade of multiple layers of nonlinear processing units for feature extraction. As higher level features are derived from lower level features, these levels form a hierarchy of concepts.

Deep learning originated from artificial neural networks, which can be viewed as cascading models of cell types inspired by biological neural systems. With the advent of backpropagation algorithm (Rumelhart et al. 1986), training deep neural networks from scratch attracted intensive attention in 1990s. In these early days, without large amounts of training data and without proper design and learning methods, during neural network training the learning signals vanish exponentially with the number of layers (or more rigorously the depth of credit assignment) when propagated from layer to layer, making it difficult to tune connection weights of deep neural networks, especially the recurrent versions. Hinton et al. (2006) initially overcame this problem by using unsupervised pretraining to first learn generally useful feature detectors. Then, the network is further trained by supervised learning to classify labeled data. As a result, it is possible to learn the distribution of a high-level representation using low-level representations. This seminal work marks the revival of neural networks. A variety of network architectures have since been proposed and developed, including deep belief networks (Hinton et al. 2006), stacked auto-encoders (Vincent et al. 2010), deep Boltzmann machines (Hinton and Salakhutdinov 2012), deep convolutional neural works (Krizhevsky et al. 2012), deep stacking networks (Deng et al. 2012), and deep Q-networks (Mnih et al. 2015). Capable of discovering intricate structures in high-dimensional data, deep learning has since 2010 been successfully applied to real-world tasks in artificial intelligence including notably speech recognition (Yu et al. 2010; Hinton et al. 2012), image classification (Krizhevsky et al. 2012; He et al. 2016), and NLP (all chapters in this book). Detailed analyses and reviews of deep learning have been provided in a set of tutorial survey articles (Deng 2014; LeCun et al. 2015; Juang 2016).

As speech recognition is one of core tasks in NLP, we briefly discuss it here due to its importance as the first industrial NLP application in real world impacted strongly by deep learning. Industrial applications of deep learning to large-scale speech recognition started to take off around 2010. The endeavor was initiated with a collaboration between academia and industry, with the original work presented at the 2009 NIPS Workshop on Deep Learning for Speech Recognition and Related Applications. The workshop was motivated by the limitations of deep generative models of speech, and the possibility that the big-compute, big-data era warrants a serious exploration of deep neural networks. It was believed then that pretraining DNNs using generative models of deep belief nets based on the contrastive divergence learning algorithm would overcome the main difficulties of neural nets encountered in the 1990s (Dahl et al. 2011; Mohamed et al. 2009). However, early into this research at Microsoft, it was discovered that without contrastive divergence pretraining, but with the use of large amounts of training data together with the deep neural networks designed with corresponding large, context-dependent output layers and with careful engineering, dramatically lower recognition errors could be obtained than then-state-of-the-art (shallow) machine learning systems (Yu et al. 2010, 2011; Dahl et al. 2012). This finding was quickly verified by several other major speech recognition research groups in North America (Hinton et al. 2012; Deng et al. 2013) and subsequently overseas. Further, the nature of recognition errors produced by the two types of systems was found to be characteristically different, offering technical insights into how to integrate deep learning into the existing highly efficient, run-time speech decoding system deployed by major players in speech recognition industry (Yu and Deng 2015; Abdel-Hamid et al. 2014; Xiong et al. 2016; Saon et al. 2017). Nowadays, backpropagation algorithm applied to deep neural nets of various forms is uniformly used in all current state-of-the-art speech recognition systems (Yu and Deng 2015; Amodei et al. 2016; Saon et al. 2017), and all major commercial speech recognition systems—Microsoft Cortana, Xbox, Skype Translator, Amazon Alexa, Google Assistant, Apple Siri, Baidu and iFlyTek voice search, and more—are all based on deep learning methods.

The striking success of speech recognition in 2010–2011 heralded the arrival of the third wave of NLP and artificial intelligence. Quickly following the success of deep learning in speech recognition, computer vision (Krizhevsky et al. 2012) and machine translation (Bahdanau et al. 2015) were taken over by the similar deep learning paradigm. In particular, while the powerful technique of neural embedding of words was developed in as early as 2011 (Bengio et al. 2001), it is not until more than 10 year later it was shown to be practically useful at a large and practically useful scale (Mikolov et al. 2013) due to the availability of big data and faster computation. In addition, a large number of other real-world NLP applications, such as image captioning (Karpathy and Fei-Fei 2015; Fang et al. 2015; Gan et al. 2017), visual question answering (Fei-Fei and Perona 2016), speech understanding (Mesnil et al. 2013), web search (Huang et al. 2013b), and recommendation systems, have been made successful due to deep learning, in addition to many non-NLP tasks including drug discovery and toxicology, customer relationship management, recommendation systems, gesture recognition, medical informatics, advertisement, medical image analysis, robotics, self-driving vehicles, board and eSports games (e.g., Atari, Go, Poker, and the latest, DOTA2), and so on. For more details, see https://en.wikipedia.org/wiki/deep_learning.

In more specific text-based NLP application areas, machine translation is perhaps impacted the most by deep learning. Advancing from the shallow statistical machine translation developed during the second wave of NLP, the current best machine translation systems in real-world applications are based on deep neural networks. For example, Google announced the first stage of its move to neural machine translation in September 2016 and Microsoft made a similar announcement 2 months later. Facebook has been working on the conversion to neural machine translation for about a year, and by August 2017 it is at full deployment. Details of the deep learning techniques in these state-of-the-art large-scale machine translation systems will be reviewed in Chap. 6.

In the area of spoken language understanding and dialogue systems, deep learning is also making a huge impact. The current popular techniques maintain and expand the statistical methods developed during second-wave era in several ways. Like the empirical, (shallow) machine learning methods, deep learning is also based on data-intensive methods to reduce the cost of hand-crafted complex understanding and dialogue management, to be robust against speech recognition errors under noise environments and against language understanding errors, and to exploit the power of Markov decision processes and reinforcement learning for designing dialogue policy, e.g., (Gasic et al. 2017; Dhingra et al. 2017). Compared with the earlier methods, deep neural network models and representations are much more powerful and they make end-to-end learning possible. However, deep learning has not yet solved the problems of interpretability and domain scalability associated with earlier empirical techniques. Details of the deep learning techniques popular for current spoken language understanding and dialogue systems as well as their challenges will be reviewed in Chaps. 2 and 3.

Two important recent technological breakthroughs brought about in applying deep learning to NLP problems are sequence-to-sequence learning (Sutskevar et al. 2014) and attention modeling (Bahdanau et al. 2015). The sequence-to-sequence learning introduces a powerful idea of using recurrent nets to carry out both encoding and decoding in an end-to-end manner. While attention modeling was initially developed to overcome the difficulty of encoding a long sequence, subsequent developments significantly extended its power to provide highly flexible alignment of two arbitrary sequences that can be learned together with neural network parameters. The key concepts of sequence-to-sequence learning and of attention mechanism boosted the performance of neural machine translation based on distributed word embedding over the best system based on statistical learning and local representations of words and phrases. Soon after this success, these concepts have also been applied successfully to a number of other NLP-related tasks such as image captioning (Karpathy and Fei-Fei 2015; Devlin et al. 2015), speech recognition (Chorowski et al. 2015), meta-learning for program execution, one-shot learning, syntactic parsing, lip reading, text understanding, summarization, and question answering and more.

Setting aside their huge empirical successes, models of neural-network-based deep learning are often simpler and easier to design than the traditional machine learning models developed in the earlier wave. In many applications, deep learning is performed simultaneously for all parts of the model, from feature extraction all the way to prediction, in an end-to-end manner. Another factor contributing to the simplicity of neural network models is that the same model building blocks (i.e., the different types of layers) are generally used in many different applications. Using the same building blocks for a large variety of tasks makes the adaptation of models used for one task or data to another task or data relatively easy. In addition, software toolkits have been developed to allow faster and more efficient implementation of these models. For these reasons, deep neural networks are nowadays a prominent method of choice for a large variety of machine learning and artificial intelligence tasks over large datasets including, prominently, NLP tasks.

Although deep learning has proven effective in reshaping the processing of speech, images, and videos in a revolutionary way, the effectiveness is less clear-cut in intersecting deep learning with text-based NLP despite its empirical successes in a number of practical NLP tasks. In speech, image, and video processing, deep learning effectively addresses the semantic gap problem by learning high-level concepts from raw perceptual data in a direct manner. However, in NLP, stronger theories and structured models on morphology, syntax, and semantics have been advanced to distill the underlying mechanisms of understanding and generation of natural languages, which have not been as easily compatible with neural networks. Compared with speech, image, and video signals, it seems less straightforward to see that the neural representations learned from textual data can provide equally direct insights onto natural language. Therefore, applying neural networks, especially those having sophisticated hierarchical architectures, to NLP has received increasing attention and has become the most active area in both NLP and deep learning communities with highly visible progresses made in recent years (Deng 2016; Manning and Socher 2017). Surveying the advances and analyzing the future directions in deep learning for NLP form the main motivation for us to write this chapter and to create this book, with the desire for the NLP researchers to accelerate the research further in the current fast pace of the progress.

1.5 Transitions from Now to the Future

Before analyzing the future dictions of NLP with more advanced deep learning, here we first summarize the significance of the transition from the past waves of NLP to the present one. We then discuss some clear limitations and challenges of the present deep learning technology for NLP, to pave a way to examining further development that would overcome these limitations for the next wave of innovations.

1.5.1 From Empiricism to Deep Learning: A Revolution

On the surface, the deep learning rising wave discussed in Sect. 1.4 in this chapter appears to be a simple push of the second, empiricist wave of NLP (Sect. 1.3) into an extreme end with bigger data, larger models, and greater computing power. After all, the fundamental approaches developed during both waves are data-driven and are based on machine learning and computation, and have dispensed with human-centric "rationalistic" rules that are often brittle and costly to acquire in practical NLP applications. However, if we analyze these approaches holistically and at a deeper level, we can identify aspects of conceptual revolution moving from empiricist machine learning to deep learning, and can subsequently analyze the future directions of the field (Sect. 1.6). This revolution, in our opinion, is no less significant than the revolution from the earlier rationalist wave to empiricist one as analyzed at the beginning (Church and Mercer 1993) and at the end of the empiricist era (Charniak 2011).

Empiricist machine learning and linguistic data analysis during the second NLP wave started in early 1990 s by crypto-analysts and computer scientists working on natural language sources that are highly limited in vocabulary and application domains. As we discussed in Sect. 1.3, surface-level text observations, i.e., words and their sequences, are counted using discrete probabilistic models without relying on deep structure in natural language. The basic representations were "one-hot" or localist, where no semantic similarity between words was exploited. With restrictions in domains and associated text content, such structure-free representations and empirical models are often sufficient to cover much of what needs to be covered. That is, the shallow, count-based statistical models can naturally do well in limited and specific NLP tasks. But when the domain and content restrictions are lifted for more realistic NLP applications in real-world, count-based models would necessarily become ineffective, no manner how many tricks of smoothing have been invented in an attempt to mitigate the problem of combinatorial counting sparseness. This is where deep learning for NLP truly shines—distributed representations of words via embedding, semantic generalization due to the embedding, longer span deep sequence modeling, and end-to-end learning methods have all contributed to beating empiricist, count-based methods in a wide range of NLP tasks as discussed in Sect. 1.4.

1.5.2 Limitations of Current Deep Learning Technology

Despite the spectacular successes of deep learning in NLP tasks, most notably in speech recognition/understanding, language modeling, and in machine translation, there remain huge challenges. The current deep learning methods based on neural networks as a black box generally lack interpretability, even further away from explainability, in contrast to the "rationalist" paradigm established during the first

NLP wave where the rules devised by experts were naturally explainable. In practice, however, it is highly desirable to explain the predictions from a seemingly "black-box" model, not only for improving the model but for providing the users of the prediction system with interpretations of the suggested actions to take (Koh and Liang 2017).

In a number of applications, deep learning methods have proved to give recognition accuracy close to or exceeding humans, but they require considerably more training data, power consumption, and computing resources than humans. Also, the accuracy results are statistically impressive but often unreliable on the individual basis. Further, most of the current deep learning models have no reasoning and explaining capabilities, making them vulnerable to disastrous failures or attacks without the ability to foresee and thus to prevent them. Moreover, the current NLP models have not taken into account the need for developing and executing goals and plans for decision-making via ultimate NLP systems. A more specific limitation of current NLP methods based on deep learning is their poor abilities for understanding and reasoning inter-sentential relationships, although huge progresses have been made for interwords and phrases within sentences.

As discussed earlier, the success of deep learning in NLP has largely come from a simple strategy thus far—given an NLP task, apply standard sequence models based on (bidirectional) LSTMs, add attention mechanisms if information required in the task needs to flow from another source, and then train the full models in an end-to-end manner. However, while sequence modeling is naturally appropriate for speech, human understanding of natural language (in text form) requires more complex structure than sequence. That is, current sequence-based deep learning systems for NLP can be further advanced by exploiting modularity, structured memories, and recursive, tree-like representations for sentences and larger text (Manning 2016).

To overcome the challenges outlined above and to achieve the ultimate success of NLP as a core artificial intelligence field, both fundamental and applied research are needed. The next new wave of NLP and artificial intelligence will not come until researchers create new paradigmatic, algorithmic, and computation (including hardware) breakthroughs. Here, we outline several high-level directions toward potential breakthroughs.

1.6 Future Directions of NLP

1.6.1 Neural-Symbolic Integration

A potential breakthrough is in developing advanced deep learning models and methods that are more effective than current methods in building, accessing, and exploiting memories and knowledge, including, in particular, common-sense knowledge. It is not clear how to best integrate the current deep learning methods, centered on distributed representations (of everything), with explicit, easily interpretable, and localist-represented knowledge about natural language and the world and with related reasoning mechanisms.

One path to this goal is to seamlessly combine neural networks and symbolic language systems. These NLP and artificial intelligence systems will aim to discover by themselves the underlying causes or logical rules that shape their prediction and decision-making processes interpretable to human users in symbolic natural language forms. Recently, very preliminary work in this direction made use of an integrated neural-symbolic representation called tensor-product neural memory cells, capable of decoding back to symbolic forms. This structured neural representation is provably lossless in the coded information after extensive learning within the neural-tensor domain (Palangi et al. 2017; Smolensky et al. 2016; Lee et al. 2016). Extensions of such tensor-product representations, when applied to NLP tasks such as machine reading and question answering, are aimed to learn to process and understand massive natural language documents. After learning, the systems will be able not only to answer questions sensibly but also to truly understand what it reads to the extent that it can convey such understanding to human users in providing clues as to what steps have been taken to reach the answer. These steps may be in the form of logical reasoning expressed in natural language which is thus naturally understood by the human users of this type of machine reading and comprehension systems. In our view, natural language understanding is not just to accurately predict an answer from a question with relevant passages or data graphs as its contextual knowledge in a supervised way after seeing many examples of matched questions–passages–answers. Rather, the desired NLP system equipped with real understanding should resemble human cognitive capabilities. As an example of such capabilities (Nguyen et al. 2017)—after an understanding system is trained well, say, in a question answering task (using supervised learning or otherwise), it should master all essential aspects of the observed text material provided to solve the question answering tasks. What such mastering entails is that the learned system can subsequently perform well on other NLP tasks, e.g., translation, summarization, recommendation, etc., without seeing additional paired data such as raw text data with its summary, or parallel English and Chinese texts, etc.

One way to examine the nature of such powerful neural-symbolic systems is to regard them as ones incorporating the strength of the "rationalist" approaches marked by expert reasoning and structure richness popular during the first wave of NLP discussed in Sect. 1.2. Interestingly, prior to the rising of deep learning (third) wave of NLP, (Church 2007) argued that the pendulum from rationalist to empiricist approaches has swung too far at almost the peak of the second NLP wave, and predicted that the new rationalist wave would arrive. However, rather than swinging back to a renewed rationalist era of NLP, deep learning era arrived in full force in just a short period from the time of writing by Church (2007). Instead of adding the rationalist flavor, deep learning has been pushing empiricism of NLP to its pinnacle with big data and big compute, and with conceptually revolutionary ways of representing a sweeping range of linguistic entities by massive parallelism and distributedness, thus drastically enhancing the generalization capability of new-generation NLP models. Only after the sweeping successes of current deep learning methods for NLP

(Sect. 1.4) and subsequent analyses of a series of their limitations, do researchers look into the next wave of NLP—not swinging back to rationalism while abandoning empiricism but developing more advanced deep learning paradigms that would organically integrate the missing essence of rationalism into the structured neural methods that are aimed to approach human cognitive functions for language.

1.6.2 Structure, Memory, and Knowledge

As discussed earlier in this chapter as well as in the current NLP literature (Manning and Socher 2017), NLP researchers at present still have very primitive deep learning methods for exploiting structure and for building and accessing memories or knowledge. While LSTM (with attention) has been pervasively applied to NLP tasks to beat many NLP benchmarks, LSTM is far from a good memory model for human cognition. In particular, LSTM lacks adequate structure for simulating episodic memory, and one key component of human cognitive ability is to retrieve and re-experience aspects of a past novel event or thought. This ability gives rise to one-shot learning skills and can be crucial in reading comprehension of natural language text or speech understanding, as well as reasoning over events described by natural language. Many recent studies have been devoted to better memory modeling, including external memory architectures with supervised learning (Vinyals et al. 2016; Kaiser et al. 2017) and augmented memory architectures with reinforcement learning (Graves et al. 2016; Oh et al. 2016). However, they have not shown general effectiveness, but have suffered from a number of of limitations including notably scalability (arising from the use of attention which has to access every stored element in the memory). Much work remains in the direction of better modeling of memory and exploitation of knowledge for text understanding and reasoning.

1.6.3 Unsupervised and Generative Deep Learning

Another potential breakthrough in deep learning for NLP is in new algorithms for unsupervised deep learning, which makes use of ideally no direct teaching signals paired with inputs (token by token) to guide the learning. Word embedding discussed in Sect. 1.4 can be viewed as a weak form of unsupervised learning, making use of adjacent words as "cost-free" surrogate teaching signals, but for real-world NLP prediction tasks, such as translation, understanding, summarization, etc., such embedding obtained in an "unsupervised manner" has to be fed into another supervised architecture which requires costly teaching signals. In truly unsupervised learning which requires no expensive teaching signals, new types of objective functions and new optimization algorithms are needed, e.g., the objective function for unsupervised learning should not require explicit target label data aligned with the input data as in cross entropy that is most popular for supervised learning. Development of unsupervised deep learning algorithms has been significantly behind that of supervised and reinforcement deep learning where backpropagation and Q-learning algorithms have been reasonably mature.

The most recent preliminary development in unsupervised learning takes the approach of exploiting sequential output structure and advanced optimization methods to alleviate the need for using labels in training prediction systems (Russell and Stefano 2017; Liu et al. 2017). Future advances in unsupervised learning are promising by exploiting new sources of learning signals including the structure of input data and the mapping relationships from input to output and vice versa. Exploiting the relationship from output to input is closely connected to building conditional generative models. To this end, the recent popular topic in deep learning—generative adversarial networks (Goodfellow et al. 2014)—is a highly promising direction where the long-standing concept of analysis-by-synthesis in pattern recognition and machine learning is likely to return to spotlight in the near future in solving NLP tasks in new ways.

Generative adversarial networks have been formulated as neural nets, with dense connectivity among nodes and with no probabilistic setting. On the other hand, probabilistic and Bayesian reasoning, which often takes computational advantage of sparse connections among "nodes" as random variables, has been one of the principal theoretical pillars to machine learning and has been responsible for many NLP methods developed during the empiricist wave of NLP discussed in Sect. 1.3. What is the right interface between deep learning and probabilistic modeling? Can probabilistic thinking help understand deep learning techniques better and motivate new deep learning methods for NLP tasks? How about the other way around? These issues are widely open for future research.

1.6.4 Multimodal and Multitask Deep Learning

Multimodal and multitask deep learning are related learning paradigms, both concerning the exploitation of latent representations in the deep networks pooled from different modalities (e.g., audio, speech, video, images, text, source codes, etc.) or from multiple cross-domain tasks (e.g., point and structured prediction, ranking, recommendation, time-series forecasting, clustering, etc.). Before the deep learning wave, multimodal and multitask learning had been very difficult to be made effective, due to the lack of intermediate representations that share across modalities or tasks. See a most striking example of this contrast for multitask learning—multilingual speech recognition during the empiricist wave (Lin et al. 2008) and during the deep learning wave (Huang et al. 2013a).

Multimodal information can be exploited as low-cost supervision. For instance, standard speech recognition, image recognition, and text classification methods make use of supervision labels within each of the speech, image, and text modalities separately. This, however, is far from how children learn to recognize speech, image, and to classify text. For example, children often get the distant "supervision" signal for speech sounds by an adult pointing to an image scene, text, or handwriting that is associated with the speech sounds. Similarly, for children learning image categories, they may exploit speech sounds or text as supervision signals. This type of learning that occurs in children can motivate a learning scheme that leverages multimodal data to improve engineering systems for multimodal deep learning. A similarity measure needs to be defined in the same semantic space, which speech, image, and text are all mapped into, via deep neural networks that may be trained using maximum mutual information across different modalities. The huge potential of this scheme has not been explored and found in the NLP literature.

Similar to multimodal deep learning, multitask deep learning can also benefit from leveraging multiple latent levels of representations across tasks or domains. The recent work on joint many-task learning solves a range of NLP tasks—from morphological to syntactic and to semantic levels, within one single, big deep neural network model (Hashimoto et al. 2017). The model predicts different levels of linguistic outputs at successively deep layers, accomplishing standard NLP tasks of tagging, chunking, syntactic parsing, as well as predictions of semantic relatedness and entailment. The strong results obtained using this single, end-to-end learned model point to the direction to solve more challenging NLP tasks in real world as well as tasks beyond NLP.

1.6.5 Meta-learning

A further future direction for fruitful NLP and artificial intelligence research is the paradigm of learning-to-learn or meta-learning. The goal of meta-learning is to learn how to learn new tasks faster by reusing previous experience, instead of treating each new task in isolation and learning to solve each of them from scratch. That is, with the success of meta-learning, we can train a model on a variety of learning tasks, such that it can solve new learning tasks using only a small number of training samples. In our NLP context, successful meta-learning will enable the design of intelligent NLP systems that improve or automatically discover new learning algorithms (e.g., sophisticated optimization algorithms for unsupervised learning), for solving NLP tasks using small amounts of training data.

The study of meta-learning, as a subfield of machine learning, started over three decades ago (Schmidhuber 1987; Hochreiter et al. 2001), but it was not until recent years when deep learning methods reasonably matured that stronger evidence of the potentially huge impact of meta-learning has become apparent. Initial progresses of meta-learning can be seen in various techniques successfully applied to deep learning, including hyper-parameter optimization (Maclaurin et al. 2015), neural network architecture optimization (Wichrowska et al. 2017), and fast reinforcement learning (Finn et al. 2017). The ultimate success of meta-learning in real world will allow the development of algorithms to solve most NLP and computer science problems to be reformulated as a deep learning problem and to be solved by a uniform infrastructure designed for deep learning today. Meta-learning is a powerful emerging artificial intelligence and deep learning paradigm, which is a fertile research area expected to impact real-world NLP applications.

1.7 Summary

In this chapter, to set up the fundamental framework for the book, we first provided an introduction to the basics of natural language processing (NLP), which is more application-oriented than computational linguistics, both belonging to a field of artificial intelligence and computer science. We survey the historical development of the NLP field, spanning over several decades, in terms of three waves of NLP—starting from rationalism and empiricism, to the current deep learning wave. The goal of the survey is to distill insights from the historical developments that serve to guide future directions.

The conclusion from our three-wave analysis is that the current deep learning technology for NLP is a conceptual and paradigmatic revolution from the NLP technologies developed from the previous two waves. The key pillars underlying this revolution consist of distributed representations of linguistic entities (sub-words, words, phrases, sentences, paragraphs, documents, etc.) via embedding, semantic generalization due to the embedding, long-span deep sequence modeling of language, hierarchical networks effective for representing linguistic levels from low to high, and end-to-end deep learning methods to jointly solve many NLP tasks. None of these were possible before the deep learning wave, not only because of the lack of big data and powerful computation in the previous waves but, equally importantly, due to missing the right framework until the deep learning paradigm emerged in recent years.

After we surveyed the prominent successes of selected NLP application areas attributed to deep learning (with a much more comprehensive coverage of the NLP successful areas in the remaining chapters of this book), we pointed out and analyzed several key limitations of current deep learning technology in general, as well as those for NLP more specifically. This investigation led us to five research directions for future advances in NLP—frameworks for neural-symbolic integration, exploration of better memory models, and better use of knowledge, as well as better deep learning paradigms including unsupervised and generative learning, multimodal and multitask learning, and meta-learning.

In conclusion, deep learning has ushered in a world that gives our NLP field a much brighter future than any time in the past. Deep learning not only provides a powerful modeling framework for representing human cognitive abilities of natural language in computer systems but, as importantly, it has already been creating superior practical results in a number of key application areas of NLP. In the remaining chapters of this book, detailed descriptions of NLP techniques developed using the deep learning framework will be provided, and where possible, benchmark results will be presented contrasting deep learning with more traditional techniques developed before the deep learning tidal wave hit the NLP shore just a few years ago. We hope this comprehensive set of material will serve as a mark along the way where NLP researchers are developing better and more advanced deep learning methods to overcome some or all the current limitations discussed in this chapter, possibly inspired by the research directions we analyzed here as well.

References

Abdel-Hamid, O., Mohamed, A., Jiang, H., Deng, L., Penn, G., & Yu, D. (2014). *Convolutional neural networks for speech recognition*. IEEE/ACM Trans. on Audio, Speech and Language Processing.

Amodei, D., Ng, A., et al. (2016). Deep speech 2: End-to-end speech recognition in English and Mandarin. In *Proceedings of ICML*.

Bahdanau, D., Cho, K., & Bengio, Y. (2015). Neural machine translation by jointly learning to align and translate. In *Proceedings of ICLR*.

Baker, J., et al. (2009a). Research developments and directions in speech recognition and understanding. *IEEE Signal Processing Magazine*, *26*(4).

Baker, J., et al. (2009b). Updated MINDS report on speech recognition and understanding. *IEEE Signal Processing Magazine*, *26*(4).

Baum, L., & Petrie, T. (1966). Statistical inference for probabilistic functions of finite state markov chains. *The Annals of Mathematical Statistics*.

Bengio, Y. (2009). *Learning Deep Architectures for AI*. Delft: NOW Publishers.

Bengio, Y., Ducharme, R., Vincent, P., & d Jauvin, C. (2001). A neural probabilistic language model. *Proceedings of NIPS*.

Bishop, C. (1995). *Neural Networks for Pattern Recognition*. Oxford: Oxford University Press.

Bishop, C. (2006). *Pattern Recognition and Machine Learning*. Berlin: Springer.

Bridle, J., et al. (1998). An investigation of segmental hidden dynamic models of speech coarticulation for automatic speech recognition. *Final Report for 1998 Workshop on Language Engineering, Johns Hopkins University CLSP*.

Brown, P. F., Della Pietra, S. A., Della Pietra, V. J., & Mercer, R. L. (1993). The mathematics of statistical machine translation: Parameter estimation. *Computational Linguistics*, *19*.

Charniak, E. (2011). The brain as a statistical inference engine—and you can too. *Computational Linguistics*, *37*.

Chiang, D. (2007). Hierarchical phrase-based translation. *Computaitional Linguistics*.

Chomsky, N. (1957). *Syntactic Structures*. The Hague: Mouton.

Chorowski, J., Bahdanau, D., Serdyuk, D., Cho, K., & Bengio, Y. (2015). Attention-based models for speech recognition. In *Proceedings of NIPS*.

Church, K. (2007). A pendulum swung too far. *Linguistic Issues in Language Technology*, *2*(4).

Church, K. (2014). The case for empiricism (with and without statistics). In *Proceedings of Frame Semantics in NLP*.

Church, K., & Mercer, R. (1993). Introduction to the special issue on computational linguistics using large corpora. *Computational Linguistics*, *9*(1).

Collins, M. (1997). *Head-driven statistical models for natural language parsing*. Ph.D. thesis, University of Pennsylvania, Philadelphia.

Collins, M. (2002). Discriminative training methods for hidden markov models: Theory and experiments with perceptron algorithms. In *Proceedings of EMNLP*.

Collobert, R., Weston, J., Bottou, L., Karlen, M., Kavukcuoglu, K., & Kuksa, P. (2011). Natural language processing (almost) from scratch. *Journal of Machine Learning Reserach*, *12*.

Dahl, G., Yu, D., & Deng, L. (2011). Large-vocabulry continuous speech recognition with context-dependent DBN-HMMs. In *Proceedings of ICASSP*.

Dahl, G., Yu, D., Deng, L., & Acero, A. (2012). Context-dependent pre-trained deep neural networks for large-vocabulary speech recognition. *IEEE Transaction on Audio, Speech, and Language Processing, 20*.

Deng, L. (1998). A dynamic, feature-based approach to the interface between phonology and phonetics for speech modeling and recognition. *Speech Communication, 24*(4).

Deng, L. (2014). A tutorial survey of architectures, algorithms, and applications for deep learning. *APSIPA Transactions on Signal and Information Processing, 3*.

Deng, L. (2016). Deep learning: From speech recognition to language and multimodal processing. *APSIPA Transactions on Signal and Information Processing, 5*.

Deng, L. (2017). Artificial intelligence in the rising wave of deep learning—The historical path and future outlook. In *IEEE Signal Processing Magazine, 35*.

Deng, L., & O'Shaughnessy, D. (2003). *SPEECH PROCESSING A Dynamic and Optimization-Oriented Approach*. New York: Marcel Dekker.

Deng, L., & Yu, D. (2007). Use of differential cepstra as acoustic features in hidden trajectory modeling for phonetic recognition. In *Proceedings of ICASSP*.

Deng, L., & Yu, D. (2014). *Deep Learning: Methods and Applications*. Delft: NOW Publishers.

Deng, L., Hinton, G., & Kingsbury, B. (2013). New types of deep neural network learning for speech recognition and related applications: An overview. In *Proceedings of ICASSP*.

Deng, L., Seltzer, M., Yu, D., Acero, A., Mohamed, A., & Hinton, G. (2010). Binary coding of speech spectrograms using a deep autoencoder. In *Proceedings of Interspeech*.

Deng, L., Yu, D., & Platt, J. (2012). Scalable stacking and learning for building deep architectures. In *Proceedings of ICASSP*.

Devlin, J., et al. (2015). Language models for image captioning: The quirks and what works. In *Proceedings of CVPR*.

Dhingra, B., Li, L., Li, X., Gao, J., Chen, Y., Ahmed, F., & Deng, L. (2017). Towards end-to-end reinforcement learning of dialogue agents for information access. In *Proceedings of ACL*.

Fang, H., et al. (2015). From captions to visual concepts and back. In *Proceedings of CVPR*.

Fei-Fei, L., & Perona, P. (2005). A Bayesian hierarchical model for learning natural scene categories. In *Proceedings of CVPR*.

Fei-Fei, L., & Perona, P. (2016). Stacked attention networks for image question answering. In *Proceedings of CVPR*.

Finn, C., Abbeel, P., & Levine, S. (2017). Model-agnostic meta-learning for fast adaptation of deep networks. In *Proceedings of ICML*.

Gan, Z., et al. (2017). Semantic compositional networks for visual captioning. In *Proceedings of CVPR*.

Gasic, M., Mrk, N., Rojas-Barahona, L., Su, P., Ultes, S., Vandyke, D., Wen, T., & Young, S. (2017). Dialogue manager domain adaptation using gaussian process reinforcement learning. *Computer Speech and Language, 45*.

Goodfellow, I., Bengio, Y., & Courville, A. (2016). *Deep Learning*. Cambridge: MIT Press.

Goodfellow, I., et al. (2014). Generative adversarial networks. In *Proceedings of NIPS*.

Graves, A., et al. (2016). Hybrid computing using a neural network with dynamic external memory. *Nature, 538*.

Hashimoto, K., Xiong, C., Tsuruoka, Y., & Socher, R. (2017). Investigation of recurrent-neural-network architectures and learning methods for spoken language understanding. In *Proceedings of EMNLP*.

He, X., & Deng, L. (2012). Maximum expected BLEU training of phrase and lexicon translation models. In *Proceedings of ACL*.

He, X., & Deng, L. (2013). Speech-centric information processing: An optimization-oriented approach. *Proceedings of the IEEE, 101*.

He, X., Deng, L., & Chou, W. (2008). Discriminative learning in sequential pattern recognition. *IEEE Signal Processing Magazine, 25*(5).

He, K., Zhang, X., Ren, S., & Sun, J. (2016). Deep residual learning for image recognition. In *Proceedings of CVPR*.

Hinton, G., & Salakhutdinov, R. (2012). A better way to pre-train deep Boltzmann machines. In *Proceedings of NIPS*.
Hinton, G., Deng, L., Yu, D., Dahl, G., Mohamed, A.-r., Jaitly, N., Senior, A., Vanhoucke, V., Nguyen, P., Kingsbury, B., & Sainath, T. (2012). Deep neural networks for acoustic modeling in speech recognition. *IEEE Signal Processing Magazine, 29*.
Hinton, G., Osindero, S., & Teh, Y. -W. (2006). A fast learning algorithm for deep belief nets. *Neural Computation, 18*.
Hochreiter, S., et al. (2001). Learning to learn using gradient descent. In *Proceedings of International Conference on Artificial Neural Networks*.
Huang, P., et al. (2013b). Learning deep structured semantic models for web search using clickthrough data. *Proceedings of CIKM*.
Huang, J. -T., Li, J., Yu, D., Deng, L., & Gong, Y. (2013a). Cross-lingual knowledge transfer using multilingual deep neural networks with shared hidden layers. In *Proceedings of ICASSP*.
Jackson, P. (1998). *Introduction to Expert Systems*. Boston: Addison-Wesley.
Jelinek, F. (1998). *Statistical Models for Speech Recognition*. Cambridge: MIT Press.
Juang, F. (2016). Deep neural networks a developmental perspective. *APSIPA Transactions on Signal and Information Processing, 5*.
Kaiser, L., Nachum, O., Roy, A., & Bengio, S. (2017). Learning to remember rare events. In *Proceedings of ICLR*.
Karpathy, A., & Fei-Fei, L. (2015). Deep visual-semantic alignments for generating image descriptions. In *Proceedings of CVPR*.
Koh, P., & Liang, P. (2017). Understanding black-box predictions via influence functions. In *Proceedings of ICML*.
Krizhevsky, A., Sutskever, I., & Hinton, G. (2012). Imagenet classification with deep convolutional neural networks. In *Proceedings of NIPS*.
Lafferty, J., McCallum, A., & Pereira, F. (2001). Conditional random fields: Probabilistic models for segmenting and labeling sequence data. In *Proceedings of ICML*.
LeCun, Y., Bengio, Y., & Hinton, G. (2015). Deep learning. *Nature, 521*.
Lee, L., Attias, H., Deng, L., & Fieguth, P. (2004). A multimodal variational approach to learning and inference in switching state space models. In *Proceedings of ICASSP*.
Lee, M., et al. (2016). Reasoning in vector space: An exploratory study of question answering. In *Proceedings of ICLR*.
Lin, H., Deng, L., Droppo, J., Yu, D., & Acero, A. (2008). Learning methods in multilingual speech recognition. In *NIPS Workshop*.
Liu, Y., Chen, J., & Deng, L. (2017). An unsupervised learning method exploiting sequential output statistics. In arXiv:1702.07817.
Ma, J., & Deng, L. (2004). Target-directed mixture dynamic models for spontaneous speech recognition. *IEEE Transaction on Speech and Audio Processing, 12*(4).
Maclaurin, D., Duvenaud, D., & Adams, R. (2015). Gradient-based hyperparameter optimization through reversible learning. In *Proceedings of ICML*.
Manning, C. (2016). Computational linguistics and deep learning. In *Computational Linguistics*.
Manning, C., & Schtze, H. (1999). *Foundations of statistical natural language processing*. Cambridge: MIT Press.
Manning, C., & Socher, R. (2017). *Lectures 17 and 18: Issues and Possible Architectures for NLP; Tackling the Limits of Deep Learning for NLP*. CS224N Course: NLP with Deep Learning.
Mesnil, G., He, X., Deng, L., & Bengio, Y. (2013). Investigation of recurrent-neural-network architectures and learning methods for spoken language understanding. In *Proceedings of Interspeech*.
Mikolov, T., Sutskever, I., Chen, K., Corrado, G., & Dean, J. (2013). Distributed representations of words and phrases and their compositionality. In *Proceedings of NIPS*.
Mnih, V., Kavukcuoglu, K., Silver, D., Rusu, A. A., Veness, J., Bellemare, M. G., Graves, A., Riedmiller, M., Fidjeland, A. K., Ostrovski, G., Petersen, S., Beattie, C., Sadik, A., Antonoglou, I., King, H., Kumaran, D., Wierstra, D., Legg, S., & Hassabis, D. (2015). Human-level control through deep reinforcement learning. *Nature, 518*.

Mohamed, A., Dahl, G., & Hinton, G. (2009). Acoustic modeling using deep belief networks. In *NIPS Workshop on Speech Recognition*.
Murphy, K. (2012). *Machine Learning: A Probabilistic Perspective*. Cambridge: MIT Press.
Nguyen, T., et al. (2017). MS MARCO: A human generated machine reading comprehension dataset. arXiv:1611,09268
Nilsson, N. (1982). *Principles of Artificial Intelligence*. Berlin: Springer.
Och, F. (2003). Maximum error rate training in statistical machine translation. In *Proceedings of ACL*.
Och, F., & Ney, H. (2002). Discriminative training and maximum entropy models for statistical machine translation. In *Proceedings of ACL*.
Oh, J., Chockalingam, V., Singh, S., & Lee, H. (2016). Control of memory, active perception, and action in minecraft. In *Proceedings of ICML*.
Palangi, H., Smolensky, P., He, X., & Deng, L. (2017). Deep learning of grammatically-interpretable representations through question-answering. arXiv:1705.08432
Parloff, R. (2016). Why deep learning is suddenly changing your life. In *Fortune Magazine*.
Pereira, F. (2017). A (computational) linguistic farce in three acts. In http://www.earningmyturns.org.
Picone, J., et al. (1999). Initial evaluation of hidden dynamic models on conversational speech. In *Proceedings of ICASSP*.
Plamondon, R., & Srihari, S. (2000). Online and off-line handwriting recognition: A comprehensive survey. *IEEE Transactions on Pattern Analysis and Machine Intelligence, 22*.
Rabiner, L., & Juang, B. -H. (1993). *Fundamentals of Speech Recognition*. USA: Prentice-Hall.
Ratnaparkhi, A. (1997). A simple introduction to maximum entropy models for natural language processing. Technical report, University of Pennsylvania.
Reddy, R. (1976). Speech recognition by machine: A review. *Proceedings of the IEEE, 64*(4).
Rumelhart, D., Hinton, G., & Williams, R. (1986). Learning representations by back-propagating errors. *Nature, 323*.
Russell, S., & Stefano, E. (2017). Label-free supervision of neural networks with physics and domain knowledge. In *Proceedings of AAAI*.
Saon, G., et al. (2017). English conversational telephone speech recognition by humans and machines. In *Proceedings of ICASSP*.
Schmidhuber, J. (1987). *Evolutionary principles in self-referential learning*. Diploma Thesis, Institute of Informatik, Technical University Munich.
Seneff, S., et al. (1991). Development and preliminary evaluation of the MIT ATIS system. In *Proceedings of HLT*.
Smolensky, P., et al. (2016). Reasoning with tensor product representations. arXiv:1601,02745
Sutskevar, I., Vinyals, O., & Le, Q. (2014). Sequence to sequence learning with neural networks. In *Proceedings of NIPS*.
Tur, G., & Deng, L. (2011). *Intent Determination and Spoken Utterance Classification; Chapter 4 in book: Spoken Language Understanding*. Hoboken: Wiley.
Turing, A. (1950). Computing machinery and intelligence. *Mind, 14*.
Vapnik, V. (1998). *Statistical Learning Theory*. Hoboken: Wiley.
Vincent, P., Larochelle, H., Lajoie, I., Bengio, Y., & Manzagol, P. -A. (2010). Stacked denoising autoencoders: Learning useful representations in a deep network with a local denoising criterion. *The Journal of Machine Learning Research, 11*.
Vinyals, O., et al. (2016). Matching networks for one shot learning. In *Proceedings of NIPS*.
Viola, P., & Jones, M. (2004). Robust real-time face detection. *International Journal of Computer Vision, 57*.
Wang, Y. -Y., Deng, L., & Acero, A. (2011). *Semantic Frame Based Spoken Language Understanding; Chapter 3 in book: Spoken Language Understanding*. Hoboken: Wiley.
Wichrowska, O., et al. (2017). Learned optimizers that scale and generalize. In *Proceedings of ICML*.
Winston, P. (1993). *Artificial Intelligence*. Boston: Addison-Wesley.

Xiong, W., et al. (2016). Achieving human parity in conversational speech recognition. In *Proceedings of Interspeech*.

Young, S., Gasic, M., Thomson, B., & Williams, J. (2013). Pomdp-based statistical spoken dialogue systems: A review. *Proceedings of the IEEE, 101*.

Yu, D., & Deng, L. (2015). *Automatic Speech Recognition: A Deep Learning Approach*. Berlin: Springer.

Yu, D., Deng, L., & Dahl, G. (2010). Roles of pre-training and fine-tuning in context-dependent dbn-hmms for real-world speech recognition. In *NIPS Workshop*.

Yu, D., Deng, L., Seide, F., & Li, G. (2011). Discriminative pre-training of deep nerual networks. In *U.S. Patent No. 9,235,799, granted in 2016, filed in 2011*.

Zue, V. (1985). The use of speech knowledge in automatic speech recognition. *Proceedings of the IEEE, 73*.

Chapter 2
Deep Learning in Conversational Language Understanding

Gokhan Tur, Asli Celikyilmaz, Xiaodong He, Dilek Hakkani-Tür and Li Deng

Abstract Recent advancements in AI resulted in increased availability of conversational assistants that can help with tasks such as seeking times to schedule an event and creating a calendar entry at that time, finding a restaurant and booking a table there at a certain time. However, creating automated agents with human-level intelligence still remains one of the most challenging problems of AI. One key component of such systems is conversational language understanding, which is a holy grail area of research for decades, as it is not a clearly defined task but relies heavily on the AI application it is used for. Nevertheless, this chapter attempts to compile the recent deep learning based literature on such goal-oriented conversational language understanding studies, starting with a historical perspective, pre-deep learning era work, moving toward most recent advances in this field.

2.1 Introduction

In the last decade, a variety of practical goal-oriented conversation language understanding (CLU) systems have been built, especially as part of the virtual personal assistants such as Google Assistant, Amazon Alexa, Microsoft Cortana, or Apple Siri.

G. Tur (✉) · D. Hakkani-Tür
Google, Mountain View, CA, USA
e-mail: gokhan.tur@ieee.org

D. Hakkani-Tür
e-mail: dilek@ieee.org

A. Celikyilmaz · X. He
Microsoft Research, Redmond, WA, USA
e-mail: asli@ieee.org

X. He
e-mail: xiaohe@microsoft.com

L. Deng
Citadel, Chicago & Seattle, USA
e-mail: l.deng@ieee.org

L. Deng and Y. Liu (eds.), *Deep Learning in Natural Language Processing*, https://doi.org/10.1007/978-981-10-5209-5_2

In contrast to speech recognition, which aims to automatically transcribe the sequence of spoken words (Deng and O'Shaughnessy 2003; Huang and Deng 2010), CLU is not a clearly defined task. At the highest level, CLU's goal is to extract "meaning" from natural language in the context of conversations, spoken or in text. In practice, this may mean any practical application allowing its users to perform some task with natural (optionally spoken) language. In the literature, CLU is often used to denote the task of understanding natural language in spoken form in conversation or otherwise. So CLU discussed in this chapter and book is closely related to and sometimes synonymous with spoken language understanding (SLU) in the literature (Tur and Mori 2011; Wang et al. 2011).

Here, we further elaborate on the connections among speech recognition, CLU/SLU, and natural language understanding in text form. Speech recognition does not concern understanding, and is responsible only for converting language from spoken form to text form (Deng and Li 2013). Errors in speech recognition can be viewed as "noise" in downstream language processing systems (He and Deng 2011). Handling this type of noisy NLP problems can be connected to the problem of noisy speech recognition where the "noise" comes from acoustic environments (as opposed to from recognition errors) (Li et al. 2014).

For SLU and CLU with spoken input, the inevitable errors in speech recognition would make understanding harder than when the input is text, free of speech recognition errors (He and Deng 2013). In the long history of SLU/CLU research, the difficulties caused by speech recognition errors forced the domains of SLU/CLU to be substantially narrower than language understanding in text form (Tur and Deng 2011). However, due to the recent huge success of deep learning in speech recognition (Hinton et al. 2012), recognition errors have been dramatically reduced, leading to increasingly broader application domains in current CLU systems.

One category of conversational understanding tasks roots in old artificial intelligence (AI) work, such as the MIT Eliza system built in 1960s (Weizenbaum 1966), mainly used for chit-chat systems, mimicking understanding. For example, if the user says "*I am depressed*", Eliza would say "*are you depressed often?*". The other extreme is building generic understanding capabilities, using deeper semantics and are demonstrated to be successful for very limited domains. These systems are typically heavily knowledge-based and rely on formal semantic interpretation defined as mapping sentences into their logical forms. In its simplest form, a logical form is a context-independent representation of a sentence covering its predicates and arguments. For example, if the sentence is *John loves Mary*, the logical form would be `love(john,mary)`. Following these ideas, some researchers worked toward building universal semantic grammars (or interlingua), which assume that all languages have a shared set of semantic features (Chomsky 1965). Such interlingua-based approaches also heavily influenced machine translation research until the late 90s, before statistical approaches began to dominate. (Allen 1995) may be consulted for more information on the artificial intelligence-based techniques for language understanding.

Having a semantic representation for CLU that is both broad coverage and simple enough to be applicable to several different tasks and domains is challenging, and

W	find	flights	to	new	york	tomorrow
	↓	↓	↓	↓	↓	↓
S	O	O	O	B-Dest	I-Dest	B-Date
D	flight					
I	find_flight					

Fig. 2.1 An example semantic parse of an utterance (W) with slot (S), domain (D), intent (I) annotations, following the IOB (in-out-begin) representation for slot values

hence most CLU tasks and approaches depend on the application and environment (such as mobile vs. TV) they have been designed for. In such a "targeted understanding" setup, three key tasks are domain classification (*what is the user talking about, e.g., "travel"*), intent determination (*what does the user want to do, e.g., "book a hotel room"*), and slot filling (*what are the parameters of this task*, e.g., "two bedroom suite near disneyland") (Tur and Mori 2011), aiming to form a semantic frame that captures the semantics of user utterances/queries. An example semantic frame is shown in Fig. 2.1 for a flight-related query: *find flights to boston tomorrow*.

In this chapter, we will review the state-of-the-art deep learning based CLU methods in detail, mainly focusing on these three tasks. In the next section, we will provide the task definitions more formally, and then present pre-deep learning era literature. Then in Sect. 2.4 we will cover the recent studies targeting this task.

2.2 A Historical Perspective

In the United States, the study of the frame-based CLU started in the 1970s at DARPA Speech Understanding Research (SUR) and then the resource management (RM) tasks. At this early stage, natural language understanding (NLU) techniques like finite state machines (FSMs) and augmented transition networks (ATNs) were applied for SLU (Woods 1983).

The study of targeted frame-based SLU surged in the 1990s, with the DARPA Air Travel Information System (ATIS) project evaluations (Price 1990; Hemphill et al. 1990; Dahl et al. 1994). Multiple research labs from both academia and industry, including AT&T, BBN Technologies (originally Bolt, Beranek and Newman), Carnegie Mellon University, MIT, and SRI, developed systems that attempted to understand users' spontaneous spoken queries for air travel information (including flight information, ground transportation information, airport service information, etc.) and then obtain the answers from a standard database. ATIS is an important milestone for frame-based SLU, largely thanks to its rigorous component-wise and end-to-end evaluation, participated by multiple institutions, with a common test set. Later, ATIS was extended to cover multi-turn dialogs, via DARPA Communicator program (Walker et al. 2001). In the meantime, the AI community had separate efforts in building a conversational planning agent, such as the TRAINS system (Allen et al. 1996), and parallel efforts were made on the other side of the Atlantic. The French EVALDA/MEDIA project aimed at designing and testing the evaluation methodology to compare and diagnose the context-dependent and context-independent SLU capability in spoken language dialogs (Bonneau-Maynard et al. 2005). Participants included both academic organizations (IRIT, LIA, LIMSI, LORIA, VALORIA, and CLIPS) and industrial institutions (FRANCE TELECOM R&D, TELIP). Like ATIS, the domain of this study was restricted to database queries for tourist and hotel information. The more recent LUNA project sponsored by the European Union focused on the problem of real-time understanding of spontaneous speech in the context of advanced telecom services (Hahn et al. 2011).

Pre-deep learning era researchers employed known sequence classification methods for filling frame slots of the application domain using the provided training dataset and performed comparative experiments. These approaches used generative models such as hidden Markov models (HMMs) (Pieraccini et al. 1992), discriminative classification methods (Kuhn and Mori 1995), knowledge-based methods, and probabilistic context-free grammars (CFGs) (Seneff 1992; Ward and Issar 1994), and finally conditional random fields (CRFs) (Raymond and Riccardi 2007; Tur et al. 2010).

Almost simultaneously with the slot filling approaches, a related CLU task, mainly used for machine-directed dialog in call center IVR (Interactive Voice Response) systems have emerged. In IVR systems, the interaction is completely controlled by the machines. Machine-initiative systems ask user-specific questions and expect the users input to be one of predetermined keywords or phrases. For example, a mail delivery system may prompt the user to say *schedule a pick-up, track a package, get rate or order supply* or a pizza delivery system may ask for *possible toppings*. Such IVR systems are typically extended to form a machine-initiative directed dialog in call centers and are now widely implemented using established and standardized platforms such as VoiceXML (VXML).

The success of these IVR systems has triggered more sophisticated versions of this very same idea of *classifying users' utterances into predefined categories* (called as *call types* or *intents*), employed by almost all major players, such as AT&T (Gorin et al. 1997, 2002; Gupta et al. 2006), Bell Labs (Chu-Carroll and Carpenter 1999), BBN (Natarajan et al. 2002), and the France Telecom (Damnati et al. 2007).

While this is a totally different perspective for the task of CLU, it is actually complementary to frame filling. For example, there are utterances in the ATIS corpus asking about ground transportation or the capacity of planes on a specific flight, and hence the users may have other intents than basically finding flight information.

A detailed survey of pre-deep learning era approaches for domain detection, intent determination, and slot filling can be found in (Tur and Mori 2011).

2.3 Major Language Understanding Tasks

In this section, we mainly cover the key tasks of targeted conversational language understanding as used in human/machine conversational systems. These include utterance classification tasks for domain detection or intent determination and slot filling.

2.3.1 Domain Detection and Intent Determination

The semantic utterance classification tasks of domain detection and intent determination aim at classifying a given speech utterance X_r into one of M semantic classes, $\hat{C}_r \in \mathcal{C} = \{C_1, \ldots, C_M\}$ (where r is the utterance index). Upon the observation of X_r, $\hat{C}_r$ is chosen so that the class-posterior probability given X_r, $P(C_r|X_r)$, is maximized. Formally,

$$\hat{C}_r = \arg\max_{C_r} P(C_r|X_r). \tag{2.1}$$

Semantic classifiers require operation with significant freedom in utterance variations. A user may say "*I want to fly from Boston to New York next week*" and another user may express the same information by saying "*I am looking to flights from JFK to Boston in the coming week*". In spite of this freedom of expression, utterances in such applications have a clear structure that binds the specific pieces of information together. Not only is there no a priori constraint on what the user can say, but the system should be able to generalize well from a tractably small amount of training data. For example, the phrase "*Show all flights*" and "*Give me flights*" should be interpreted as variants of a single semantic class "Flight". On the other hand, the command "*Show me fares*" should be interpreted as an instance of another semantic class, "Fare". Traditional text categorization techniques devise learning methods to maximize the probability of C_r given the text W_r, i.e., the class-posterior probability $P(C_r|W_r)$. Other semantically motivated features like domain gazetteers (lists of entities), named entities (like organization names or time/date expressions) and contextual features (such as the previous dialog turn), can be used to enrich the feature set.

2.3.2 Slot Filling

The semantic structure of an application domain is defined in terms of the *semantic frames*. Each semantic frame contains several typed components called "*slots*". For example, in Fig. 2.1, the *Flights* domain may contain slots like *Departure_City*, *Arrival_City*, *Departure_Date*, *Airline_Name*, etc. The task of slot filling is then to instantiate the slots in semantic frames.

Some SLU systems have adopted a hierarchical representation as that is more expressive and allows the sharing of substructures. This is mainly motivated by syntactic constituency trees.

In statistical frame-based conversational language understanding, the task is often formalized as a pattern recognition problem. Given the word sequence W, the goal of slot filling is to find the semantic tag sequence, S, that has the maximum *a posteriori* probability $P(S \mid W)$:

$$\hat{\mathbf{S}} = \arg\max_{\mathbf{S}} P(\mathbf{S} \mid \mathbf{W}). \tag{2.2}$$

2.4 Elevating State of the Art: From Statistical Modeling to Deep Learning

In this section, we review the recent deep learning based efforts for conversational language understanding, both task by task, and also covering joint, multitask approaches.

2.4.1 *Domain Detection and Intent Determination*

The first applications of deep learning for utterance classification started as deep belief networks (DBNs) (Hinton et al. 2006) gained popularity in various areas of information processing applications. DBNs are stacks of restricted Boltzmann machines (RBMs) followed by fine-tuning. RBM is a two-layer network, which can be trained reasonably efficiently in an unsupervised fashion. Following the introduction of this RBM learning and layer-by-layer construction of deep architectures, DBNs have been successfully used for numerous tasks in speech and language processing, and finally for intent determination in a call routing setup (Sarikaya et al. 2011). This work has been extended in (Sarikaya et al. 2014), where additional unlabeled data is exploited for better pretraining.

Following the success of DBN, Deng and Yu proposed the use of deep convex net (DCN), which directly attacks the scalability issue of DBN-like deep learning techniques (Deng and Yu 2011). DCN is shown to be superior to DBN, not only in terms of accuracy but also in training scalability and efficiency. A DCN is a regular feed-forward neural network, but the input vector is also considered at each hidden layer.

Figure 2.2 shows the conceptual structure of a DCN, where **W** denotes input, and **U** denotes weights. In this study, mean square error is used as the loss function, given the target vectors, **T**. However, the network is pretrained using DBN as described above.

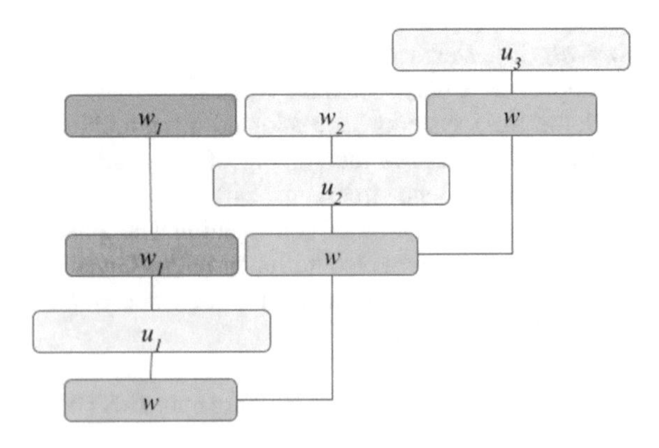

Fig. 2.2 A typical DCN architecture

In this early work, since vocabulary size was too big for the input vector, instead of feature transformation, a boosting (Freund and Schapire 1997)-based feature selection was employed to find the salient phrases for the classification task and results were compared with this boosting baseline.

After this early work, DBNs have been used more rarely for pretraining, and the state of the art is using convolutional neural networks (CNNs), and its varieties (Collobert and Weston 2008; Kim 2014; Kalchbrenner et al. 2014 among others).

Figure 2.3 shows a typical CNN architecture for sentence or utterance classification. A convolution operation involves a filter, $\mathbf{U}$, which is applied to a window of h words in the input sentence to produce a new feature, c_i. For example,

$$c_i = \tanh(\mathbf{U}.W_{i:i+h-1} + b),$$

where b is the bias, $\mathbf{W}$ is the input vector of words, and c_i is the new feature. Then, max-over-time pooling operation is applied over $\mathbf{c} = [c_1, c_2, \ldots, c_{n-h+1}]$ to take the maximum valued feature, $\hat{c} = max\mathbf{c}$. These features are passed to a fully connected softmax layer whose output is the probability distribution over labels:

$$P(y = j|\mathbf{x}) = \frac{e^{\mathbf{x}^{\mathrm{T}}\mathbf{w}_j}}{\sum_{k=1}^{k} e^{\mathbf{x}^{\mathrm{T}}\mathbf{w}_k}}.$$

There are few studies trying to use methods for domain detection inspired from recurrent neural networks (RNNs), and combining with CNNs, trying to get the best out of two worlds. Lee and Dernoncourt (2016) tried to build an RNN encoder, which can then be fed into a feed-forward network and compared that with a regular CNN. Figure 2.4 shows the conceptual model of the RNN-based encoder employed.

One notable work which is not using a feed-forward or convolutional neural network for utterance classification is by Ravuri and Stolcke (2015). They have

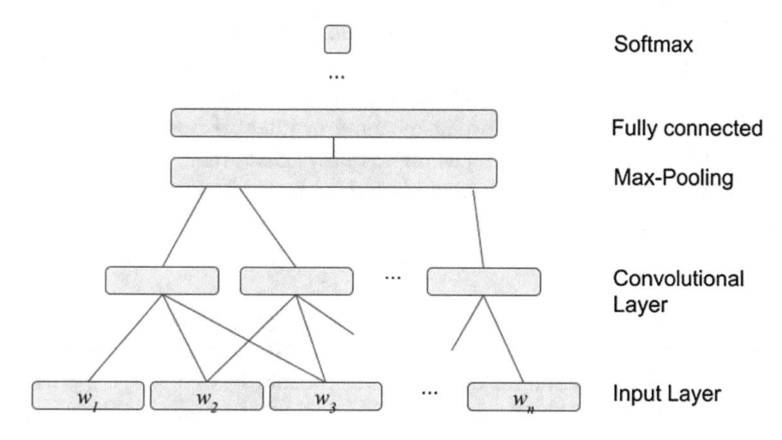

Fig. 2.3 A typical CNN architecture

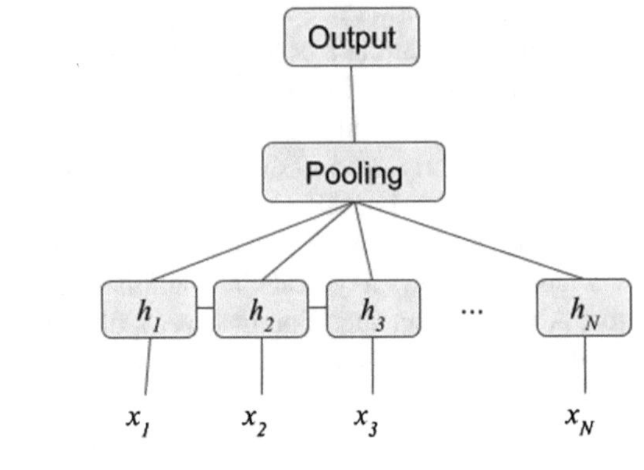

Fig. 2.4 An RNN–CNN-based encoder for sentence classification

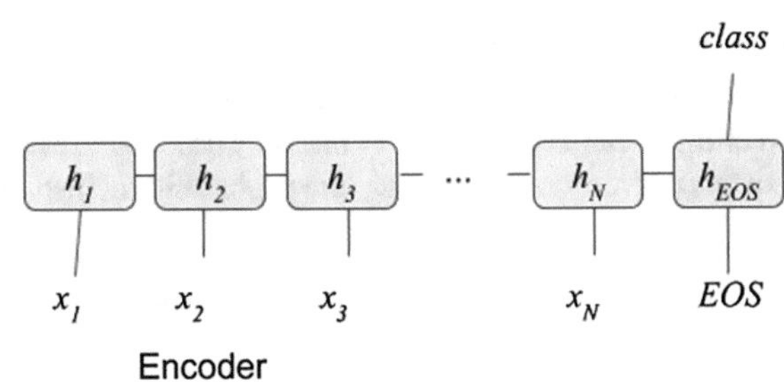

Fig. 2.5 An RNN-only-based encoder for sentence classification

simply used an RNN encoder to model the utterance where the end of sentence token decodes the class as shown in Fig. 2.5. While they have not compared their results with CNN or simple DNN, this work is significant because one can simply extend this architecture to a bidirectional RNN, and also load the begin-of-sentence token as the class, and as presented in (Hakkani-Tür et al. 2016), support not only utterance intent but also slot filling towards a joint semantic parsing model, which is covered in the next section.

Aside from these representative modeling studies, one approach worth mentioning is the unsupervised utterance classification work by Dauphin et al. (2014). This approach relies on search queries associated with their clicked URLs. The assumption is that the queries will have a similar meaning or intent if they result in clicks to similar URLs. Figure 2.6 shows an example query-click graph. This data is used to train a simple deep network with multiple hidden layers, last of which is supposed to capture the latent intent of a given query. Note that this is different from other word embedding training methods and can directly provide an embedding to a given query.

The zero-shot classifier then simply finds the category whose embedding is the semantically closest to the query, assuming that the class names (e.g., restaurants or sports) are given in a meaningful way. Then, the probability of belonging to a class is a simply softmax over all classes, based on the Euclidean distance of the embeddings of the query and the class name.

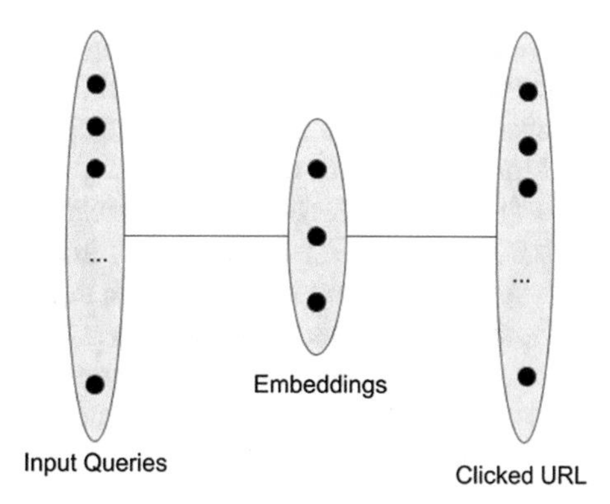

Fig. 2.6 A bi-partite query-click graph from queries to clicked URLs

2.4.2 *Slot Filling*

The state of the art in slot filling relies on RNN-based approaches and its variations. Pre-RNN approaches include neural network Markov models (NN-MMs) or DNN with conditional random fields (CRFs). In one of the pre-RNN era works, among several approaches, Deoras and Sarikaya (2013) have investigated deep belief networks for slot filling. They propose discriminative embedding technique which projects the sparse and large input layer onto a small, dense, and real-valued feature vector, which is then subsequently used for pretraining the network and then to do discriminative classification using local classification. They apply it to the well-studied spoken language understanding task of ATIS and obtained new state-of-the-art performances, outperforming the best CRF-based system.

CNNs are used for feature extraction and have been shown to perform well for learning sentence semantics (Kim 2014). CNNs have also been used for learning hidden features for slot tagging as well. Xu and Sarikaya (2013) have investigated using CNN as a lower layer that extracts features for each word in relation to its neighboring words, capturing the utterance local semantics. A CRF layer sits on top of the CNN layer which produces hidden features for the CRF. The entire network is trained end-to-end with backpropagation and applied on personal assistant domains. Their results showed significant improvements over the standard CRF models while providing a flexibility in feature engineering for the domain expert.

With the advances in recurrent neural network (RNN)-based models, they have first been used for slot filling by Yao et al. (2013) and Mesnil et al. (2013) simultaneously. For example, Mesnil et al. implemented and compared several important architectures of the RNN, including the Elman-type (Elman 1990) and Jordan-type (Jordan 1997) recurrent networks and their variants. Experimental results show that both Elman- and Jordan-type networks, while giving similar performance, outperform the widely used CRF baseline substantially. Moreover, the results also show that the bidirectional RNN that take into account both past and future dependencies among slots gave the best performance. The effectiveness of word embeddings for initializing the RNNs for slot filling is studied in both papers, too. The work is further extended in (Mesnil et al. 2015), where the authors performed a comprehensive evaluation of the standard RNN architectures, and hybrid, bidirectional, and CRF extensions, and set a new state of the art in this area.

More formally, to estimate the sequence of tags $Y = y_1, \ldots, y_n$ in the form of IOB labels as in (Raymond and Riccardi 2007) (with three outputs corresponding to "B", "I", and "O"), and as shown in Fig. 2.1 corresponding to an input sequence of tokens $X = x_1, \ldots, x_n$, the Elman RNN architecture (Elman 1990) consists of an input layer, a number of hidden layers, and an output layer. The input, hidden, and output layers consist of a set of neurons representing the input, hidden, and output at each time step t, x_t, h_t, and y_t, respectively. The input is typically represented by one-hot vectors or word-level embeddings. Given the input layer x_t at time t, and hidden state from the previous time step h_{t-1}, the hidden and output layers for the current time step are computed as follows:

$$h_t = \phi(W_{xh}\left[\begin{smallmatrix} h_{t-1} \\ x_t \end{smallmatrix}\right])$$

$$p_t = \texttt{softmax}(W_{hy}h_t)$$

$$\hat{y}_t = \texttt{argmax}\ p_t,$$

where W_{xh} and W_{hy} are the matrices that denote the weights between the input and hidden layers and hidden and output layers, respectively. ϕ denotes the activation function, i.e., tanh or $sigm$.

In contrast, the Jordan RNN computes the recurrent hidden layer for the current time step from the output layer of the previous time step plus input layer at the current time step, i.e.,

$$h_t = \phi(W_{xp}\left[\begin{smallmatrix} p_{t-1} \\ x_t \end{smallmatrix}\right]).$$

The architectures of the feed-forward NN, the Elman RNN, and the Jordan RNN are illustrated in Fig. 2.7.

An alternative approach would be augmenting these with explicit sequence-level optimization. This is important as, for example, the model can model an I tag cannot follow an O tag. Liu and Lane (2015) propose such an architecture where the hidden state also uses the previous prediction as shown in Fig. 2.8:

$$h_t = f(Ux_t + Wh_{t-1} + Qy\_out_{t-1}),$$

where y_out_{t-1} is the vector representing output label at time $t-1$, and Q is the weight matrix connecting output label vector and the hidden layer.

A recent paper by Dupont et al. (2017) should also be mentioned here for proposing a new variant RNN architecture where the output label is also concatenated into the next input.

Especially with the re-discovery of LSTM cells (Hochreiter and Schmidhuber 1997) for RNNs, this architecture has started to emerge (Yao et al. 2014). LSTM cells are shown to have superior properties, such as faster convergence and elimination of

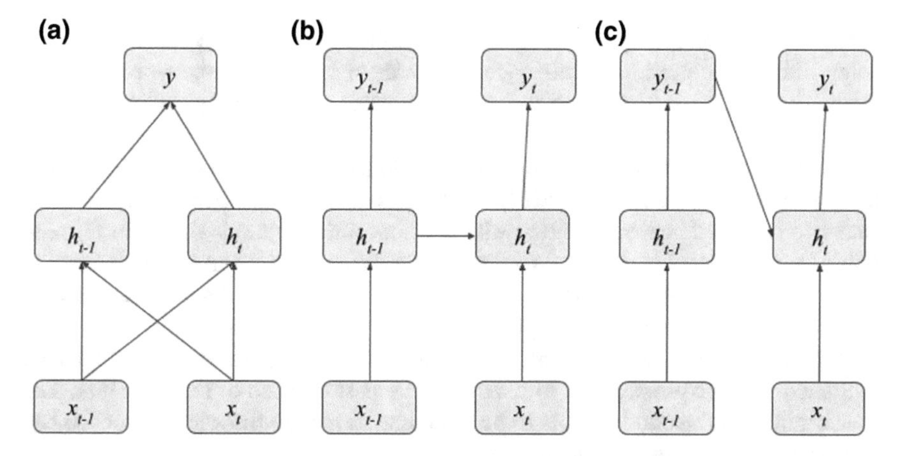

Fig. 2.7 **a** Feed-forward NN; **b** Elman RNN; **c** Jordan RNN

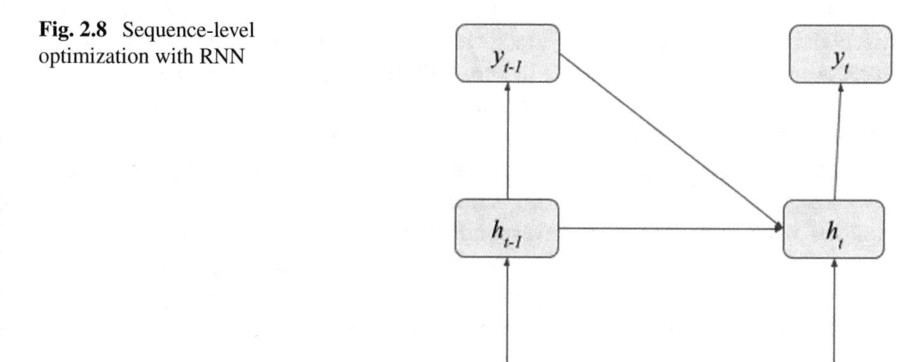

Fig. 2.8 Sequence-level optimization with RNN

the problem of vanishing or exploding gradients in sequences via self-regularization. As a result, LSTM is shown to be more robust than RNN in capturing long-span dependencies.

We have compiled a comprehensive review of RNN-based slot filling approaches in (Mesnil et al. 2015). While pre-LSTM/GRU RNN studies focused on look-ahead and look-back features (e.g., Mesnil et al. 2013; Vu et al. 2016), nowadays, state-of-the-art slot filling methods usually rely on bidirectional LSTM/GRU models (Hakkani-Tür et al. 2016; Mesnil et al. 2015; Kurata et al. 2016a; Vu et al. 2016; Vukotic et al. 2016) among others.

Extensions include encoder–decoder models (Liu and Lane 2016; Zhu and Yu 2016a among others) or memory (Chen et al. 2016) as we will describe below. In this respect, common sentence encoders include sequence-based recurrent neural networks with LSTMs or GRU units, which accumulate information over the sentence sequentially; convolutional neural networks, which accumulate information using filters over short local sequences of words or characters; and tree-structured recursive neural networks (RecNNs), which propagate information up a binary parse tree (Socher et al. 2011; Bowman et al. 2016).

Related to recursive neural networks (RecNNs), two papers are worth mentioning here. The first is by Guo et al. (2014) where the syntactic parse structure of an input sentence is tagged instead of the words. The conceptual figure is shown in Fig. 2.9. Every word is associated with a word vector, and these vectors are given as input to the bottom of the network. Then, the network propagates the information upward by repeatedly applying a neural network at each node until the root node outputs a single vector. This vector is then used as the input to a semantic classifier, and the network is trained via backpropagation to maximize the performance of this classifier. The nonterminals correspond to slots to be filled and at the top the whole sentence can be classified for intent or domain.

While this architecture is very elegant and expensive, it did not result in superior performance due to various reasons: (i) the underlying parse trees can be noisy, and the model cannot jointly train a syntactic and semantic parser, (ii) the phrases do not necessarily correspond to slots one to one, and (iii) the high-level tag sequence is not considered, hence needs a final Viterbi layer. Hence, an ideal architecture would be a hybrid RNN/RecNN model.

A more promising approach is presented by Andreas et al. (2016) for question answering. As shown in Fig. 2.10, a semantic parse is built bottom up using the composition of neural modules corresponding to six key logical functions in the task, namely, *lookup*, *find*, *relate*, *and*, *exists*, and *describe*. An advantage over RecNNs is that the model jointly learns the structure or layout of the parse during training using these primitives, starting from an existing syntactic parser.

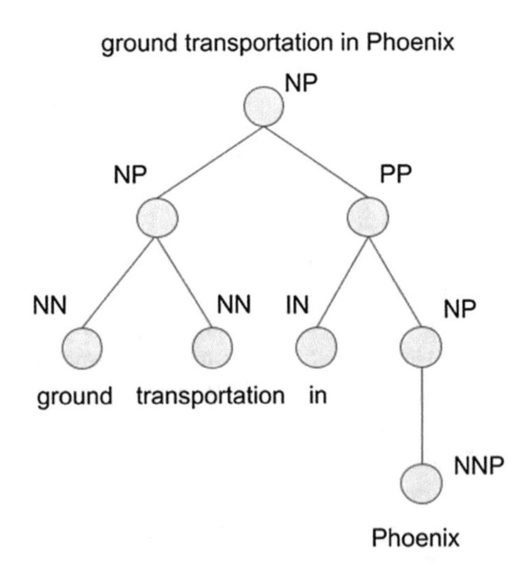

Fig. 2.9 Recursive neural networks building on top of a given syntactic parse tree

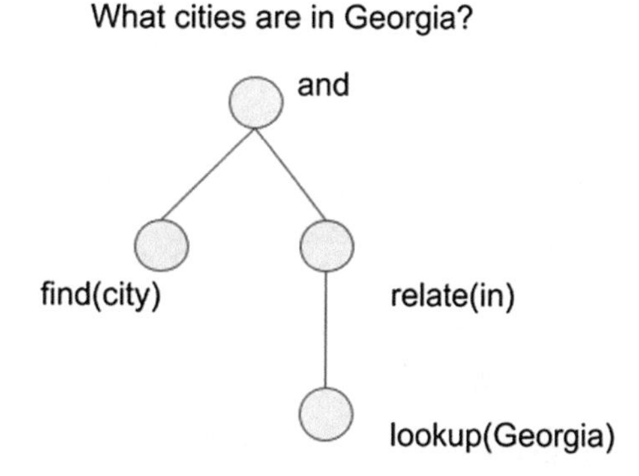

Fig. 2.10 Composition of neural modules for semantic parsing

Vu et al. (2016) have proposed to use a ranking loss function, instead of the conventional cross entropy loss. One benefit of this is that it does not force the model to learn a pattern for the artificial class O (which might not exist). It learns to maximize the distance between the true label y and the best competitive label c given a data point x. The objective function is

$$L = \log(1 + \exp(\gamma(m_{cor} s_\theta(x)y))) + \log(1 + \exp(\gamma(m_{inc} + s_\theta(x)c))),$$

where $s_\theta(x)y$ and $s_?(x)c$ being the scores for the classes y and c, respectively. The parameter ? controls the penalization of the prediction errors, and m_{cor} and m_{inc} are margins for the correct and incorrect classes. ?, m_{cor}, and m_{inc} are hyperparameters which can be tuned on the development set. For the class O, only the second summand of equation is calculated. By doing this, the model does not learn a pattern for class O but nevertheless increase its difference to the best competitive label. During testing, the model will predict class O if the score for all the other classes is lower than 0.

Besides tagger LSTM models, there are few studies focusing on encoder/decoder RNN architectures after advances in similar studies (Sutskever et al. 2014; Vinyals and Le 2015). Kurata et al. (2016b) proposed using an architecture like in Fig. 2.11, where the input sentence is encoded into a fixed length vector by the encoder LSTM. Then, the slot label sequence is predicted by the labeler LSTM whose hidden state is initialized with the encoded vector by the encoder LSTM. With this encoder-labeler LSTM, the label sequence can be predicted using the whole sentence embedding explicitly.

Note that in this model, since the output is the usual tag sequence, the words are also fed into the tagger (reversed as usually done by other encoder/decoder studies) in addition to the previous prediction.

Another benefit of such an approach comes with the attention mechanism (Simonnet et al. 2015), where the decoder can attend longer distance dependencies while tagging. The attention is another vector, c, which is a weighted sum of all the hidden state embeddings on the encoder side. There are multiple ways to determine these weights:

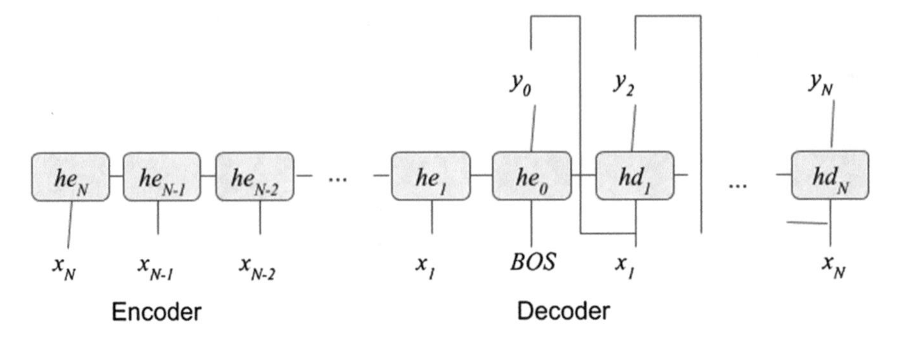

Fig. 2.11 Encoder/decoder RNN using both words and labels

$$c_t = \sum_{i=1}^{T} \alpha_{t_i} h_i.$$

Consider the example sentence *flights departing from london no later than next Saturday afternoon*, the tag for the word *afternoon* is *departure_time* and only evident by the head verb which is eight words away. In such cases, an attention mechanism can be useful.

Zhu and Yu (2016b) have further extended this encoder/decoder architecture using "focus" (or direct attention) mechanism, which is emphasizing the aligned encoders hidden states. In other words, attention is no longer learned but simply assigned to the corresponding hidden state:

$$c_t = h_t.$$

Zhai et al. (2017) have later extended the encoder/decoder architecture using pointer networks (Vinyals et al. 2015) on the chunked outputs of the input sentence. The main motivation is that RNN models still need to treat each token independently using the IOB scheme, instead of a complete unit. If we can eliminate this drawback, it could result in more accurate labeling, especially for multiword chunks. Sequence chunking is a natural solution to overcome this problem. In sequence chunking, the original sequence labeling task is divided into two subtasks: (1) Segmentation, to identify scope of the chunks explicitly; and (2) Labeling, to label each chunk as a single unit based on the segmentation results. Hence, the authors have proposed a joint model which chunks the input sentence during the encoding phase and the decoder simply tags those chunks as shown in Fig. 2.12.

Regarding unsupervised training of slot filling models, one paper worth mentioning is by Bapna et al. (2017) proposing an approach that can utilize only the slot description in context without the need for any labeled or unlabeled in-domain examples, to quickly bootstrap a new domain. The main idea of this work is to leverage the encoding of the slot names and descriptions within a multitask deep learned slot filling model, to implicitly align slots across domains, assuming an already trained background model.

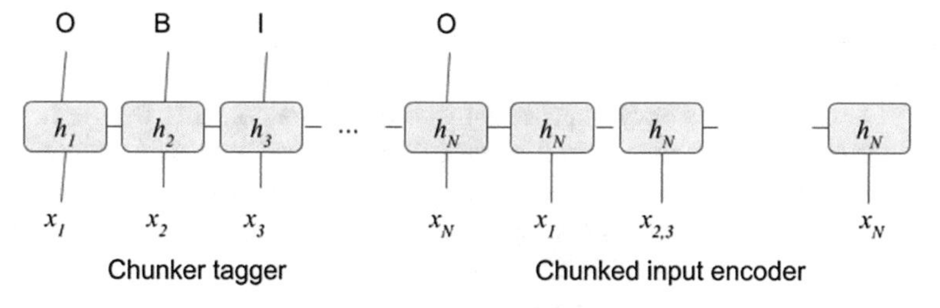

Fig. 2.12 Pointer encoder/decoder RNN using chunked input

If one of the already covered domains contains a similar slot, a continuous representation of the slot obtained from shared pretrained embeddings can be leveraged in a domain agnostic model. An obvious example would be adding United Airlines when the multitask model can already parse queries for American Airlines and Turkish Airlines. While the slot names may be different, the concept of departure city or arrival city should persist and can be transferred to the new task of United Airlines using their natural language descriptions. Such an approach is promising for solving the domain scaling problem and eliminating the need for any manually annotated data or explicit schema alignment.

2.4.3 Joint Multitask Multi-domain Modeling

Historically, intent determination has been seen as an example classification problem and slot filling as sequence classification problem, and in the pre-deep-learning era the solutions for these two tasks are typically not the same, they have been modeled separately. For example, SVMs are used for intent determination and CRFs are used for slot filling. With the advances in deep learning, it is now possible to get the whole semantic parse using a single model in a multitask fashion. This allows the slot decisions to help intent determination and vice versa.

Furthermore, domain classification is often completed first, serving as a top-level triage for subsequent processing. Intent determination and slot filling are then run for each domain to fill a domain-specific semantic template. This modular design approach (i.e., modeling semantic parsing as three separate tasks) has the advantage of flexibility; specific modifications (e.g., insertions, deletions) to a domain can be implemented without requiring changes to other domains. Another advantage is that, in this approach, one can use task-/domain-specific features, which often significantly improve the accuracy of these task-/domain-specific models. Also, this approach often yields more focused understanding in each domain since the intent determination only needs to consider a relatively small set of intent and slot classes over a single (or limited set) of domains, and model parameters could be optimized for the specific set of intent and slots.

However, this approach also has disadvantages: First of all, one needs to train these models for each domain. This is an error-prone process, requiring careful engineering to ensure consistency in processing across domains. Also, during run-time, such pipelining of tasks results in transfer of errors from one task to the following task. Furthermore, there is no data or feature sharing between the individual domain models, resulting in data fragmentation, whereas some semantic intents (such as finding or buying a domain-specific entity) and slots (such as dates, times, and locations) could actually be common to many domains (Kim et al. 2015; Chen et al. 2015a). Finally, the users may not know which domains are covered by the system and to what extent, so this issue results in interactions where the users do not know what to expect and hence resulting in user dissatisfaction (Chen et al. 2013, 2015b).

To this end, Hakkani-Tür et al. (2016) proposed a single RNN architecture that integrates the three tasks of domain detection, intent detection, and slot filling for multiple domains in a single RNN model. This model is trained using all available utterances from all domains, paired with their semantic frames. The input of this RNN is the input sequence of words (e.g., user queries) and the output is the full semantic frame, including domain, intent, and slots, as shown in Fig. 2.13. This is similar to the multitask parsing and entity extraction work by Tafforeau et al. (2016).

For joint modeling of domain, intent, and slots, an additional token is inserted at the beginning and end of each input utterance k, *<BOS>* and *<EOS>*, and associate a combination of domain and intent tags d_k and i_k to this sentence initial and final tokens by concatenating these tags. Hence, the new input and output sequence are

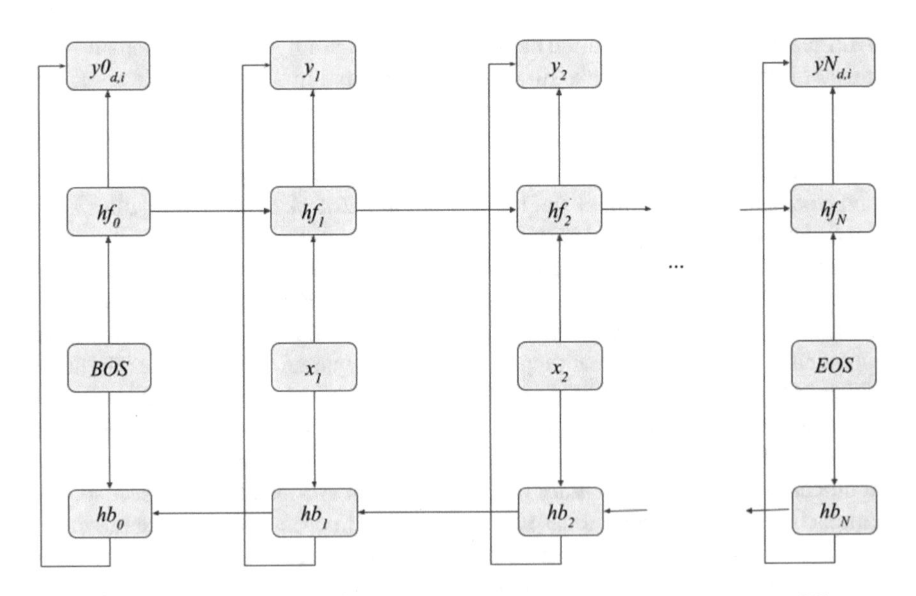

Fig. 2.13 Bidirectional RNN for joint domain detection, intent determination, and slot filling

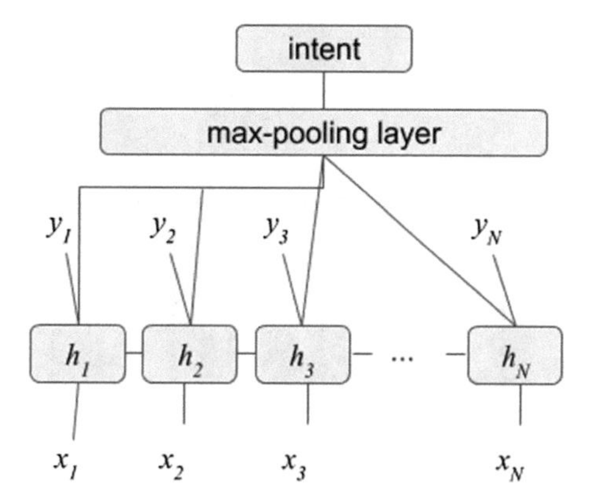

Fig. 2.14 Joint slot filling and intent determination model with max-pooling layer

$$X = \text{<}BOS\text{>}, x_1, \ldots, x_n, \text{<}EOS\text{>}$$

$$Y = d_k\_i_k, s_1, \ldots, s_n, d_k\_i_k,$$

where X is the input and Y is the output (Fig. 2.13).

The main rationale of this idea is similar to the sequence-to-sequence modeling approach, as used in machine translation (Sutskever et al. 2014) or chit-chat (Vinyals and Le 2015) systems approaches: The last hidden layer of the query (in each direction) is supposed to contain a latent semantic representation of the whole input utterance, so that it can be utilized for domain and intent prediction (d_k, i_k).

Zhang and Wang (2016) extended this architecture so as to add a max-pooling layer is employed to capture global features of a sentence for intent classification (Fig. 2.14). A united loss function, which is a weighted sum of cross entropy for slot filling and intent determination, is used while training.

Liu and Lane (2016) proposed a joint slot filling and intent determination model based on an encoder/decoder architecture as shown in Fig. 2.15. It is basically a multi-headed model sharing the sentence encoder with task-specific attention, c_i.

Note that such a joint modeling approach can be extremely useful for scaling to new domains, starting from the larger background model trained from multiple domains, analogous to language model adaptation (Bellegarda 2004). Jaech et al. (2016) presented such a study where the multitask approach is exploited for scalable CLU model training via transfer learning. The key to scalability is reducing the amount of training data needed to learn a model for a new task. The proposed multitask model delivers better performance with less data by leveraging patterns that it learns from the other tasks. The approach supports an open vocabulary, which allows the models to generalize to unseen words, which is particularly important when very little training data is used.

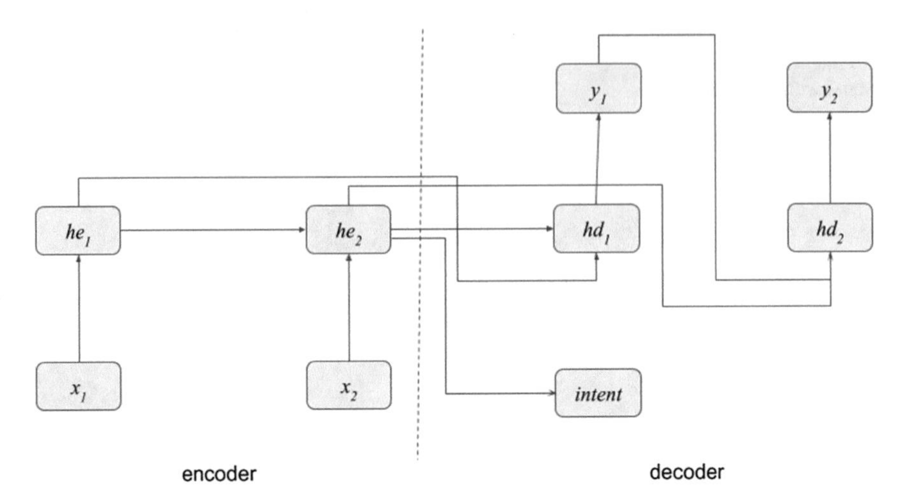

Fig. 2.15 Joint slot filling and intent determination model using encoder/decoder model

2.4.4 Understanding in Context

Natural language understanding involves understanding the context in which the language is used. But understanding context involves multiple challenges. First, in many languages certain words can be used in multiple senses. That makes it important to eliminate the ambiguity of all such words so that their usage in a particular document can be accurately detected. Word-sense disambiguation is an ongoing research area in natural language processing and specifically important when building natural language understanding systems. Second, understanding task involves documents from different domains such as travel reservation, understanding legal documents, news articles, arxiv articles, and the like. Each of these domains carries a certain property, hence domain-specific context, that the natural language understanding models should learn to capture. Third, in spoken and written text, many words are used as proxies for other concepts. For instance, most commonly, "Xerox" is used for "copy" or "fedex" for "overnight courier", and so on. Finally, documents contain words or phrases which refer to knowledge which is not explicitly included in the text. Only with intelligent methods, we can learn to use "prior" knowledge to be able to understand such information that exist in text.

Recently, deep learning architectures have been applied to various natural language processing tasks and have shown the advantages to capture the relevant semantic and syntactic aspects of units in context. As word distributions are composed to form the meanings of phrases or multiword expressions, the goal is to extend distributed phrase-level representations to single- and multi-sentence (discourse) levels, and produce hierarchical structure of entire texts.

With the goal of learning *context* in natural language text, Hori et al. (2014) proposed an efficient context-sensitive spoken language understanding approach using role-based LSTM layers. Specifically, to understand speaker intentions accurately in a dialog, it is important to consider the sentence in the context of the surrounding sequence of dialog turns. In their work, LSTM recurrent neural networks are used to train a context-sensitive model to predict sequences of dialog concepts from the spoken word sequences. Thus, to capture long-term characteristics over an entire dialog, they implemented LSTMs representing intention using consequent word sequences of each concept tag. To train such a model, they build LSTMs from a human-to-human dialog corpus annotated with concept tags which represent client and agent intentions for hotel reservation. The expressions are characterized by each role of agent and client.

As shown in Fig. 2.16, there are two LSTM layers that have different parameters depending on the speaker roles. The input vector is thus processed differently by the left layer for the clients' utterances, and by the right layer for the agents' utterances representing these different roles. The recurrent LSTM inputs thus receive the output from the role-dependent layer active at the previous frame, allowing for transitions between roles. This approach can learn the model *context* from an intelligent language understanding system by characterizing expressions of utterances varied among each different role.

In (Chen et al. 2016), one of the first end-to-end neural network based conversational understanding models is proposed that uses memory networks to extract the prior information as context knowledge for the encoder in understanding natural language utterances of conversational dialogs. As shown in Fig. 2.17, their approach

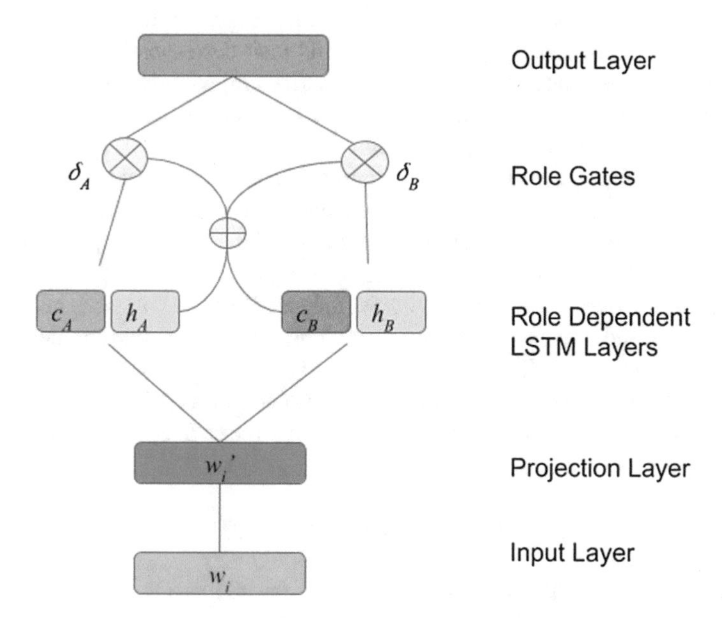

Fig. 2.16 LSTM with role-dependent layers. Layer (A) corresponds to client utterance states and Layer (B) corresponds to agent utterance states. Role gates control which role is active

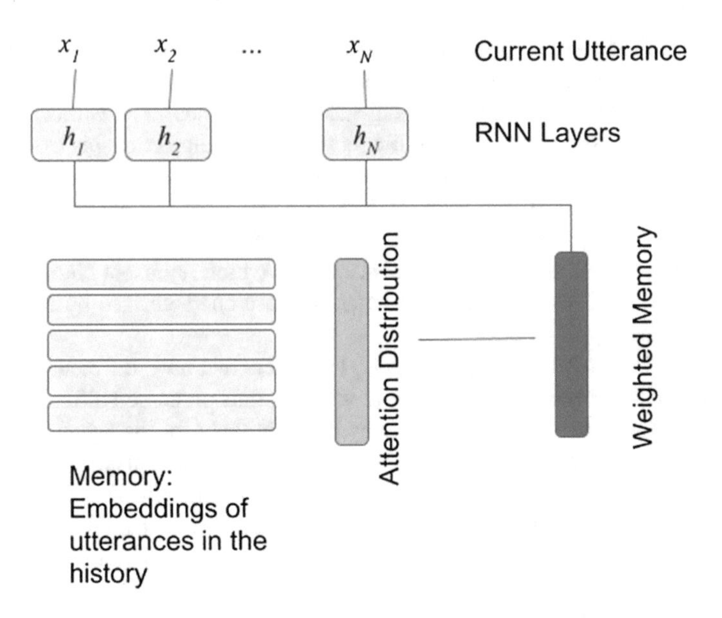

Fig. 2.17 The illustration of the proposed end-to-end memory network model for multi-turn spoken language understanding

is combined with an RNN-based encoder which learns to encode prior information from a possibly large external memory before parsing the utterance from the dialog. Provided that there are input utterances and their corresponding semantic tags, their model can be trained end-to-end directly from input–output pairs. employing an end-to-end neural network model to model long-term knowledge carryover for multi-turn spoken language understanding.

citeankur:arxiv17 have extended this approach using hierarchical dialog encoders, an extension of hierarchical recurrent encoder–decoders (HRED) proposed by Sordoni et al. (2015), where the query level encodings are combined with a representation of the current utterance, before feeding it into the session level encoder. In the proposed architecture, instead of a simple cosine-based memory network, the encoder employed a feed-forward network whose input is the current and previous utterances in context which is then feeding into an RNN as shown in Fig. 2.18. More formally, the current utterance encoding c is combined with each memory vector m_k, for $1, \ldots, n_k$, by concatenating and passing them through a feed-forward (FF) layer to produce context encodings, denoted by $g_1, g_2, \ldots g_{t-1}$

$$g_k = sigmoid(FF(m_k, c))$$

for $k = 0, \ldots, t-1$. These context encodings are fed as token-level inputs into the bidirectional GRU RNN session encoder. The final state of the session encoder represents the dialog context encoding h_t.

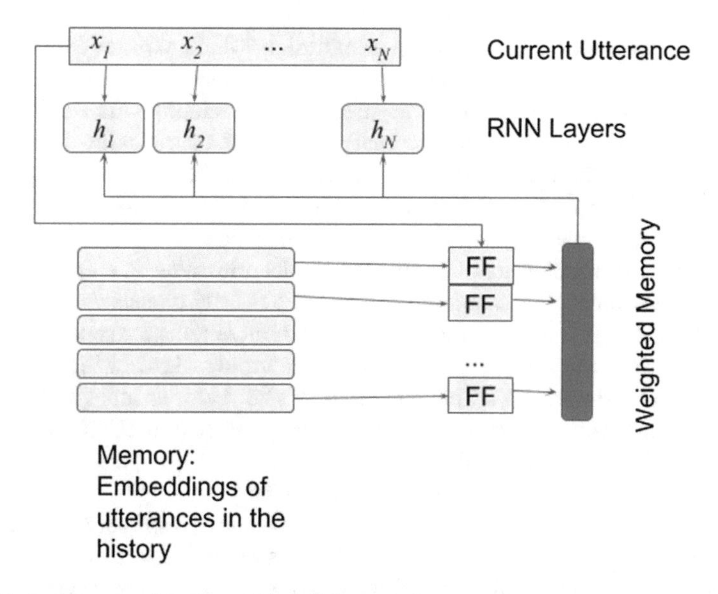

Fig. 2.18 Architecture of the hierarchical dialog encoder network

As mentioned earlier in the chapter, CNNs have mainly used in natural language understanding tasks to learn latent features that are otherwise impossible to learn. Celikyilmaz et al. (2016) introduced a pretraining method for deep neural network models, using CNNs in particular to jointly learn the *context* as the network structure from large unlabeled data, while learning to predict task-specific contextual information from labeled sequences. Extending the supervised CNN with CRF architecture of (Xu and Sarikaya 2013), they use CNN as the bottom layer to learn the feature representations from labeled and unlabeled sequences by the way of semi-supervised learning. At the top layer, they use two CRF structures to decode the output sequences as semantic slot tags as well as latent class labels per each word as output sequences. This allows the network to simultaneously learn the transition and emission weights for slot tagging and class labeling of the words in utterances in a single model.

2.5 Summary

Advances in deep learning based approaches lead the CLU field in two dimensions. The first dimension is end-to-end learning. Conversational language understanding is one of the many subsystems in a complete conversation system. For example, it usually takes the speech recognition results as the input and its output will be fed into the dialog manager for state tracking and response generation. Therefore, an end-to-end optimal design of the whole conversational system usually leads to better user experience. He and Deng (2013) discussed an optimization-oriented statistical framework for the overall system design, which exploits the uncertainty in each subsystem output and the interactions between the subsystems. In the framework, parameters of all subsystems are treated as correlating with each other and are trained end-to-end to optimize the final performance metric of the whole conversation system. Furthermore, more recently, reinforcement learning based methods combined with user simulators also started to invade the CLU task, providing seamless end-to-end natural language dialog (see next chapter).

The second dimension in CLU enabled by deep learning is efficient encoders implemented without RNN unrolling. RNNs are powerful models that are uniquely capable of dealing with sequential data, like natural language, speech, video, etc. With RNNs, we can now understand sequential data and make decisions. Traditional neural networks are stateless. They take a fixed size vector as input and produce a vector as output. Having this unique property of being stateful RNNs has been the most used tools in language understanding systems today.

Networks without hidden layers are very limited in the input–output mappings that they can model. Adding a layer of hand-coded features (as in perceptrons) makes them much more powerful, but the hard bit is designing these features. We would like to find good features without requiring insights into the task or repeated trial and error of different features. We need to automate the trial-and-error feature designing loop. Reinforcement learning can learn such structures by perturbing weights. How reinforcement learning help in deep learning is actually not that complex. They randomly perturb one weight and see if it improves performance—if so, save the change. This could be inefficient and hence the machine learning community, especially in deep reinforcement learning has been focusing on this in recent years.

Since meaning in natural language sentences is known to be constructed recursively according to a tree structure, more efficient encoders study tree-structured neural network encoders, specifically TreeLSTMs (Socher et al. 2011; Bowman et al. 2016). The idea is to be able to encode faster and efficiently while maintaining the efficiency. On the other hand, models that can learn a network structure predictor jointly with module parameters themselves have shown to improve natural language understanding while reducing the issues that come with longer text sequences, thus the backpropagation in RNNs. Andreas et al. (2016) present such a model that uses natural language strings to automatically assemble neural networks from a collection of composable modules. Parameters for these modules are learned jointly with network-assembly parameters via reinforcement learning, with only (world, question, answer) triples as supervision.

To conclude, we believe advances in deep learning have led to exciting new research frontiers for human/machine conversational systems, especially for CLU. The studies mentioned here would be considered as scratching the surface over the next decade, tackling toy tasks with manually annotated data. The future research includes transfer learning, unsupervised learning, and reinforcement learning more than ever for any high-quality scalable CLU solution.

References

Allen, J. (1995). *Natural language understanding*, chapter 8. Benjamin/Cummings.

Allen, J. F., Miller, B. W., Ringger, E. K., & Sikorski, T. (1996). A robust system for natural spoken dialogue. In *Proceedings of the Annual Meeting of the Association for Computational Linguistics*, pp. 62–70.

Andreas, J., Rohrbach, M., Darrell, T., & Klein, D. (2016). Learning to compose neural networks for question answering. In *Proceedings of NAACL*.

Bapna, A., Tur, G., Hakkani-Tur, D., & Heck, L. (2017). Towards zero-shot frame semantic parsing for domain scaling. In *Proceedings of the Interspeech*.

Bellegarda, J. R. (2004). Statistical language model adaptation: Review and perspectives. *Speech Communication Special Issue on Adaptation Methods for Speech Recognition*, *42*, 93–108.

Bonneau-Maynard, H., Rosset, S., Ayache, C., Kuhn, A., & Mostefa, D. (2005). Semantic annotation of the French MEDIA dialog corpus. In *Proceedings of the Interspeech*, Lisbon, Portugal.

Bowman, S. R., Gauthier, J., Rastogi, A., Gupta, R., & Manning, C. D. (2016). A fast unified model for parsing and sentence understanding. In *Proceedings of ACL*.

Celikyilmaz, A., Sarikaya, R., Hakkani, D., Liu, X., Ramesh, N., & Tur, G. (2016). A new pre-training method for training deep learning models with application to spoken language understanding. In *Proceedings of The 17th Annual Meeting of the International Speech Communication Association (INTERSPEECH 2016)*.

Chen, Y.-N., Hakkani-Tur, D., & He, X. (2015a). Zero-shot learning of intent embeddings for expansion by convolutional deep structured semantic models. In *Proceedings of the IEEE ICASSP*.

Chen, Y.-N., Hakkani-Tür, D., Tur, G., Gao, J., & Deng, L. (2016). End-to-end memory networks with knowledge carryover for multi-turn spoken language understanding. In *Proceedings of the Interspeech*, San Francisco, CA.

Chen, Y.-N., Wang, W. Y., Gershman, A., & Rudnicky, A. I. (2015b). Matrix factorization with knowledge graph propagation for unsupervised spoken language understanding. In *Proceedings of the ACLIJCNLP*.

Chen, Y.-N., Wang, W. Y., & Rudnicky, A. I. (2013). Unsupervised induction and filling of semantic slots for spoken dialogue systems using frame-semantic parsing. In *Proceedings of the IEEE ASRU*.

Chomsky, N. (1965). *Aspects of the theory of syntax*. Cambridge, MA: MIT Press.

Chu-Carroll, J., & Carpenter, B. (1999). Vector-based natural language call routing. *Computational Linguistics*, *25*(3), 361–388.

Collobert, R., & Weston, J. (2008). A unified architecture for natural language processing: Deep neural networks with multitask learning. In *Proceedings of the ICML*, Helsinki, Finland.

Dahl, D. A., Bates, M., Brown, M., Fisher, W., Hunicke-Smith, K., Pallett, D., et al. (1994). Expanding the scope of the ATIS task: the ATIS-3 corpus. In *Proceedings of the Human Language Technology Workshop*. Morgan Kaufmann.

Damnati, G., Bechet, F., & de Mori, R. (2007). Spoken language understanding strategies on the france telecom 3000 voice agency corpus. In *Proceedings of the ICASSP*, Honolulu, HI.

Dauphin, Y., Tur, G., Hakkani-Tür, D., & Heck, L. (2014). Zero-shot learning and clustering for semantic utterance classification. In *Proceedings of the ICLR*.

Deng, L., & Li, X. (2013). Machine learning paradigms for speech recognition: An overview. *IEEE Transactions on Audio, Speech, and Language Processing*, *21*(5), 1060–1089.

Deng, L., & O'Shaughnessy, D. (2003). *Speech processing: A dynamic and optimization-oriented approach*. Marcel Dekker, New York: Publisher.

Deng, L., & Yu, D. (2011). Deep convex nets: A scalable architecture for speech pattern classification. In *Proceedings of the Interspeech*, Florence, Italy.

Deoras, A., & Sarikaya, R. (2013). Deep belief network based semantic taggers for spoken language understanding. In *Proceedings of the IEEE Interspeech*, Lyon, France.

Dupont, Y., Dinarelli, M., & Tellier, I. (2017). Label-dependencies aware recurrent neural networks. arXiv preprint arXiv:1706.01740.

Elman, J. L. (1990). Finding structure in time. *Cognitive science, 14*(2), 179–211.
Freund, Y., & Schapire, R. E. (1997). A decision-theoretic generalization of on-line learning and an application to boosting. *Journal of Computer and System Sciences, 55*(1), 119–139.
Gorin, A. L., Abella, A., Alonso, T., Riccardi, G., & Wright, J. H. (2002). Automated natural spoken dialog. *IEEE Computer Magazine, 35*(4), 51–56.
Gorin, A. L., Riccardi, G., & Wright, J. H. (1997). How may I help you? *Speech Communication, 23*, 113–127.
Guo, D., Tur, G., Yih, W.-t., & Zweig, G. (2014). Joint semantic utterance classification and slot filling with recursive neural networks. In *In Proceedings of the IEEE SLT Workshop*.
Gupta, N., Tur, G., Hakkani-Tür, D., Bangalore, S., Riccardi, G., & Rahim, M. (2006). The AT&T spoken language understanding system. *IEEE Transactions on Audio, Speech, and Language Processing, 14*(1), 213–222.
Hahn, S., Dinarelli, M., Raymond, C., Lefevre, F., Lehnen, P., Mori, R. D., et al. (2011). Comparing stochastic approaches to spoken language understanding in multiple languages. *IEEE Transactions on Audio, Speech, and Language Processing, 19*(6), 1569–1583.
Hakkani-Tür, D., Tur, G., Celikyilmaz, A., Chen, Y.-N., Gao, J., Deng, L., & Wang, Y.-Y. (2016). Multi-domain joint semantic frame parsing using bi-directional RNN-LSTM. In *Proceedings of the Interspeech*, San Francisco, CA.
He, X., & Deng, L. (2011). Speech recognition, machine translation, and speech translation a unified discriminative learning paradigm. In *IEEE Signal Processing Magazine, 28*(5), 126–133.
He, X. & Deng, L. (2013). Speech-centric information processing: An optimization-oriented approach. In *Proceedings of the IEEE, 101*(5), 1116–1135.
Hemphill, C. T., Godfrey, J. J., & Doddington, G. R. (1990). The ATIS spoken language systems pilot corpus. In *Proceedings of the Workshop on Speech and Natural Language, HLT'90*, pp. 96–101, Morristown, NJ, USA. Association for Computational Linguistics.
Hinton, G., Deng, L., Yu, D., Dahl, G., Rahman Mohamed, A., Jaitly, N., et al. (2012). Deep neural networks for acoustic modeling in speech recognition. *IEEE Signal Processing Magazine, 29*(6), 82–97.
Hinton, G. E., Osindero, S., & Teh, Y. W. (2006). A fast learning algorithm for deep belief nets. *Advances in Neural Computation, 18*(7), 1527–1554.
Hochreiter, S., & Schmidhuber, J. (1997). Long short-term memory. *Neural computation, 9*(8), 1735–1780.
Hori, C., Hori, T., Watanabe, S., & Hershey, J. R. (2014). Context sensitive spoken language understanding using role dependent lstm layers. In *Proceedings of the Machine Learning for SLU Interaction NIPS 2015 Workshop*.
Huang, X., & Deng, L. (2010). An overview of modern speech recognition. In *Handbook of Natural Language Processing, Second Edition, Chapter 15*.
Jaech, A., Heck, L., & Ostendorf, M. (2016). Domain adaptation of recurrent neural networks for natural language understanding. In *Proceedings of the Interspeech*, San Francisco, CA.
Jordan, M. (1997). Serial order: A parallel distributed processing approach. Technical Report 8604, University of California San Diego, Institute of Computer Science.
Kalchbrenner, N., Grefenstette, E., & Blunsom, P. (2014). A convolutional neural network for modelling sentences. In *Proceedings of the ACL*, Baltimore, MD.
Kim, Y. (2014). Convolutional neural networks for sentence classification. In *Proceedings of the EMNLP*, Doha, Qatar.
Kim, Y.-B., Stratos, K., Sarikaya, R., & Jeong, M. (2015). New transfer learning techniques for disparate label sets. In *Proceedings of the ACL-IJCNLP*.
Kuhn, R., & Mori, R. D. (1995). The application of semantic classification trees to natural language understanding. *IEEE Transactions on Pattern Analysis and Machine Intelligence, 17*, 449–460.
Kurata, G., Xiang, B., Zhou, B., & Yu, M. (2016a). Leveraging sentence-level information with encoder LSTM for semantic slot filling. In *Proceedings of the EMNLP*, Austin, TX.
Kurata, G., Xiang, B., Zhou, B., & Yu, M. (2016b). Leveraging sentence-level information with encoder lstm for semantic slot filling. arXiv preprint arXiv:1601.01530.

Lee, J. Y., & Dernoncourt, F. (2016). Sequential short-text classification with recurrent and convolutional neural networks. In *Proceedings of the NAACL.*

Li, J., Deng, L., Gong, Y., & Haeb-Umbach, R. (2014). An overview of noise-robust automatic speech recognition. *IEEE/ACM Transactions on Audio, Speech, and Language Processing*, *22*(4), 745–777.

Liu, B., & Lane, I. (2015). *Recurrent neural network structured output prediction for spoken language understanding*. In Proc: NIPS Workshop on Machine Learning for Spoken Language Understanding and Interactions.

Liu, B., & Lane, I. (2016). Attention-based recurrent neural network models for joint intent detection and slot filling. In *Proceedings of the Interspeech*, San Francisco, CA.

Mesnil, G., Dauphin, Y., Yao, K., Bengio, Y., Deng, L., Hakkani-Tür, D., et al. (2015). Using recurrent neural networks for slot filling in spoken language understanding. *IEEE Transactions on Audio, Speech, and Language Processing*, *23*(3), 530–539.

Mesnil, G., He, X., Deng, L., & Bengio, Y. (2013). Investigation of recurrent-neural-network architectures and learning methods for spoken language understanding. In *Proceedings of the Interspeech*, Lyon, France.

Natarajan, P., Prasad, R., Suhm, B., & McCarthy, D. (2002). Speech enabled natural language call routing: BBN call director. In *Proceedings of the ICSLP*, Denver, CO.

Pieraccini, R., Tzoukermann, E., Gorelov, Z., Gauvain, J.-L., Levin, E., Lee, C.-H., et al. (1992). A speech understanding system based on statistical representation of semantics. In *Proceedings of the ICASSP*, San Francisco, CA.

Price, P. J. (1990). Evaluation of spoken language systems: The ATIS domain. In *Proceedings of the DARPA Workshop on Speech and Natural Language*, Hidden Valley, PA.

Ravuri, S., & Stolcke, A. (2015). Recurrent neural network and lstm models for lexical utterance classification. In *Proceedings of the Interspeech.*

Raymond, C., & Riccardi, G. (2007). Generative and discriminative algorithms for spoken language understanding. In *Proceedings of the Interspeech*, Antwerp, Belgium.

Sarikaya, R., Hinton, G. E., & Deoras, A. (2014). Application of deep belief networks for natural language understanding. *IEEE Transactions on Audio, Speech, and Language Processing*, *22*(4).

Sarikaya, R., Hinton, G. E., & Ramabhadran, B. (2011). Deep belief nets for natural language call-routing. In *Proceedings of the ICASSP*, Prague, Czech Republic.

Seneff, S. (1992). TINA: A natural language system for spoken language applications. *Computational Linguistics*, *18*(1), 61–86.

Simonnet, E., Camelin, N., Deleglise, P., & Esteve, Y. (2015). Exploring the use of attention-based recurrent neural networks for spoken language understanding. In *Proceedings of the NIPS Workshop on Machine Learning for Spoken Language Understanding and Interaction.*

Socher, R., Lin, C. C., Ng, A. Y., & Manning, C. D. (2011). Parsing natural scenes and natural language with recursive neural networks. In *Proceedings of ICML.*

Sordoni, A., Bengio, Y., Vahabi, H., Lioma, C., Simonsen, J. G., & Nie, J.-Y. (2015). A hierarchical recurrent encoder-decoder for generative context-aware query suggestion. In *Proceedings of the ACM CIKM.*

Sutskever, I., Vinyals, O., & Le, Q. V. (2014). *Advances in neural information processing systems 27*, chapter Sequence to sequence learning with neural networks.

Tafforeau, J., Bechet, F., Artiere1, T., & Favre, B. (2016). Joint syntactic and semantic analysis with a multitask deep learning framework for spoken language understanding. In *Proceedings of the Interspeech*, San Francisco, CA.

Tur, G., & Deng, L. (2011). *Intent determination and spoken utterance classification, Chapter 4 in book: Spoken language understanding*. New York, NY: Wiley.

Tur, G., Hakkani-Tür, D., & Heck, L. (2010). What is left to be understood in ATIS? In *Proceedings of the IEEE SLT Workshop*, Berkeley, CA.

Tur, G., & Mori, R. D. (Eds.). (2011). *Spoken language understanding: Systems for extracting semantic information from speech*. New York, NY: Wiley.

Vinyals, O., Fortunato, M., & Jaitly, N. (2015). Pointer networks. In *Proceedings of the NIPS.*

Vinyals, O., & Le, Q. V. (2015). A neural conversational model. In *Proceedings of the ICML.*
Vu, N. T., Gupta, P., Adel, H., & Schütze, H. (2016). Bi-directional recurrent neural network with ranking loss for spoken language understanding. In *Proceedings of the IEEE ICASSP*, Shanghai, China.
Vukotic, V., Raymond, C., & Gravier, G. (2016). A step beyond local observations with a dialog aware bidirectional gru network for spoken language understanding. In *Proceedings of the Interspeech*, San Francisco, CA.
Walker, M., Aberdeen, J., Boland, J., Bratt, E., Garofolo, J., Hirschman, L., et al. (2001). DARPA communicator dialog travel planning systems: The June 2000 data collection. In *Proceedings of the Eurospeech Conference.*
Wang, Y., Deng, L., & Acero, A. (2011). *Semantic frame based spoken language understanding, Chapter 3*. New York, NY: Wiley.
Ward, W., & Issar, S. (1994). Recent improvements in the CMU spoken language understanding system. In *Proceedings of the ARPA HLT Workshop*, pages 213–216.
Weizenbaum, J. (1966). Eliza—A computer program for the study of natural language communication between man and machine. *Communications of the ACM*, *9*(1), 36–45.
Woods, W. A. (1983). *Language processing for speech understanding*. Prentice-Hall International, Englewood Cliffs, NJ: In Computer Speech Processing.
Xu, P., & Sarikaya, R. (2013). Convolutional neural network based triangular crf for joint intent detection and slot filling. In *Proceedings of the IEEE ASRU.*
Yao, K., Peng, B., Zhang, Y., Yu, D., Zweig, G., & Shi, Y. (2014). Spoken language understanding using long short-term memory neural networks. In *Proceedings of the IEEE SLT Workshop*, South Lake Tahoe, CA. IEEE.
Yao, K., Zweig, G., Hwang, M.-Y., Shi, Y., & Yu, D. (2013). Recurrent neural networks for language understanding. In *Proceedings of the Interspeech*, Lyon, France.
Zhai, F., Potdar, S., Xiang, B., & Zhou, B. (2017). Neural models for sequence chunking. In *Proceedings of the AAAI.*
Zhang, X., & Wang, H. (2016). A joint model of intent determination and slot filling for spoken language understanding. In *Proceedings of the IJCAI.*
Zhu, S., & Yu, K. (2016a). Encoder-decoder with focus-mechanism for sequence labelling based spoken language understanding. In *submission.*
Zhu, S., & Yu, K. (2016b). Encoder-decoder with focus-mechanism for sequence labelling based spoken language understanding. arXiv preprint arXiv:1608.02097.

Chapter 3
Deep Learning in Spoken and Text-Based Dialog Systems

Asli Celikyilmaz, Li Deng and Dilek Hakkani-Tür

Abstract Last few decades have witnessed substantial breakthroughs on several areas of speech and language understanding research, specifically for building human to machine conversational dialog systems. Dialog systems, also known as interactive conversational agents, virtual agents or sometimes chatbots, are useful in a wide range of applications ranging from technical support services to language learning tools and entertainment. Recent success in deep neural networks has spurred the research in building data-driven dialog models. In this chapter, we present state-of-the-art neural network architectures and details on each of the components of building a successful dialog system using deep learning. Task-oriented dialog systems would be the focus of this chapter, and later different networks are provided for building open-ended non-task-oriented dialog systems. Furthermore, to facilitate research in this area, we have a survey of publicly available datasets and software tools suitable for data-driven learning of dialog systems. Finally, appropriate choice of evaluation metrics are discussed for the learning objective.

3.1 Introduction

In the past decade, virtual personal assistants (VPAs) or conversational chatbots have been the most exciting technological developments. Spoken Dialog Systems (SDS) are considered the brain of these VPAs. For instance Microsoft's Cortana,[1]

[1]https://www.microsoft.com/en-us/mobile/experiences/cortana/.

A. Celikyilmaz (✉)
Microsoft Research, Redmond, WA, USA
e-mail: asli@ieee.org

L. Deng
Citadel, Chicago & Seattle, USA
e-mail: l.deng@ieee.org

D. Hakkani-Tür
Google, Mountain View, CA, USA
e-mail: dilek@ieee.org

L. Deng and Y. Liu (eds.), *Deep Learning in Natural Language Processing*, https://doi.org/10.1007/978-981-10-5209-5_3

Table 3.1 Type of tasks that dialog systems are currently used

Types of tasks	Examples
Information consumption	*"what is the conference schedule"* *'which room is the talk in?"*
Task completion	*"set my alarm for 3pm tomorrow"* *"find me kid-friendly vegetarian restaurant in downtown Seattle"* *"schedule a meeting with sandy after lunch."*
Decision support	*"why are sales in south region far behind?"*
Social interaction (chit-chat)	*"how is your day going"* *"i am as smart as human?"* *"i love you too."*

Apple's Siri,[2] Amazon Alexa,[3] Google Home,[4] and Facebook's M,[5] have incorporated SDS modules in various devices, which allow users to speak naturally in order to finish tasks more efficiently. The traditional conversational systems have rather complex and/or modular pipelines. The advance of deep learning technologies has recently risen the applications of neural models to dialog modeling.

Spoken dialog systems have nearly 30 years of history, which can be divided into three generations: symbolic rule or template based (before late 90s), statistical learning based, and deep learning based (since 2014). This chapter briefly surveys the history of conversational systems, and analyzes why and how the underlying technology moved from one generation to the next. Strengths and weaknesses of these three largely distinct types of bot technology are examined and future directions are discussed.

Current dialog systems are trying to help users on several tasks to complete daily activities, play interactive games, and even be a companion (see examples in Table 3.1). Thus, conversational dialog systems have been built for many purposes, however, a meaningful distinction can be made between goal-oriented dialogs (e.g., for personal assistant systems or other task completion dialogs such as purchasing or technical support services) and non-goal-oriented dialog systems such as chit-chat, computer game characters (avatars), etc. Since they serve for different purses, structurally their dialog system designs and the components they operate on are different. In this chapter, we will provide details on the components of dialog systems for task (goal)-oriented dialog tasks. Details of the non-goal-oriented dialog systems (chit-chat) will also be provided.

As shown in Fig. 3.1, the classic spoken dialog systems incorporate several components including Automatic Speech Recognition (ASR), Language Understanding Module, State Tracker and Dialog Policy together forming the Dialog Manager, the

---

[2]http://www.apple.com/ios/siri/.

[3]https://developer.amazon.com/alexa.

[4]https://madeby.google.com/home.

[5]https://developers.facebook.com/blog/post/2016/04/12/bots-for-messenger/.

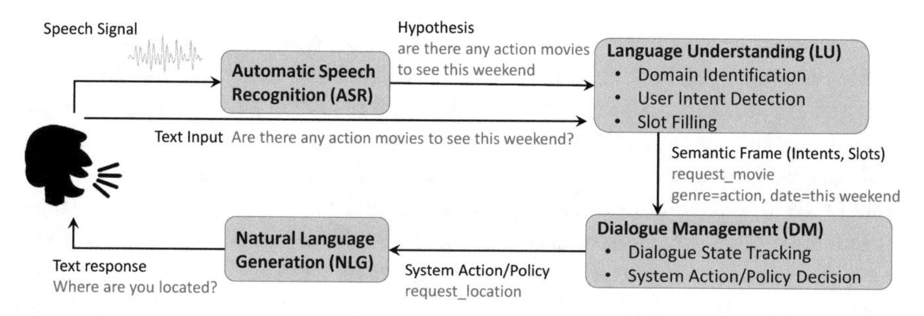

Fig. 3.1 Pipeline framework of spoken dialog system

Natural Language Generator (NLG), also known as Response Generator. In this chapter, we focus on data-driven dialog systems as well as interactive dialog systems in which human or a simulated human is involved in learning dialog system components using deep learning on real-world conversational dialogs.

The spoken language or speech recognition have huge impact on the success of the overall spoken dialog system. This front-end component involves several factors that make it difficult for machines to recognize speech. The analysis of continuous speech is a difficult task as there is huge variability in the speech signal and there are no clear boundaries between words. For technical details of such and many other difficulties in building spoken language systems, we refer readers to Huang and Deng (2010), Deng and Li (2013), Li et al. (2014), Deng and Yu (2015), Hinton et al. (2012), He and Deng (2011).

The speech recognition component of the spoken dialog systems is often speaker independent and does not take into account that it is the same user during the whole dialog. In an end-to-end spoken dialog system, the inevitable errors in speech recognition would make the language understanding component harder than when the input is text—free of speech recognition errors (He and Deng 2013). In the long history of spoken language understanding research, the difficulties caused by speech recognition errors forced the domains of spoken language understanding to be substantially narrower than language understanding in text form (Tur and Deng 2011). However, due to the huge success of deep learning in speech recognition in recent years (Yu and Deng 2015; Deng 2016), recognition errors have been dramatically reduced, leading to increasingly broader application domains in the current conversational understanding systems.[6]

Most early goal-driven dialog systems were primarily based on handcrafted rules (Aust et al. 1995; Simpson and Eraser 1993) which immediately followed machine learning techniques for all components of the dialog system (Tur and De Mori 2011; Gorin et al. 1997). Most of these work formulate dialog as a sequential decision-making problem based on Markov Decision Processes. With the deep neural networks, especially the research in speech recognition, spoken language

[6]We refer the reader to the "Deep Learning in Conversational Language Understanding" chapter in this book for more details in discussing this issue.

understanding (e.g., Feed-forward neural networks) (Hastie et al. 2009), RNNs (Goller and Kchler 1996) including LSTMs (Graves and Schmidhuber 2005), and dialog modeling (e.g., deep reinforcement learning methods) have showed incredible success in robustness and coherency of the dialog systems (Wen et al. 2016b; Dhingra et al. 2016a; Lipton et al. 2016). On the other hand, most earlier non-goal-oriented systems have used simple rules, topic models, and modeled dialog as a stochastic sequence of discrete symbols (words) using higher order Markov chains. Only recently, deep neural network architectures trained on large-scale corpora have been investigated and promising results have been observed (Ritter et al. 2011; Vinyals and Le 2015; Lowe et al. 2015a; Sordoni et al. 2015a; Serban et al. 2016b, 2017). One of the biggest challenges of non-goal-oriented systems that use deep neural networks is that they require substantially large corpora in order to achieve good results.

This chapter is structured as follows. In the next in Sect. 3.2, a high-level overview of the deep learning tools that are used in building subcomponents of the current dialog systems are provided. Section 3.3 describes the individual system components of the goal-oriented neural dialog systems and provides the examples of recently presented research work. In Sect. 3.4, types of user simulators that are use deep learning technologies are discussed. Later methods on how deep learning methods are utilized in natural language generation are presented in Sect. 3.5. Later section delves into the deep learning methods that are relevant for building end-to-end dialog systems in Sect. 3.6. In Sect. 3.7, the open-domain non-goal-oriented dialog systems are presented followed by the current datasets used to building deep dialog models and provide links to the each corpus in turn while emphasizing how the dialogs were generated and collected. Section 3.9 briefly touches on open source neural dialog system modeling software. Evaluating dialog systems and the measures used to evaluate them are presented in Sect. 3.10. Finally in Sect. 3.11, this chapter concludes with a survey of projections into the future of dialog modeling.

3.2 Learning Methodology for Components of a Dialog System

In this section, we summarize some of the deep learning techniques that are used in building conversational agents. Deep learning technologies have been used to model nearly all of the components of the dialog systems. We investigate such methods below under three different categories: *discriminative*, *generative*, and *decision-making based*, specifically reinforcement learning.

3.2.1 Discriminative Methods

Deep learning methods that model the posterior $p(y|x)$ directly with abundance of supervised data have been one of the most investigated approaches in dialog modeling research. Most advanced and prominent approaches have been investigated for the Spoken Language Understanding (SLU) tasks such as goal estimation and intention identification from users commands, which are essential components in spoken dialog systems and they are modeled as multi-output classification tasks. Most research work in this area use Deep Neural Networks for classification specifically multilayered feed-forward neural networks or multilayered perceptrons (Hastie et al. 2009). These models are called feed-forward because information flows through the function being evaluated from x, through the intermediate computations used to define f, and finally to the output y.

Deep Structured Semantic Models (DSSM), or more general, Deep Semantic Similarity Models, are one of the approaches in deep learning research which is commonly used for multi/single class text classification which intrinsically learns similarities between two text while discovering latent features. In dialog system modeling, DSSM approaches are mainly for SLU's classification tasks (Huang et al. 2013). DSSMs are a Deep Neural Network (DNN) modeling technique for representing text strings (sentences, queries, predicates, entity mentions, etc.) in a continuous semantic space and modeling semantic similarity between two text strings (e.g., Sent2Vec). Also commonly used are the Convolutional Neural Networks (CNN) which utilize layers with convolving filters that are applied to local features (LeCun et al. 1998). Originally invented for computer vision, CNN models have subsequently been shown to be effective for SLU models mainly for learning latent features that are otherwise impossible to extract with standard (non-)linear machine learning approaches.

Semantic slot filling is one of the most challenging problems in SLU and is considered as a sequence learning problem. Similarly, belief tracking or dialog state tacking are also considered sequential learning problems for the reasons that they mainly maintain the state of the dialog through each conversation in the dialog. Although CNNs are a great way to pool local information, they do not really capture the sequentiality of the data and not the first choice when it comes to sequential modeling. Hence to tackle sequential information in modeling user utterances in dialog systems, most research has focused on using Recurrent Neural Networks (RNN) which help tackle sequential information.

Memory networks (Weston et al. 2015; Sukhbaatar et al. 2015; Bordes et al. 2017) are a recent class of models that have been applied to a range of natural language processing tasks, including question answering (Weston et al. 2015), language modeling (Sukhbaatar et al. 2015), etc. Memory networks in general work by first writing and then iteratively reading from a memory component (using hops) that can store historical dialogs and short-term context to reason about the required response. They have been shown to perform well on those tasks and to outperform some other end-to-end architectures based on Recurrent Neural Networks. Also, attention-based RNN networks such as Long Short-Term-Memory Networks (LSTM) take different approach to keep the memory component and learn to attend dialog context (Liu and Lane 2016a).

Obtaining large corpora for every new applications may not be feasible to build deep supervised learning models. For this reason, the use of other related datasets can effectively bootstrap the learning process. Particularly in deep learning, the use of related datasets in pre-training a model is an effective method of scaling up to complex environments (Kumar et al. 2015). This is crucial in open-domain dialog systems, as well as multi-task dialog systems (e.g., travel domain comprising of several tasks from different domains such as hotel, flight, restaurants, etc.). Dialog modeling researchers have already proposed various deep learning approaches for applying transfer learning to build data-driven dialog systems such as learning sub-components of the dialog system (e.g., intent and dialog act classification) or learning end-to-end dialog system using transfer learning.

3.2.2 Generative Methods

Deep generative models have recently become popular due to their ability to model input data distributions and generate realistic examples from those distributions and in turn has recently entered in the dialog system modeling research field. Such approaches are largely considered in clustering objects and instances in the data, extracting latent features from unstructured text, or dimensionality reduction. A large portion of the category of dialog modeling systems that use deep generative models investigate open-domain dialog systems specifically focusing on neural generative models for response generation. Common to these work are encoder–decoder based neural dialog models (see Fig. 3.5) (Vinyals and Le 2015; Lowe et al. 2015b; Serban et al. 2017; Shang et al. 2015), in which the encoder network used the entire history to encode the dialog semantics and the decoder generates natural language utterance (e.g., sequence of words representing systems' response to user's request). Also used are RNN-based systems that map an abstract dialog act into an appropriate surface text (Wen et al. 2015a).

Generative Adversarial Networks (GANs) (Goodfellow et al. 2014) is one topic in generative modeling which has very recently appeared in the dialog field as neural dialog modeling tasks specifically for dialog response generation. While Li et al. (2017) use deep generative adversarial networks for response generation, Kannan and Vinyals (2016) investigate the use of an adversarial evaluation method for dialog models.

3.2.3 Decision-Making

The key to a dialog system is its decision-making module, which is also known as the dialog manager or also referred to as dialog policy. The dialog policy chooses system actions at each step of the conversation to guide the dialog to successful task completion. The system actions include interacting with the user for getting specific requirements for accomplishing the task, as well as negotiating and offering alternatives. Optimization of statistical dialog managers using *Reinforcement Learning*

(RL) methods is an active and promising area of research (Fatemi et al. 2016a, b; Su et al. 2016; Lipton et al. 2016; Shah et al. 2016; Williams and Zweig 2016a; Dhingra et al. 2016a). The RL setting fits the dialog setting quite well because RL is meant for situations when feedback may be delayed. When a conversational agent carries a dialog with a user, it will often know whether or not the dialog was successful and the task was achieved only after the dialog is ended.

Aside from the above categories, deep dialog systems have also been introduced with novel solutions involving applications of transfer learning and domain adaptation for next generation dialog systems, specifically focusing on domain transfer in spoken language understanding (Kim et al. 2016a, b, 2017a, b) and dialog modeling (Gai et al. 2015, 2016; Lipton et al. 2016).

3.3 Goal-Oriented Neural Dialog Systems

The most useful applications of dialog systems can be considered to be the goal-oriented and transactional, in which the system needs to understand a user request and complete a related task with a clear goal within a limited number of dialog turns. We will provide description and recent related work for each component of goal-oriented dialog systems in detail.

3.3.1 Neural Language Understanding

With the power of deep learning, there is increasing research work focusing on applying deep learning for language understanding. In the context of goal-oriented dialog systems, language understanding is tasked with interpreting user utterances according to a semantic meaning representation, in order to enable with the back-end action or knowledge providers. Three key tasks in such targeted understanding applications are domain classification, intent determination, and slot filling (Tur and De Mori 2011), aiming to form a semantic frame to capture the semantics of user utterances/queries. Domain classification is often completed first in spoken language understanding (SLU) systems, serving as a top-level triage for subsequent processing. Intent determination and slot filling are then executed for each domain to fill a domain-specific semantic template. An example semantic frame for a movie-related utterance, "*find recent action movies by Jackie Chan*", is shown in Fig. 3.2.

With the advances on deep learning, Deep Belief Networks (DBNs) with Deep Neural Networks (DNNs) have been applied to domain and intent classification tasks (Sarikaya et al. 2011; Tur et al. 2012; Sarikaya et al. 2014). More recently, Ravuri and Stolcke (2015) proposed an RNN architecture for intent determination, where an encoder network first predicts a representation for the input utterance, and then a single step decoder predicts a domain/intent class for the input utterance using a single step decoder network.

		find	recent	action	movies	by	jackie	chan
Slots		O	B-date	B-genre	B-type	O	B-director	I-director
Intent	movies							
Domain	find_movie							

Fig. 3.2 An example utterance with annotations of semantic slots in IOB format, domain, and intent, B-dir and I-dir denote the director name

For slot filling task, deep learning has been mostly used as a feature generator. For instance (Xu and Sarikaya, 2013) extracted features using convolutional neural networks to feed into a CRF model. Yao et al. (2013) and Mesnil et al. (2015) later used RNNs for sequence labeling in order to perform slot filling. More recent work focus on sequence-to-sequence models (Kurata et al. 2016), sequence-to-sequence models with attention (Simonnet et al. 2015), multi-domain training (Jaech et al. 2016), multi-task training (Tafforeau et al. 2016), multi-domain joint semantic frame parsing (Hakkani-Tür et al. 2016; Liu and Lane 2016b), and context modeling using end-to-end memory networks (Chen et al. 2016; Bapna et al. 2017). These will be described in more detail in the language understanding chapter.

3.3.2 *Dialog State Tracker*

The next step in spoken dialog systems pipeline is Dialog State Tracking (DST), which aims to track system's belief on user's goal through the course of a conversation. The dialog state is used for querying the back-end knowledge or information sources and for determining the next state action by the dialog manager. At each turn in a dialog, DST gets as input the estimated dialog state from the previous user turn, s_{t-1}, and the most recent system and user utterances and estimates the dialog state s_t for the current turn. In the past few years, the research on dialog state tracking has accelerated owing to the data sets and evaluations performed by the dialog state tracking challenges (Williams et al. 2013; Henderson et al. 2014). The state-of-the-art dialog managers focus on monitoring the dialog progress by neural dialog state tracking models. Among the initial models are the RNN based dialog state tracking approaches (Henderson et al. 2013) that has shown to outperform Bayesian networks (Thomson and Young 2010). More recent work on Neural Dialog Managers that provide conjoint representations between the utterances, slot-value pairs as well as knowledge graph representations (Wen et al. 2016b; Mrkšić et al. 2016) demonstrates that using neural dialog models can overcome current obstacles of deploying dialog systems in larger dialog domains.

3.3.3 *Deep Dialog Manager*

A dialog manager is a component of a conversational dialog system, which interacts in a natural way to help the user complete the tasks that the system is designed to support. It is responsible for the state and flow of the conversation, hence determines what policy should be used. The input to the dialog manager is the human utterance, which is converted to some system-specific semantic representation by the natural language understanding component. For example, in a flight-planning dialog system, the input may look like "ORDER(from = SFO, to = SEA, date = 2017-02-01)". The dialog manager usually maintains state variables, such as the dialog history, the latest unanswered question, the recent user intent and entities, etc., depending on the domain of the dialog. The output of the dialog manager is a list of instructions to other parts of the dialog system, usually in a semantic representation, for example "Inform (flight-num = 555,flight-time = 18:20)". This semantic representation is converted into natural language by the natural language generation component.

Typically, an expert manually designs a dialog management policy and incorporates several dialog design choices. Manual dialog policy design is intractable and does not scale as the performance of the dialog policy depends on several factors including domain-specific features, robustness of the automatic speech recognizer (ASR) system, the task difficulty, to name a few. Instead of letting a human expert write a complex set of decision rules, it is more common to use reinforcement learning. The dialog is represented as a Markov Decision Process (MDP)—a process where, in each state, the dialog manager has to select an action, based on the state and the possible rewards from each action. In this setting, the dialog author should only define the reward function, for example: in restaurant reservation dialogs, the reward is the user success in reserving a table successfully; in information seeking dialogs, the reward is positive if the human receives the information, but there is also a negative reward for each dialog step. Reinforcement learning techniques are then used to learn a policy, for example, what type of confirmation should the system use in each state (Lemon and Rieserr 2009). A different way to learn dialog policies is to try to imitate humans, using Wizard of Oz experiments, in which a human sits in a hidden room and tells the computer what to say (Passonneau et al. 2011).

For complex dialog systems, it is often impossible to specify a good policy a priori and the dynamics of an environment may change over time. Thus, learning policies online and interactively via reinforcement learning have emerged as a popular approach (Singh et al. 2016; Gasic et al. 2010; Fatemi et al. 2016b). For instance, the ability to compute an accurate reward function is essential for optimizing a dialog policy via reinforcement learning. In real-world applications, using explicit user feedback as the reward signal is often unreliable and costly to collect. Su et al. (2016) propose an online learning framework in which the dialog policy is jointly trained alongside the reward model via active learning with a Gaussian process model. They propose three main system components which include dialog policy, dialog embedding creation, and reward modeling based on user feedback (see Fig. 3.3). They use episodic turn-level features extracted from a dialog and build a Bidirectional Long Short-Term Memory network (BLSTM) for their dialog embedding creation.

Efficient dialog policy learning with deep learning technologies has recently been the focus of dialog researcher with the recent advancements in deep reinforcement learning. For instance, Lipton et al. (2016) investigate understanding boundaries of the deep neural network structure of the dialog policy model to efficiently explore different trajectories via Thompson sampling, drawing Monte Carlo samples from a Bayesian neural network (Blundell et al. 2015). They use deep Q-network to optimize the policy. They explore a version of their approach that incorporates the intrinsic reward from Variational Information Maximizing Exploration (VIME) (Blundell et al. 2015). Their Bayesian approach addresses uncertainty in the Q-value given the current policy, whereas VIME addresses uncertainty of the dynamics of under-explored parts of the environment. Thus, there is a synergistic effect of combining the approaches. On the domain extension task, the combined exploration method proved especially promising, outperforming all other methods.

There are several other aspects that affect the policy optimization for dialog managers. Some of which include learning policies under multi-domain systems (Gasic et al. 2015; Ge and Xu 2016), committee-based learning for multi-domain systems (Gasic et al. 2015), learning domain-independent policies (Wang et al. 2015), adapting to grounded word meanings (Yu et al. 2016), adapting to new user behaviors (Shah et al. 2016), to name a few. Among these systems, Peng et al. (2017) investigate hierarchal policy learning for task-oriented systems that have composite subtasks. This domain is particularly challenging and the authors tackle with the issue of reward sparsity, satisfying slot constraints across subtasks. This requirement makes most of the existing methods of learning multi-domain dialog agents (Cuayahuitl et al. 2016; Gasic et al. 2015) inapplicable: these methods train a collection of policies, one for each domain, and there are no cross-domain constraints required to successfully complete a dialog. As shown in Fig. 3.4, their composite task completion dialog agent consists of four components: (1) an LSTM-based language understanding module for identifying user intents and extracting associated slots; (2) a dialog state tracker; (3) a dialog policy which selects the next action based on the current state; and (4) a model-based natural language generator for converting agent actions to natural language responses. Following the options over MDP's formalism (Sutton and Singh 1999), they build their agent to learn a composite tasks such as travel planning, subtasks like book flight ticket and reserve hotel which can be modeled as options.

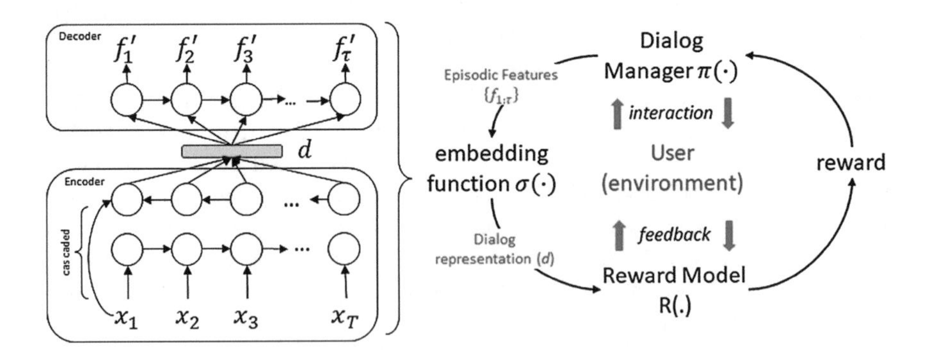

Fig. 3.3 Schematic of the dialog policy learning with deep encoder–decoder networks. The three main system components: dialog policy, dialog embedding creation, and reward modeling based on user feedback

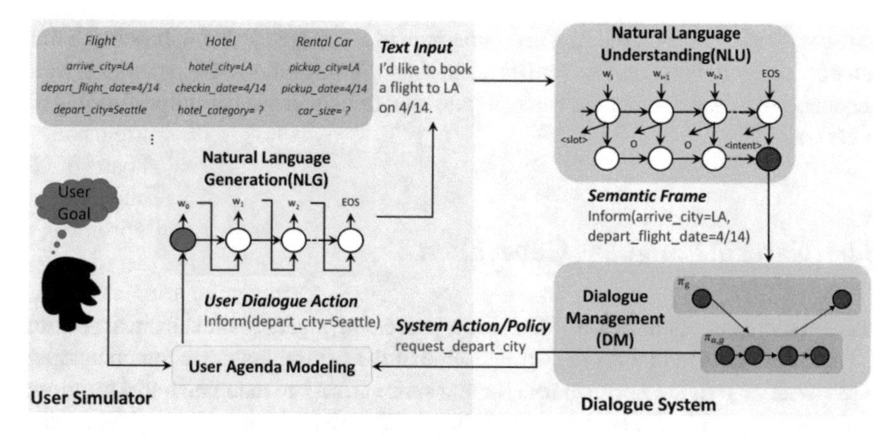

Fig. 3.4 Illustration of the composite task completion dialog system

3.4 Model-Based User Simulators

User simulators for spoken dialog systems aim at generating artificial interactions supposed to be representative of what would be an actual dialog between a human user and a given dialog system. Model-based simulated users for building dialog models are not as common as the other components of the dialog systems; detailed reviews of some of these methods are presented in Schatzmann et al. (2006), Georgila et al. (2005, 2006). In this section, we only investigate deep learning methods for user simulation, that is, methods purely based on data and deep learning approaches models.

The early spoken dialog systems optimization required a lot of data because of inefficiency of reinforcement learning algorithms, justifying the use of simulation. In recent years, sample efficient reinforcement learning methods were applied to spoken dialog systems optimization. With this, models can be trained to learn optimal dialog strategies directly from large amounts of data collected even from suboptimal systems with actual users (Li et al. 2009; Pietquin et al. 2011b) but also from online interactions (Pietquin et al. 2011a). This makes it much more appealing for the dialog systems to be trained using a simulated user with user feedback and corrected as the process continues.

There are several reasons that make learning parameters of a user simulation model hard to optimize because most of the system features are hidden (e.g., user goal, mental states, dialog history, etc.). Focusing on this problem, Asri et al. (2016) presented a sequence-to-sequence base user simulator on non-goal-oriented domains (e.g., chit-chat) that takes into account the entire dialog history. Their user simulator does not rely on any external data structure to ensure coherent user behavior, and it does not require mapping to a summarized action space, which makes it able to model user behavior with finer granularity.

Crook and Marin (2017) explore sequence-to-sequence learning approach for NL-to-NL simulated user models for goal-oriented dialog systems. They present several extensions to their architecture to incorporate context in different ways and investigate the efficacy of each method in comparison to language modeling baseline simulator on a personal assistant system domain. Their findings showed that context-based sequence-to-sequence method can generate human like utterances outperforming all other baselines.

3.5 Natural Language Generation

Natural Language Generation (NLG) is the process of generating text from a meaning representation. It can be taken as the reverse of the natural language understanding. NLG systems provide a critical role for text summarization, machine translation, and dialog systems. While several general-purpose rule-based generation systems have been developed (Elhadad and Robin 1996), they are often quite difficult to adapt to small, task-oriented applications because of their generality. To overcome this, several people have proposed different solutions. Bateman and Henschel (1999) have described a lower cost and more efficient generation system for a specific application using an automatically customized sub-grammar. Busemann and Horacek (1998) describe a system that mixes templates and rule-based generation. This approach takes advantage of templates and rule-based generation as needed by specific sentences or utterances. Stent (1999) has also proposed a similar approach for a spoken dialog system. Although such approaches are conceptually simple and tailored to the domain, they lack generality (e.g., repeatedly encode linguistic rules such as subject–verb agreement), have little variation in style and difficult to grow and maintain (e.g., usually each new utterance is added by hand). Such approaches impose the requirement of writing grammar rules and acquiring the appropriate lexicon, which requires a specialist activity.

Machine learning based (trainable) NLG systems are more common in today's dialog systems. Such NLG systems use several sources as input such as: *content plan*, representing meaning representation of what to communicate with the user (e.g., describe a particular restaurant), *knowledge base*, structured database to return domain-specific entities, (e.g., database of restaurants), *user model*, a model that imposes constraints on output utterance (e.g., user wants short utterances), *dialog history*, the information from previous turns to avoid repetitions, referring expressions, etc. The goal is to use these meaning representations that indicate what to say (e.g., entities described by features in an ontology) to output natural language string describing the input (e.g., *zucca's food is delicious.*).

Trainable NLG systems can produce various candidate utterances (e.g., stochastically or rule base) and use a statistical model to rank them (Dale and Reiter 2000). The statistical model assigns scores to each utterance and is learnt based on textual data. Most of these systems use bigram and trigram language models to generate utterances. The trainable generator approach exemplified by the HALOGEN (Langkilde and Knight 1998) and SPaRKy system (Stent et al. 2004) are among the most notable trainable approaches. These systems include various trainable modules within their framework to allow the model to adapt to different domains (Walker et al. 2007), or reproduce certain style (Mairesse and Walker 2011). However, these approaches still require a handcrafted generator to define the decision space. The resulting utterances are therefore constrained by the predefined syntax and any domain-specific colloquial responses must be added manually. In addition to these approaches, corpus-based methods (Oh and Rudnicky 2000; Mairesse and Young 2014; Wen et al. 2015a) have been shown to have flexible learning structures with the goal of learning generation directly from data by adopting an over-generation and re-ranking paradigm (Oh and Rudnicky 2000), in which final responses are obtained by re-ranking a set of candidates generated from a stochastic generator.

With the advancement of deep neural network systems, more sophisticated NLG systems can be developed that can be trained from un-aligned data or produce longer utterances. Recent study has shown that especially with the RNN methods (e.g., LSTMs, GRUs, etc.), more coherent, realistic, and proposer answers can be generated. Among these studies, the work by Vinyals and Le (2015), on Neural Conversational Model has opened a new chapter in using encoder–decoder based models for generation. Their model is based on two LSTM layers. One for encoding the input sentence into a "thought vector", and another for decoding that vector into a response. This model is called sequence-to-sequence or seq2seq. The model only gives simple and short answers to questions.

Sordoni et al. (2015b) propose three neural models to generate a response (r) based on a context and message pair (c, m). The context is defined as a single message. They propose several models, the first one of which is a basic Recurrent Language Model that is fed the whole (c, m, r) triple. The second model encodes context and message into a BoW representation, puts it through a feed-forward neural network encoder, and then generates the response using an RNN decoder. The last model is similar but keeps the representations of context and message separate instead of encoding them into a single BoW vector. The authors train their models on 29M triple data set from Twitter and evaluate using BLEU, METEOR, and human evaluator scores. Because (c, m) is very long on average the authors expect their first model to perform poorly. Their model generates responses degrade with length after eight tokens.

Li et al. (2016b) present a method which adds coherency to the response generated by sequence-to-sequence models such as the Neural Conversational Model (Vinyals and Le 2015). They define persona as the character that an agent performs during conversational interactions. Their model combines identity, language, behavior, and interaction style. Their model may be adapted during the conversation itself. Their proposed models yield performance improvements in both perplexity and BLEU scores over baseline sequence-to-sequence models. Compared to Persona based Neural Conversational Model, the baseline Neural Conversational Model fails to maintain a consistent persona throughout the conversation resulting in incoherent responses. A similar approach in Li et al. (2016a) uses a Maximum Mutual Information (MMI) objective function to generate conversational responses. They still train their models with maximum likelihood, but use MMI to generate responses during decoding. The idea behind MMI is that it promotes more diversity and penalizes trivial responses. The authors evaluate their method using BLEU scores, human evaluators, and qualitative analysis and find that the proposed metric indeed leads to more diverse responses.

Serban et al. (2017) presents a hierarchical latent variable encoder–decoder model for generating dialogs. Their goal is to generate natural language dialog responses. Their model assumes that each output sequence can be modeled in a two-level hierarchy: sequences of subsequences, and subsequences of tokens. For example, a dialog may be modeled as a sequence of utterances (subsequences), with each utterance modeled as a sequence of words. Given this, their model consists of three RNN modules: an encoder RNN, a context RNN and a decoder RNN. Each subsequence of tokens is deterministically encoded into a real-valued vector by the encoder RNN. This is given as input to the context RNN, which updates its internal hidden state to reflect all information up to that point in time. The context RNN deterministically outputs a real-valued vector, which the decoder RNN conditions on to generate the next subsequence of tokens (see Fig. 3.5).

Recent work in natural language generation has focused on using reinforcement learning strategies to explore different learning signals (He et al. 2016; Williams and Zweig 2016b; Wen et al. 2016a; Cuayahuitl 2016). The motivation for this renewed interest in reinforcement learning stems from issues of using teacher forcing for learning. Text generation systems trained using word-by-word cross-entropy loss with gold sequences as supervision have produced locally coherent generations, but generally fail to capture the contextual dynamics of the domain they are modeling. Recipe generation systems that are conditioned on their ingredients and recipe title, for example, do not manage to combine the starting ingredients into their end dish in a successful way. Similarly, dialog generation systems often fail to condition their responses on previous utterances in the conversation. Reinforcement learning allows models to be trained with rewards that go beyond predicting the correct word. Mixing reward schemes using teacher forcing and other more "global" metrics has recently become popular for producing more domain-relevant generations.

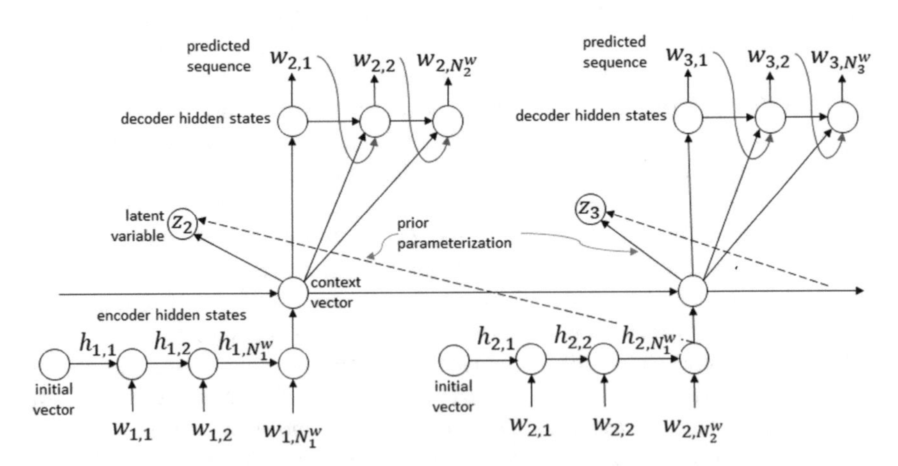

Fig. 3.5 Hierarchal Encoder–Decoder Model computational graph. Diamond boxes represent deterministic variables and rounded boxes represent stochastic variables. Full lines represent the generative model and dashed lines represent the approximate posterior model

3.6 End-to-End Deep Learning Approaches to Building Dialog Systems

End-to-end dialog systems are considered a cognitive system, which has to carry out natural language understanding, reasoning, decision-making, and natural language generation within the same network in order to replicate or emulate the behavior of the agents in the training corpus. This has not been fully investigated before the deep learning technologies have started to be used for dialog system building. Building such systems with today's deep learning technologies are much easier because of the fact that with the deep learning systems and backpropagation all parameters can be trained jointly. In the next, we will briefly investigate the recent end-to-end dialog models for goal- and non-goal-oriented systems.

One of the major obstacles in building end-to-end goal-oriented dialog systems is that the database calls made by the system to retrieve the information requested by the user are not differentiable. Specifically, the query generated by the system and sent to knowledge base is done in a manual way, which means that part of the system is not trained and no function is learnt. This cripples the deep learning model into incorporating the knowledge base response and the information it receives. Also, the neural response generation part is trained and run as separate from the dialog policy network. Putting all this together, training the whole cycle end-to-end has not been fully investigated until recently.

Recently, there has been a growing body of literature focusing on building end-to-end dialog systems, which combine feature extraction and policy optimization using deep neural networks. Wen et al. (2015b) introduced a modular neural dialog agent, which uses a hard knowledge base lookup, thus breaking the differentiability of the whole system. As a result, training of various components of the dialog system is performed separately. The intent network and belief trackers are trained using supervised labels specifically collected for them; while the policy network and generation network are trained separately on the system utterances.

Dhingra et al. (2016b) introduce a modular approach, consisting of: a belief tracker module for identifying user intents, extracting associated slots, and tracking the dialog state; an interface with the database to query for relevant results (Soft-KB lookup); a summary module to summarize the state into a vector; a dialog policy which selects the next system action based on current state and a easily configurable template-based Natural Language Generator (NLG) for converting dialog acts into natural language (see Fig. 3.6). The main contribution of their work is that it retains modularity of the end-to-end network by keeping the belief trackers separate, but replaces the hard lookup with a differentiable one. They propose a differentiable probabilistic framework for querying a database given the agents' beliefs over its fields (or slots) showing that the downstream reinforcement learner can discover better dialog policies by providing it more information.

The non-goal-oriented end-to-end dialog systems investigate the task of building open-domain, conversational dialog systems based on large dialog corpora. Serban et al. (2015) incorporate generative models to produce system responses that are

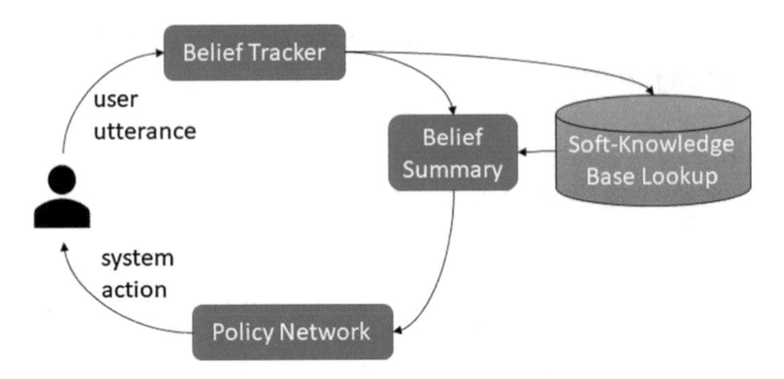

Fig. 3.6 High-level overview of the end-to-end Knowledge-Base-InfoBot: a multi-turn dialog agent which helps users search knowledge bases without composing complicated queries. Such goal-oriented dialog agents typically need to interact with an external database to access real-world knowledge. This model replaces symbolic queries with an induced soft posterior distribution over the knowledge base that indicates which entities the user is interested in. The components with trainable parameters are highlighted in gray

autonomously generated word by word, opening up the possibility for realistic, flexible interactions. They demonstrate that a hierarchical recurrent neural network generative model can outperform both n-gram based models and baseline neural network models on the task of modeling utterances and speech acts.

3.7 Deep Learning for Open Dialog Systems

Open-domain dialog systems, also known as non-task-oriented systems, do not have a stated goal to work towards. These types of dialog systems are mainly useful for interactions in social environments (e.g., social bots) as well as many other useful scenarios (e.g., keeping elderly people company) (Higashinaka et al. 2014), or entertaining users (Yu et al. 2015), to name a few. Open-domain spoken dialog systems support a natural conversation about any topic within a wide coverage Knowledge Graph (KG). The KG can contain not only ontological information about entities but also the operations that might be applied to those entities (e.g., find flight information, book a hotel room, buy an ebook, etc.)

The non-task-oriented systems do not have a goal, nor have a set of states or slots to follow but they do have intentions. Due to this, there have been several work on non-goal-oriented dialog systems that focus preliminarily on response generation which use dialog history (human–agent conversations) as input to propose a response to the user. Among these work are machine translation (Ritter et al. 2011), retrieval-based response selection (Banchs and Li 2012), and sequence-to-sequence models with different structures, such as, vanilla recurrent neural networks (Vinyals and Le 2015), hierarchical neural models (Serban et al. 2015, 2016a; Sordoni et al. 2015b; Shang et al. 2015), and memory neural networks (Dodge et al. 2015). There are several motivations for developing non-goal-driven systems. They may be deployed directly for tasks which do not naturally exhibit a directly measurable goal (e.g., language learning) or simply for entertainment. Also if they are trained on corpora related to the task of a goal-driven dialog system (e.g., corpora which cover conversations on similar topics) then these models can be used to train a user simulator, which can then train the policy strategy.

Until very recently, there has been no research on combining the goal-oriented and non-goal-oriented dialog systems. In a recent work, a first attempt to create a framework that combines these two types of conversations in a natural and smooth manner for the purpose of improving conversation task success and user engagement is presented (Yu et al. 2017). Such a framework is especially useful to handle users who do not have explicit intentions.

3.8 Datasets for Dialog Modeling

In the last years, there has been several publicly available conversational dialog dataset released. Dialog corpora may vary based on several characteristics of the conversational dialog systems. Dialog corpora can be classified based on written, spoken or multi-model properties, or human-to-human or human-to-machine conversations, or natural or unnatural conversations (e.g., in a Wizard-of-Oz system, a human thinks (s)he is speaking to a machine, but a human operator is in fact controlling the dialog system). In this section, we provide a brief overview of these publicly available datasets that are used by the community, for spoken language understanding, state tracking, dialog policy learning, etc., specifically for task completion task. We leave out for open-ended non-task completion datasets in this section.

3.8.1 The Carnegie Mellon Communicator Corpus

This corpus contains human–machine interactions with a travel booking system. It is a medium-sized dataset of interactions with a system providing up-to-the-minute flight information, hotel information, and car rentals. Conversations with the system were transcribed, along with the users comments at the end of the interaction.

3.8.2 ATIS—Air Travel Information System Pilot Corpus

The Air Travel Information System (ATIS) Pilot Corpus (Hemphill et al. 1990) is one of the first human–machine corpora. It consists of interactions, lasting about 40 min each, between human participants and a travel-type booking system, secretly operated by humans. Unlike the Carnegie Mellon Communicator Corpus, it only contains 1041 utterances.

3.8.3 Dialog State Tracking Challenge Dataset

The Dialog State Tracking Challenge (DSTC) is an ongoing series of research community challenge tasks. Each task released dialog data labeled with dialog state information, such as the users desired restaurant search query given all of the dialog history up to the current turn. The challenge is to create a "tracker" that can predict the dialog state for new dialogs. In each challenge, trackers are evaluated using held-out dialog data. Williams et al. (2016) provide an overview of the challenge and datasets which we summarize below:

DSTC1.[7] This dataset consists of human–computer dialogs in the bus timetable domain. Results were presented in a special session at SIGDIAL 2013.

DSTC2 and DSTC3.[8] DSTC2 consists of human–computer dialogs in the restaurant information domain. DSTC2 comprises of large number of training dialog related to restaurant search. It has changing user goals, tracking "requested slots". Results were presented in special sessions at SIGDIAL 2014 and IEEE SLT 2014. DSTC3 is in tourist information domain which addressed the problem of adaptation to a new domain. DSTC2 and 3 were organized by Matthew Henderson, Blaise Thomson, and Jason D. Williams.

DSTC4.[9] The focus of this challenge is on a dialog state tracking task on human–human dialogs. In addition to this main task, a series of pilot tracks is introduced for the core components in developing end-to-end dialog systems based on the same dataset. Results were presented at IWSDS 2015. DSTC4 was organized by Seokhwan Kim, Luis F. DHaro, Rafael E Banchs, Matthew Henderson, and Jason D. Williams.

DSTC5.[10] DSTC5 consists of human–human dialogs in the tourist information domain, where training dialogs were provided in one language, and test dialogs were in a different language. Results are presented in a special session at IEEE SLT 2016. DSTC5 was organized by Seokhwan Kim, Luis F. DHaro, Rafael E Banchs, Matthew Henderson, Jason D. Williams, and Koichiro Yoshino.

3.8.4 Maluuba Frames Dataset

Frames[11] is presented to for research in conversational agents which can support decision-making in complex settings, i.e., booking a vacation including flights and a hotel. With this dataset the goal is to teach conversational agents that can help users explore a database, compare items, and reach a decision. The human–human conversation frames data is collected using Wizard-of-Oz, which is designed for composite task completion dialog setting. we consider an important type of complex task, called

[7] https://www.microsoft.com/en-us/research/event/dialog-state-tracking-challenge/.

[8] http://camdial.org/~mh521/dstc/.

[9] http://www.colips.org/workshop/dstc4/.

[10] http://workshop.colips.org/dstc5/.

[11] https://datasets.maluuba.com/Frames.

composite task, which consists of a set of subtasks that need to be fulfilled collectively. For example, in order to make a travel plan, the user first needs to book air tickets, reserve a hotel, rent a car, etc., in a collective way so as to satisfy a set of cross-subtask constraints, which are called *slot constraints*. Examples of slot constraints for travel planning are: hotel check-in time should be later than the departure flight time, hotel check-out time may be earlier than the return flight depart time, the number of flight tickets equals to that of hotel check-in people, and so on.

3.8.5 Facebook's Dialog Datasets

In the last year, Facebook AI and Research (FAIR) has released task oriented dialog datasets to be used by the dialog research community (Bordes et al. 2017).[12] The objective of their project is to explore neural network architectures for question answering and goal-oriented dialog systems. They designed a set of five tasks within the goal-oriented context of restaurant reservation (see example in Fig. 3.7). Grounded with an underlying KB of restaurants and their properties (location, type of cuisine, etc.), these tasks cover several dialog stages and test if models can learn various abilities such as performing dialog management, querying KBs, interpreting the output of such queries to continue the conversation or dealing with new entities not appearing in dialogs from the training set.

3.8.6 Ubuntu Dialog Corpus

The Ubuntu Dialog Corpus Lowe et al. (2015b)[13] consists of almost one million two-person conversations extracted from the Ubuntu chat logs about technical support for various Ubuntu-related problems. The dataset targets a specific technical support domain. Therefore, it can be used as a case study for the development of AI agents in targeted applications, in contrast to chatbox systems. All conversations are carried out in text form (not audio). The dataset is orders of magnitude larger than structured corpora such as those of the DSTC. Each conversation in their dataset includes several turns, as well as long utterances.

3.9 Open Source Dialog Software

Conversational dialog systems have been the focus of many leading companies and researchers in the field have been building systems to improve several components of

---

[12]https://github.com/facebookresearch/ParlAI.

[13]https://github.com/rkadlec/ubuntu-ranking-dataset-creator.

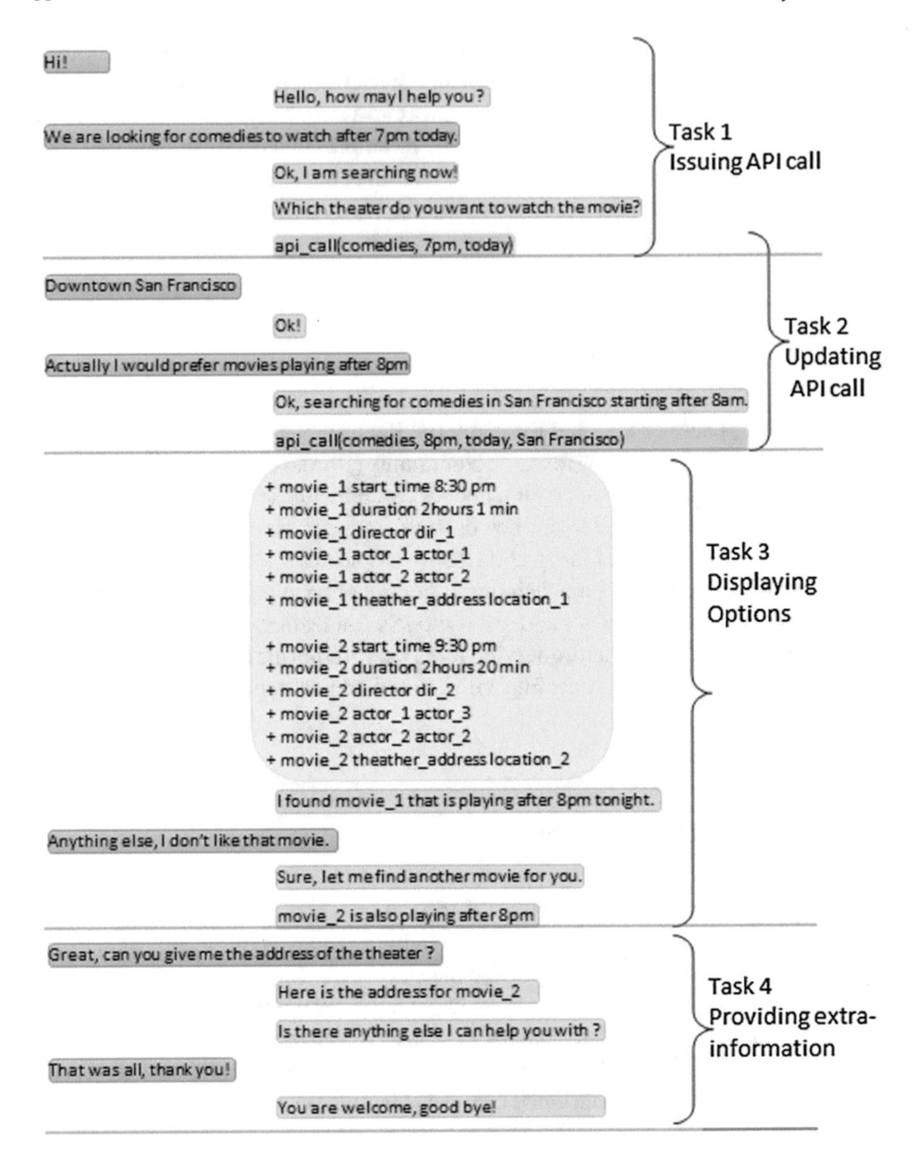

Fig. 3.7 A sample dialog between a virtual agent and a customer, in restaurant domain

the conversational dialog systems. Some work just focus on proving trainable datasets and labeling platforms, or machine learning algorithms that can learn through interaction, others provide environment (simulators) to train interactive dialog systems. Below, we briefly summarize the open source software/platforms that are readily accessible for dialog researchers.

- **OpenDial**[14]: The toolkit has been originally developed by the Language Technology Group of the University of Oslo (Norway), with Pierre Lison as main developer. It is a Java-based, domain-independent toolkit for developing spoken dialog systems. OpenDial provides a tool to build full-fledged, end-to-end dialog system, integrating speech recognition, language understanding, generation and speech synthesis. The purpose of OpenDial is to combine the benefits of logical and statistical approaches to dialog modeling into a single framework. The toolkit relies on probabilistic rules to represent the domain models in a compact and human-readable format. Supervised or reinforcement learning techniques can be applied to automatically estimate unknown rule parameters from relatively small amounts of data (Lison 2013). The tool also enables to incorporate expert knowledge and domain-specific constraints in a robust, probabilistic framework.
- **ParlAI**: Along with the datasets, Facebook AI and Research (FAIR) have released a platform entitled ParlAI[15] with the goal of providing researchers a unified framework for training and testing dialog models, multitask training over many datasets at once as well as seamless integration of Amazon Mechanical Turk for data collection and human evaluation.
- **Alex Dialog Systems Framework**[16]: This is a dialog systems framework that facilitates research into and development of spoken dialog system. It is provided by a group at UFAL[17]—the Institute of Formal and Applied Linguistics, Faculty of Mathematics and Physics, Charles University in Prague, Czech Republic. The tool provides baseline components that are required for a building spoken dialog systems as well as provides additional tools for processing dialog system interactions logs, e.g., for audio transcription, semantic annotation, or spoken dialog system evaluation.
- **SimpleDS**: This is a simple deep reinforcement learning dialog system[18] that enables training dialog agents with as little human intervention as possible. It includes the Deep Q-Learning with experience replay (Mnih et al. 2013) and provides support for multi-threaded and client–server processing, and fast learning via constrained search spaces.
- **Cornell Movie Dialogs Corpus**: This corpus contains a large metadata-rich collection of fictional conversations extracted from raw movie scripts (Mizil and Lee 2011). It contains several conversational exchanges between pairs of movie characters.
- **Others**: There are numerous software applications (some open sourced) that also provide non-task-oriented dialog systems, e.g., chit-chat dialog systems. Such systems provide machine learning tools and conversational dialog engine for creating chat bots. Examples include **Chatterbot**,[19] a conversational dialog engine for

[14]https://github.com/plison/opendial.

[15]https://github.com/facebookresearch/ParlAI.

[16]https://github.com/UFAL-DSG/alex.

[17]http://ufal.mff.cuni.cz/.

[18]https://github.com/cuayahuitl/SimpleDS.

[19]https://github.com/gunthercox/ChatterBot.

creating chat bots, **chatbot-rnn**,[20] a toy chatbot powered by deep learning and trained on data from Reddit, to name a few. In metaguide.com,[21] top 100 chatbots are listed.

3.10 Dialog System Evaluation

Throughout this chapter, we have been investigated several types of dialog models, i.e., task oriented, which are considered domain dependent as well as open-domain dialog software, which are semi-domain dependent which can open ended or can switch back and froth between task-oriented and open-domain conversational dialogs.

The task-oriented dialog systems, which are typically component base, are evaluated based on the performance of each individual component. For instance, the CLU is evaluated based on the performance of the intent detection model, the slot sequence tagging models (Hakkani-Tür et al. 2016; Celikyilmaz et al. 2016; Tur and De Mori 2011; Chen et al. 2016), etc., whereas the dialog state tracker is evaluated based on the accuracy of the state changes discovered during the dialog turns. The dialog policy for task-oriented systems is typically evaluated based on the success rate of the completed task judged by either user or the real human. Typically, evaluation is done using human-generated supervised signals, such as a task completion test or a user satisfaction score. Also the length of the dialog has played role in shaping the dialog policy (Schatzmann et al. 2006).

The real problem in evaluating the dialog models performance arises when the dialog systems are open domain. Most approaches focus on evaluating the dialog response generation systems, which are trained to produce a reasonable utterance given a conversational context. This is a very challenging task since automatically evaluating language generation models is intractable to the availability of possibly very large set of correct answers. Nevertheless, today, several performance measures are used to automatically evaluate how appropriate the proposed response is to the conversation (Liu et al. 2016). Most of these metrics compare the generated response to the ground truth response of the conversation using word based similarity metrics and word-embedding based similarity metrics. Below, we will summarize some of the metrics that are most commonly used in the dialog systems:

BLEU (Papineni et al. 2002) is an algorithm for evaluating the quality of text by investigating the co-occurrences of n-grams in the ground truth sequence (text) and the generated responses. BLEU uses a modified form of precision to compare a candidate translation against multiple reference translations:

$$P_n(r, \hat{r}) = \frac{\sum_k \min(h(k, r), h(k, \hat{r}_i))}{\sum_k h(k, r_i)},$$

[20] https://github.com/pender/chatbot-rnn.

[21] http://meta-guide.com/software-meta-guide/100-best-github-chatbot.

where k represents all possible n-grams and $h(k, r)$ is the number of n-grams k in r. The metric modifies simple precision since text generation systems have been known to generate more words than are in a reference text. Such a score would favor shorter sequences. To remedy that, in Papineni et al. (2002) a brevity score is used which yields BLUE-N score, where N is the maximum length of the n-grams and is defined as :

$$BLEU\text{-}N = b(r, \hat{r}) \exp(\sum_{n=1}^{N})\beta_n \log P_n(r, \hat{r})),$$

where β_n is the weight factor and $b(\cdot)$ is the brevity penalty.

METEOR (Banerjee and Lavie 2005) is another method which is based on BLEU and is introduced to address several weaknesses of BLEU. As with BLEU, the basic unit of evaluation is the sentence, the algorithm first creates an alignment between the reference and candidate generated sentences. The alignment is a set of mappings between unigrams and has to comply with several constraints including the fact that every unigram in the candidate translation must map to zero or one unigram in the reference followed by WordNet synonym matching, stemmed tokens and paraphrases of text. The METEOR score is calculated as the harmonic mean of precision and recall between the proposed and ground truth sentence given the set of alignments.

ROUGE (Lin 2004) is another evaluation metric mainly used to evaluate the automatic summarization systems. There are five different extensions of ROUGE available: ROUGE-N, on N-gram based co-occurrence statistics; ROUGE-L, Longest Common Subsequence (LCS) based statistics (Longest common subsequence problem takes into account sentence-level structure similarity naturally and identifies longest co-occurring in sequence n-grams automatically.); ROUGE-W, weighted LCS-based statistics that favors consecutive LCSes; ROUGE-S, skip-bigram based co-occurrence statistics (Skip-bigram is any pair of words in their sentence order.); and ROUGE-SU, skip-bigram plus unigram-based co-occurrence statistics. In text generation, ROUGE-L is the most commonly used metric in text generation tasks because the LCS is easy to measure the similarity between two sentences in the same order.

Embedding-Based approaches consider the meaning of each word as defined by a word embedding, which assigns a vector to each word as opposed to the rest of the above metrics that consider n-gram matching scenarios. A word embedding learning method such as the one from Mikolov et al. (2013) is used to calculate these embeddings using distributional semantics; that is, they approximate the meaning of a word by considering how often it co-occurs with other words in the corpus. These embedding-based metrics usually approximate sentence-level embeddings using some heuristic to combine the vectors of the individual words in the sentence. The sentence-level embeddings between the generated and reference response are compared using a measure such as cosine distance.

RUBER (Tao et al. 2017) is a Referenced metric and Unreferenced metric Blended Evaluation Routine for open-domain dialog systems. RUBER has the following dis-

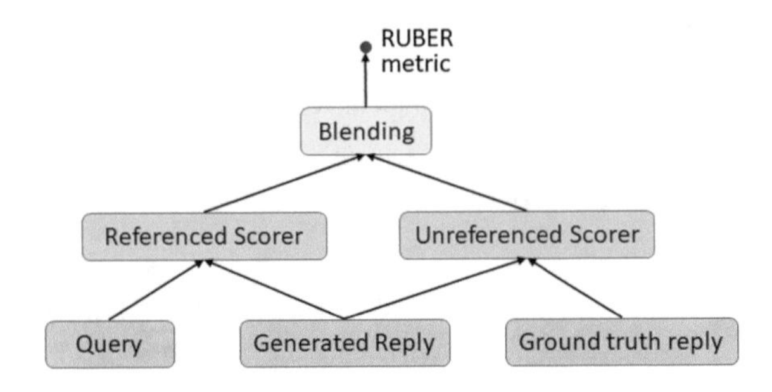

Fig. 3.8 Overview of RUBER metric

tinct features: (i) An embedding-based scorer named *referenced metric*, which measures the similarity between a generated reply and the ground truth. Instead of using word-overlapping information (as in BLEU and ROUGE), RUBER's reference metric measures the similarity by pooling of word embeddings (Forgues et al. 2014) which is more suited to dialog systems due to the diversity of replies. (ii) A neural network-based scorer named *unreferenced metric* that measures the relatedness between the generated reply and its query. This scorer is unreferenced because it does not refer to ground truth and requires no manual annotation labels. (iii) The referenced and unreferenced metrics are combined with strategies like averaging further improves the performance (see Fig. 3.8).

3.11 Summary

This chapter presents an extensive survey on current approaches in data-driven dialog modeling that use deep learning technologies, after some detailed introduction to various components of a spoken dialog system including speech recognition, language understanding (spoken or text-based), dialog manager, and language generation (spoken or text-based) . The chapter also describes available deep dialog modeling software and datasets suitable for research, development, and evaluation.

Deep learning technologies have yielded recent improvements in dialog systems as well as new research activities. Most of the current dialog systems and research on them are moving towards large-scale data-driven and specifically end-to-end trainable models. In addition to the current new approaches and datasets, also highlighted in this chapter are potential future directions in building conversational dialog systems including hierarchical structures, multi-agent systems as well as domain adaptation.

Dialog systems, especially the spoken version, are a representative instance of multiple-stage information processing exemplified in NLP. The multiple stages include speech recognition, language understanding (Chap. 2), decision-making (via dialog manager), and language/speech generation. Such multiple-stage processing schemes suit ideally well deep learning methodology, which is based on end-to-end learning in multiple-layered (or deep) systems. The current progress in applying deep learning to dialog systems as reviewed, in this chapter, has largely been limited to using deep learning to modeling and optimizing each individual processing stage in the overall system. The future progress is expected to broaden such a scope and to succeed in the fully end-to-end systems.

References

Asri, L. E., He, J., & Suleman, K. (2016). A sequence-to-sequence model for user simulation in spoken dialogue systems. *Interspeech.*

Aust, H., Oerder, M., Seide, F., & Steinbiss, V. (1995). The philips automatic train timetable information system. *Speech Communication, 17*, 249–262.

Banchs, R. E., & Li., H. (2012). Iris: A chat-oriented dialogue system based on the vector space model. *ACL.*

Banerjee, S., & Lavie, A. (2005). Meteor: An automatic metric for mt evaluation with improved correlation with human judgments. In *ACL Workshop on Intrinsic and Extrinsic Evaluation Measures for Machine Translation and/or Summarization.*

Bapna, A., Tur, G., Hakkani-Tur, D., & Heck, L. (2017). Improving frame semantic parsing with hierarchical dialogue encoders.

Bateman, J., & Henschel, R. (1999). From full generation to near-templates without losing generality. In *KI'99 Workshop, "May I Speak Freely?".*

Blundell, C., Cornebise, J., Kavukcuoglu, K., & Wierstra, D. (2015). Weight uncertainty in neural networks. *ICML.*

Bordes, A., Boureau, Y.-L., & Weston, J. (2017). Learning end-to-end goal-oriented dialog. In *ICLR 2017*

Busemann, S., & Horacek, H. (1998). A flexible shallow approach to text generation. In *International Natural Language Generation Workshop, Niagara-on-the-Lake, Canada*

Celikyilmaz, A., Sarikaya, R., Hakkani-Tur, D., Liu, X., Ramesh, N., & Tur, G. (2016). A new pre-training method for training deep learning models with application to spoken language understanding. In *Proceedings of Interspeech* (pp. 3255–3259).

Chen, Y.-N., Hakkani-Tür, D., Tur, G., Gao, J., & Deng, L. (2016). End-to-end memory networks with knowledge carryover for multi-turn spoken language understanding. In *Proceedings of The 17th Annual Meeting of the International Speech Communication Association (INTERSPEECH)*, San Francisco, CA. ISCA.

Crook, P., & Marin, A. (2017). Sequence to sequence modeling for user simulation in dialog systems. *Interspeech.*

Cuayahuitl, H. (2016). Simpleds: A simple deep reinforcement learning dialogue system. In *International Workshop on Spoken Dialogue Systems (IWSDS).*

Cuayahuitl, H., Yu, S., Williamson, A., & Carse, J. (2016). Deep reinforcement learning for multi-domain dialogue systems. arXiv:1611.08675.

Dale, R., & Reiter, E. (2000). *Building natural language generation systems.* Cambridge, UK: Cambridge University Press.

Deng, L. (2016). Deep learning from speech recognition to language and multi-modal processing. In *APSIPA Transactions on Signal and Information Processing.* Cambridge University Press.

Deng, L., & Yu, D. (2015). *Deep learning: Methods and applications.* NOW Publishers.

Deng, L., & Li, X. (2013). Machine learning paradigms for speech recognition: An overview. *IEEE Transactions on Audio, Speech, and Language Processing, 21*(5), 1060–1089.

Dhingra, B., Li, L., Li, X., Gao, J., Chen, Y.-N., Ahmed, F., & Deng, L. (2016a). End-to-end reinforcement learning of dialogue agents for information access. arXiv:1609.00777.

Dhingra, B., Li, L., Li, X., Gao, J., Chen, Y.-N., Ahmed, F., & Deng, L. (2016b). Towards end-to-end reinforcement learning of dialogue agents for information access. *ACL.*

Dodge, J., Gane, A., Zhang, X., Bordes, A., Chopra, S., Miller, A., Szlam, A., & Weston, J. (2015). Evaluating prerequisite qualities for learning end-to-end dialog systems. arXiv:1511.06931.
Elhadad, M., & Robin, J. (1996). An overview of surge: A reusable comprehensive syntactic realization component. *Technical Report 96-03, Department of Mathematics and Computer Science, Ben Gurion University, Beer Sheva, Israel.*
Fatemi, M., Asri, L. E., Schulz, H., He, J., & Suleman, K. (2016a). Policy networks with two-stage training for dialogue systems. arXiv:1606.03152.
Fatemi, M., Asri, L. E., Schulz, H., He, J., & Suleman, K. (2016b). Policy networks with two-stage training for dialogue systems. arXiv:1606.03152.
Forgues, G., Pineau, J., Larcheveque, J.-M., & Tremblay, R. (2014). Bootstrapping dialog systems with word embeddings. *NIPS ML-NLP Workshop*.
Gai, M., Mrki, N., Su, P.-H., Vandyke, D., Wen, T.-H., & Young, S. (2015). Policy committee for adaptation in multi-domain spoken dialogue sytems. *ASRU*.
Gai, M., Mrki, N., Rojas-Barahona, L. M., Su, P.-H., Ultes, S., Vandyke, D., et al. (2016). Dialogue manager domain adaptation using Gaussian process reinforcement learning. *Computer Speech and Language*, *45*, 552–569.
Gasic, M., Jurcicek, F., Keizer, S., Mairesse, F., Thomson, B., Yu, K., & Young, S. (2010). Gaussian processes for fast policy optimisation of POMDP-based dialogue managers. In *SIGDIAL*.
Gasic, M., Mrksic, N., Su, P.-H., Vandyke, D., & Wen, T.-H. (2015). Multi-agent learning in multi-domain spoken dialogue systems. *NIPS workshop on Spoken Language Understanding and Interaction*.
Ge, W., & Xu, B. (2016). Dialogue management based on multi-domain corpus. In *Special Interest Group on Discourse and Dialog*.
Georgila, K., Henderson, J., & Lemon, O. (2005). Learning user simulations for information state update dialogue systems. In *9th European Conference on Speech Communication and Technology (INTERSPEECH—EUROSPEECH)*.
Georgila, K., Henderson, J., & Lemon, O. (2006). User simulation for spoken dialogue systems: Learning and evaluation. In *INTERSPEECH—EUROSPEECH*.
Goller, C., & Kchler, A. (1996). Learning task-dependent distributed representations by backpropagation through structure. *IEEE*.
Goodfellow, I., Pouget-Abadie, J., Mirza, M., Xu, B., Warde-Farley, D., Ozair, S., Courville, A., & Bengio, Y. (2014). Generative adversarial nets. In *NIPS*.
Gorin, A. L., Riccardi, G., & Wright, J. H. (1997). How may i help you? *Speech Communication*, *23*, 113–127.
Graves, A., & Schmidhuber, J. (2005). Framewise phoneme classification with bidirectional lstm and other neural network architectures. *Neural Networks*, *18*, 602–610.
Hakkani-Tür, D., Tur, G., Celikyilmaz, A., Chen, Y.-N., Gao, J., Deng, L., & Wang, Y.-Y. (2016). Multi-domain joint semantic frame parsing using bi-directional rnn-lstm. In *Proceedings of Interspeech* (pp. 715–719).
Hastie, T., Tibshirani, R., & Friedman, J. (2009). *The Elements of Statistical Learning: Data Mining, Inference, and Prediction*. Berlin: Springer.
He, X., & Deng, L. (2011). Speech recognition, machine translation, and speech translation a unified discriminative learning paradigm. In *IEEE Signal Processing Magazine*.
He, X., & Deng, L. (2013). Speech-centric information processing: An optimization-oriented approach. In *IEEE*.
He, J., Chen, J., He, X., Gao, J., Li, L., Deng, L., & Ostendorf, M. (2016). Deep reinforcement learning with a natural language action space. *ACL*.
Hemphill, C. T., Godfrey, J. J., & Doddington, G. R. (1990). The ATIS spoken language systems pilot corpus. In *DARPA Speech and Natural Language Workshop*.
Henderson, M., Thomson, B., & Williams, J. D. (2014). The third dialog state tracking challenge. In *2014 IEEE, Spoken Language Technology Workshop (SLT)* (pp. 324–329). IEEE.
Henderson, M., Thomson, B., & Young, S. (2013). Deep neural network approach for the dialog state tracking challenge. In *Proceedings of the SIGDIAL 2013 Conference* (pp. 467–471).

Higashinaka, R., Imamura, K., Meguro, T., Miyazaki, C., Kobayashi, N., Sugiyama, H., et al. (2014). Towards an open-domain conversational system fully based on natural language processing. *COLING*.

Hinton, G., Deng, L., Yu, D., Dahl, G., Rahman Mohamed, A., Jaitly, N., et al. (2012). Deep neural networks for acoustic modeling in speech recognition. *IEEE Signal Processing Magazine*, *29*(6), 82–97.

Huang, X., & Deng, L. (2010). An overview of modern speech recognition. In *Handbook of Natural Language Processing* (2nd ed., Chapter 15).

Huang, P.-S., He, X., Gao, J., Deng, L., Acero, A., & Heck, L. (2013). Learning deep structured semantic models for web search using click-through data. In *ACM International Conference on Information and Knowledge Management (CIKM)*.

Jaech, A., Heck, L., & Ostendorf, M. (2016). *Domain adaptation of recurrent neural networks for natural language understanding*.

Kannan, A., & Vinyals, O. (2016). Adversarial evaluation of dialog models. In *Workshop on Adversarial Training, NIPS 2016, Barcelona, Spain*.

Kim, Y.-B., Stratos, K., & Kim, D. (2017a). Adversarial adaptation of synthetic or stale data. *ACL*.

Kim, Y.-B., Stratos, K., & Kim, D. (2017b). Domain attention with an ensemble of experts. *ACL*.

Kim, Y.-B., Stratos, K., & Sarikaya, R. (2016a). Domainless adaptation by constrained decoding on a schema lattice. *COLING*.

Kim, Y.-B., Stratos, K., & Sarikaya, R. (2016b). Frustratingly easy neural domain adaptation. *COLING*.

Kumar, A., Irsoy, O., Su, J., Bradbury, J., English, R., Pierce, B., et al. (2015). Ask me anything: Dynamic memory networks for natural language processing. In *Neural Information Processing Systems (NIPS)*.

Kurata, G., Xiang, B., Zhou, B., & Yu, M. (2016). Leveraging sentence level information with encoder lstm for natural language understanding. arXiv:1601.01530.

Langkilde, I., & Knight, K. (1998). Generation that exploits corpus-based statistical knowledge. *ACL*.

LeCun, Y., Bottou, L., Bengio, Y., & Haffner, P. (1998). Gradient-based learning applied to document recognition. *IEEE*, *86*, 2278–2324.

Lemon, O., & Rieserr, V. (2009). Reinforcement learning for adaptive dialogue systems—tutorial. *EACL*.

Li, L., Balakrishnan, S., & Williams, J. (2009). Reinforcement learning for dialog management using least-squares policy iteration and fast feature selection. *InterSpeech*.

Li, J., Galley, M., Brockett, C., Gao, J., & Dolan, B. (2016a). A diversity-promoting objective function for neural conversation models. *NAACL*.

Li, J., Galley, M., Brockett, C., Spithourakis, G. P., Gao, J., & Dolan, B. (2016b). A persona based neural conversational model. *ACL*.

Li, J., Monroe, W., Shu, T., Jean, S., Ritter, A., & Jurafsky, D. (2017). Adversarial learning for neural dialogue generation. arXiv:1701.06547.

Li, J., Deng, L., Gong, Y., & Haeb-Umbach, R. (2014). An overview of noise-robust automatic speech recognition. *IEEE/ACM Transactions on Audio, Speech, and Language Processing*, *22*(4), 745–777.

Lin, C.-Y. (2004). Rouge: A package for automatic evaluation of summaries. *In Text summarization branches out: ACL-04 Workshop*.

Lipton, Z. C., Li, X., Gao, J., Li, L., Ahmed, F., & Deng, L. (2016). Efficient dialogue policy learning with bbq-networks. arXiv.org.

Lison, P. (2013). *Structured probabilistic modelling for dialogue management*. Department of Informatics Faculty of Mathematics and Natural Sciences University of Osloe.

Liu, B., & Lane, I. (2016a). Attention-based recurrent neural network models for joint intent detection and slot filling. *Interspeech*.

Liu, B., & Lane, I. (2016b). Attention-based recurrent neural network models for joint intent detection and slot filling. In *SigDial*.

Liu, C.-W., Lowe, R., Serban, I. V., Noseworthy, M., Charlin, L., & Pineau, J. (2016). How not to evaluate your dialogue system: An empirical study of unsupervised evaluation metrics for dialogue response generation. *EMNLP*.

Lowe, R., Pow, N., Serban, I. V., and Pineau, J. (2015b). The ubuntu dialogue corpus: A large dataset for research in unstructure multi-turn dialogue systems. In *SIGDIAL 2015*.

Lowe, R., Pow, N., Serban, I. V., Charlin, L., and Pineau, J. (2015a). Incorporating unstructured textual knowledge sources into neural dialogue systems. In *Neural Information Processing Systems Workshop on Machine Learning for Spoken Language Understanding*.

Mairesse, F., & Young, S. (2014). Stochastic language generation in dialogue using factored language models. *Computer Linguistics*.

Mairesse, F. and Walker, M. A. (2011). Controlling user perceptions of linguistic style: Trainable generation of personality traits. *Computer Linguistics*.

Mesnil, G., Dauphin, Y., Yao, K., Bengio, Y., Deng, L., Hakkani-Tur, D., et al. (2015). Using recurrent neural networks for slot filling in spoken language understanding. *IEEE/ACM Transactions on Audio, Speech, and Language Processing*, *23*(3), 530–539.

Mikolov, T., Sutskever, I., Chen, K., Corrado, G. S., & Dean, J. (2013). Distributed representations of words and phrases and their compositionality. In *Advances in neural information processing systems* (pp. 3111–3119).

Mizil, C. D. N. & Lee, L. (2011). Chameleons in imagined conversations: A new approach to understanding coordination of linguistic style in dialogs. In *Proceedings of the Workshop on Cognitive Modeling and Computational Linguistics, ACL 2011*.

Mnih, V., Kavukcuoglu, K., Silver, D., Graves, A., Antonoglou, I., Wierstra, D., & Riedmiller, M. (2013). Playing Atari with deep reinforcement learning. *NIPS Deep Learning Workshop*.

Mrkšić, N., Séaghdha, D. Ó., Wen, T.-H., Thomson, B., & Young, S. (2016). Neural belief tracker: Data-driven dialogue state tracking. arXiv:1606.03777.

Oh, A. H., & Rudnicky, A. I. (2000). Stochastic language generation for spoken dialogue systems. *ANLP/NAACL Workshop on Conversational Systems*.

Papineni, K., Roukos, S., Ward, T., & Zhu, W. (2002). Bleu: A method for automatic evaluation of machine translation. In *40th annual meeting on Association for Computational Linguistics (ACL)*.

Passonneau, R. J., Epstein, S. L., Ligorio, T., & Gordon, J. (2011). Embedded wizardry. In *SIGDIAL 2011 Conference*.

Peng, B., Li, X., Li, L., Gao, J., Celikyilmaz, A., Lee, S., & Wong, K.-F. (2017). *Composite task-completion dialogue system via hierarchical deep reinforcement learning*. arxiv:1704.03084v2.

Pietquin, O., Geist, M., & Chandramohan, S. (2011a). Sample efficient on-line learning of optimal dialogue policies with kalman temporal differences. In *IJCAI 2011, Barcelona, Spain*.

Pietquin, O., Geist, M., Chandramohan, S., & FrezzaBuet, H. (2011b). Sample-efficient batch reinforcement learning for dialogue management optimization. *ACM Transactions on Speech and Language Processing*.

Ravuri, S., & Stolcke, A. (2015). Recurrent neural network and LSTM models for lexical utterance classification. In *Sixteenth Annual Conference of the International Speech Communication Association*.

Ritter, A., Cherry, C., & Dolan., W. B. (2011). Data-driven response generation in social media. *Empirical Methods in Natural Language Processing*.

Sarikaya, R., Hinton, G. E., & Ramabhadran, B. (2011). Deep belief nets for natural language call-routing. In *2011 IEEE International Conference on Acoustics, Speech and Signal Processing (ICASSP)* (pp. 5680–5683). IEEE.

Sarikaya, R., Hinton, G. E., & Deoras, A. (2014). Application of deep belief networks for natural language understanding. *IEEE/ACM Transactions on Audio, Speech, and Language Processing*, *22*(4), 778–784.

Schatzmann, J., Weilhammer, K., & Matt Stutle, S. Y. (2006). A survey of statistical user simulation techniques for reinforcement-learning of dialogue management strategies. *The Knowledge Engineering Review*.

Serban, I., Klinger, T., Tesauro, G., Talamadupula, K., Zhou, B., Bengio, Y., & Courville, A. (2016a). Multiresolution recurrent neural networks: An application to dialogue response generation. arXiv:1606.00776v2
Serban, I., Sordoni, A., & Bengio, Y. (2017). A hierarchical latent variable encoder-decoder model for generating dialogues. *AAAI*.
Serban, I. V., Sordoni, A., Bengio, Y., Courville, A., & Pineau, J. (2015). Building end-to-end dialogue systems using generative hierarchical neural network models. *AAAI*.
Serban, I. V., Sordoni, A., Bengio, Y., Courville, A., & Pineau, J. (2016b). Building end-to-end dialogue systems using generative hierarchical neural networks. *AAAI*.
Shah, P., Hakkani-Tur, D., & Heck, L. (2016). Interactive reinforcement learning for task-oriented dialogue management. *SIGDIAL*.
Shang, L., Lu, Z., & Li, H. (2015). Neural responding machine for short text conversation. *ACL-IJCNLP*.
Simonnet, E., Camelin, N., Deléglise, P., & Estève, Y. (2015). Exploring the use of attention-based recurrent neural networks for spoken language understanding. In *Machine Learning for Spoken Language Understanding and Interaction NIPS 2015 Workshop (SLUNIPS 2015)*.
Simpson, A. & Eraser, N. M. (1993). Black box and glass box evaluation of the sundial system. In *Third European Conference on Speech Communication and Technology*.
Singh, S. P., Kearns, M. J., Litman, D. J., & Walker, M. A. (2016). Reinforcement learning for spoken dialogue systems. *NIPS*.
Sordoni, A., Galley, M., Auli, M., Brockett, C., Ji, Y., Mitchell, M., et al. (2015a). A neural network approach to context-sensitive generation of conversational responses. In *North American Chapter of the Association for Computational Linguistics (NAACL-HLT 2015)*.
Sordoni, A., Galley, M., Auli, M., Brockett, C., Ji, Y., Mitchell, M., Nie, J.-Y., et al. (2015b). A neural network approach to context-sensitive generation of conversational responses. In *Proceedings of the 2015 Conference of the North American Chapter of the Association for Computational Linguistics: Human Language Technologies* (pp. 196–205), Denver, Colorado. Association for Computational Linguistics.
Stent, A. (1999). Content planning and generation in continuous-speech spoken dialog systems. In *KI'99 workshop, "May I Speak Freely?"*.
Stent, A., Prasad, R., & Walker, M. (2004). Trainable sentence planning for complex information presentation in spoken dialog systems. *ACL*.
Su, P.-H., Gasic, M., Mrksic, N., Rojas-Barahona, L., Ultes, S., Vandyke, D., et al. (2016). On-line active reward learning for policy optimisation in spoken dialogue systems. arXiv:1605.07669.
Sukhbaatar, S., Weston, J., Fergus, R., et al. (2015). End-to-end memory networks. In *Advances in neural information processing systems* (pp. 2440–2448).
Sutton, R. S., & Singh, S. P. (1999). Between mdps and semi-MDPs: A framework for temporal abstraction in reinforcement learning. *Artificial Intelligence*, *112*, 181–211.
Tafforeau, J., Bechet, F., Artières, T., & Favre, B. (2016). Joint syntactic and semantic analysis with a multitask deep learning framework for spoken language understanding. In *Interspeech* (pp. 3260–3264).
Tao, C., Mou, L., Zhao, D., & Yan, R. (2017). *Ruber: An unsupervised method for automatic evaluation of open-domain dialog systems*. ArXiv2017.
Thomson, B., & Young, S. (2010). Bayesian update of dialogue state: A POMDP framework for spoken dialogue systems. *Computer Speech and Language*, *24*(4), 562–588.
Tur, G., Deng, L., Hakkani-Tür, D., & He, X. (2012). Towards deeper understanding: Deep convex networks for semantic utterance classification. In *2012 IEEE International Conference on Acoustics, Speech and Signal Processing (ICASSP)* (pp. 5045–5048). IEEE.
Tur, G., & Deng, L. (2011). *Intent determination and spoken utterance classification, Chapter 4 in Book: Spoken language understanding*. New York, NY: Wiley.
Tur, G., & De Mori, R. (2011). *Spoken language understanding: Systems for extracting semantic information from speech*. New York: Wiley.
Vinyals, O., & Le, Q. (2015). A neural conversational model. arXiv:1506.05869.

Walker, M., Stent, A., Mairesse, F., & Prasad, R. (2007). Individual and domain adaptation in sentence planning for dialogue. *Journal of Artificial Intelligence Research.*

Wang, Z., Stylianou, Y., Wen, T.-H., Su, P.-H., & Young, S. (2015). Learning domain-independent dialogue policies via ontology parameterisation. In *SIGDAIL.*

Wen, T.-H., Gasic, M., Mrksic, N., Rojas-Barahona, L. M., Pei-Hao, P., Ultes, S., et al. (2016a). *A network-based end-to-end trainable task-oriented dialogue system.* arXiv.

Wen, T.-H., Gasic, M., Mrksic, N., Rojas-Barahona, L. M., Su, P.-H., Ultes, S., et al. (2016b). A network-based end-to-end trainable task-oriented dialogue system. arXiv:1604.04562.

Wen, T.-H., Gasic, M., Mrksic, N., Su, P.-H., Vandyke, D., & Young, S. (2015a). Semantically conditioned LSTM-based natural language generation for spoken dialogue systems. *EMNLP.*

Wen, T.-H., Gasic, M., Mrksic, N., Su, P.-H., Vandyke, D., & Young, S. (2015b). Semantically conditioned LSTM-based natural language generation for spoken dialogue systems. arXiv:1508.01745

Weston, J., Chopra, S., & Bordesa, A. (2015). Memory networks. In *International Conference on Learning Representations (ICLR).*

Williams, J. D., & Zweig, G. (2016a). End-to-end LSTM-based dialog control optimized with supervised and reinforcement learning. arXiv:1606.01269.

Williams, J. D., & Zweig, G. (2016b). *End-to-end LSTM-based dialog control optimized with supervised and reinforcement learning.* arXiv.

Williams, J. D., Raux, A., Ramachandran, D., & Black, A. W. (2013). The dialog state tracking challenge. In *SIGDIAL Conference* (pp. 404–413).

Williams, J., Raux, A., & Handerson, M. (2016). The dialog state tracking challenge series: A review. *Dialogue and Discourse*, *7*(3), 4–33.

Xu, P., & Sarikaya, R. (2013). Convolutional neural network based triangular CRF for joint intent detection and slot filling. In *2013 IEEE Workshop on Automatic Speech Recognition and Understanding (ASRU)* (pp. 78–83). IEEE.

Yao, K., Zweig, G., Hwang, M.-Y., Shi, Y., & Yu, D. (2013). Recurrent neural networks for language understanding. In *INTERSPEECH* (pp. 2524–2528).

Yu, Z., Black, A., & Rudnicky, A. I. (2017). Learning conversational systems that interleave task and non-task content. arXiv:1703.00099v1.

Yu, Y., Eshghi, A., & Lemon, O. (2016). Training an adaptive dialogue policy for interactive learning of visually grounded word meanings. *SIGDIAL.*

Yu, Z., Papangelis, A., & Rudnicky, A. (2015). Ticktock: A non-goal-oriented multimodal dialog system with engagement awareness. In *AAAI Spring Symposium.*

Yu, D., & Deng, L. (2015). *Automatic speech recognition: A deep learning approach.* Berlin: Springer.

Chapter 4
Deep Learning in Lexical Analysis and Parsing

Wanxiang Che and Yue Zhang

Abstract Lexical analysis and parsing tasks model the deeper properties of the words and their relationships to each other. The commonly used techniques involve word segmentation, part-of-speech tagging and parsing. A typical characteristic of such tasks is that the outputs are structured. Two types of methods are usually used to solve these structured prediction tasks: graph-based methods and transition-based methods. Graph-based methods differentiate output structures based on their characteristics directly, while transition-based methods transform output construction processes into state transition processes, differentiating sequences of transition actions. Neural network models have been successfully used for both graph-based and transition-based structured prediction. In this chapter, we give a review of applying deep learning in lexical analysis and parsing, and compare with traditional statistical methods.

4.1 Background

The properties of a word include its syntactic word categories (also known as **part of speech**, POS), **morphologies**, and so on (Manning and Schütze 1999). Obtaining these information is also known as **lexical analysis**. For languages like Chinese, Japanese, and Korean that do not separate words with whitespace, lexical analysis also includes the task of **word segmentation**, i.e., splitting a sequence of characters into words. Even in English, although whitespace is a strong clue for word boundaries, it is neither necessary nor sufficient. For example, in some situations, we might wish to treat *New York* as a single word. This is regarded as a **named entity recognition** (NER) problem (Shaalan 2014). On the other hand, punctuation marks are always adjacent to words. We also need to judge whether to segment them or not.

W. Che (✉)
Harbin Institute of Technology, Harbin, Heilongjiang, China
e-mail: wxche@ir.hit.edu.cn

Y. Zhang
Singapore University of Technology and Design, Singapore, Singapore
e-mail: yue_zhang@sutd.edu.sg

L. Deng and Y. Liu (eds.), *Deep Learning in Natural Language Processing*, https://doi.org/10.1007/978-981-10-5209-5_4

For languages like English, this is often called **tokenization** which is more a matter of convention than a serious research problem.

Once we know some properties of words, we may be interested in the relationships between them. The **parsing** task is to find and label words (or sequences of words) that are related to each other compositionally or recursively (Jurafsky and Martin 2009). There are two commonly used parses: **phrase-structure** (or **constituency**) parsing and **dependency** parsing.

All of these tasks can be regarded as structured prediction problems which is a term for supervised machine learning, i.e., the outputs are structured and influenced each other. Traditionally, huge amounts of human-designed handcrafted features are fed into a linear classifier to predict a score for each decision unit and then combine all of these scores together with satisfying some structured constraints. With the help of deep learning, we can employ an end-to-end learning paradigm which does not need costly feature engineering. The technology can even find more implicit features which are difficult to be designed by humans. Nowadays, deep learning has dominated these natural language processing tasks.

However, because of the pervasive problem of **ambiguity**, none of these tasks is trivial to predict. Some ambiguities may not even be noticed by humans.

This chapter is organized as follows. We will first select some typical tasks as examples to see where these ambiguities come from (Sect. 4.2). Then, we will review two typical structured prediction methods (Sect. 4.3): graph-based method (Sect. 4.3.1) and transition-based method (Sect. 4.3.2). Sections 4.4 and 4.5 are devoted to neural networks for graph-based and transition-based methods respectively. The chapter closes with a conclusion (Sect. 4.6).

4.2 Typical Lexical Analysis and Parsing Tasks

A natural language processing (lexical analysis and parsing here) pipeline usually includes three stages: word segmentation, POS tagging, and syntactic parsing.

4.2.1 Word Segmentation

As mentioned above, some languages, such as Chinese, are written in contiguous characters (Wong et al. 2009). Even though there are dictionaries to list all words, we cannot simply match words in a sequence of characters because ambiguity exists. For example, a Chinese sentence

- yanshouyibashoujiguanle (Shouyi Yan turned off the mobile phone)

can match words

- yanshouyi (Shouyi Yan)/ba (NA)/shouji (mobile phone)/guan (turn off)/le (NA)

which is a correct word segmentation result. However,

- yanshou (strictly)/yibashou (leader)/jiguan (office)/le (NA)
- yanshou (strictly)/yiba (handful)/shouji (mobile phone)/guan (turn off)/le (NA)
- yanshouyi (Shouyi Yan)/bashou (handle)/jiguan (office)/le (NA)

are also valid matching results but the sentence becomes meaningless with the segmentations. Obviously, the word matching method cannot distinguish which segmentation result is better than others. We need some kinds of scoring functions to assess the results.

4.2.2 POS Tagging

POS tagging is one of the most basic tasks in NLP, and it is useful in many natural language applications.[1] For example, the word *loves* can be a noun (plural of love) or a verb (third person present form of love). We can determine that *loves* is a verb but not noun in the following sentence.

- *The boy loves a girl*

The determent can be made independently without knowing the tags assigned to other words. Better POS taggers, however, take the word tags into consideration, because the tags of nearby a word can help to disambiguate its POS tag. In the example above, the following determiner *a* can help to indicate that *loves* is a verb.

Therefore, the complete POS tagging output of above sentence is a tag sequence, for example.

- *D N V D N*

(here we use *D* for a determiner, *N* for noun, and *V* for verb). The tag sequence has the same length as the input sentence, and therefore specifies a single tag for each word in the sentence (in this example *D* for *the*, *N* for *boy*, *V* for *loves*, and so on). Usually, the output of POS tagging can be written into a tagged sentence, where each word in the sentence is annotated with its corresponding POS tag, i.e., *The/D boy/N loves/V a/D girl/N*.

Like word segmentation, some sentences may have different meanings, if they are assigned with different POS tag sequences. For instance, two interpretations are possible for the sentence "Teacher strikes idle kids", depending on the POS assignments of the words in the sentence,

4.2.3 Syntactic Parsing

Phrase structures are very often constrained to correspond to the derivations of context-free grammars (CFGs) (Carnie 2012). In such a derivation, each phrase that

[1] https://en.wikipedia.org/wiki/Part-of-speech_tagging.

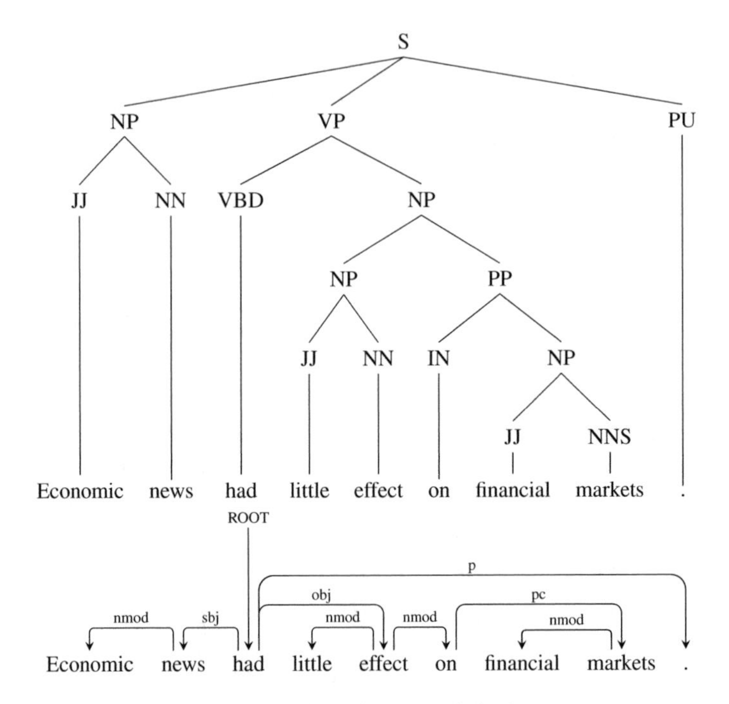

Fig. 4.1 Constituent tree (above) versus dependency tree (below)

is longer than one word is made of a sequence of non-overlapping "child" phrases or words, such that the children "cover" the yield of the parent phrase.

Another syntactic structure widely used in NLP is **dependency parse** tree (Kbler et al. 2009). A dependency parse is a directed tree where the words are vertices. Edges (also know as arcs) correspond to syntactic relations between two words and may be labeled with relation types. One extra pseudo word is the root of the tree, and each other word has a single in-bound edge from its syntactic head. For example, Fig. 4.1 shows constituent and dependency trees for the sentence, *Economic news had little effect on financial markets.*[2]

Dependency parsing can be classified into two categories: **projective** parsing (if there are no crossing arcs in the trees) and **non-projective** parsing (if there are crossing arcs in the trees). English and Chinese structures are predominantly projective.

A primary reason for using dependency structures instead of more informative constituent structures is that they are usually easier to be understood. For example, in Fig. 4.1, it is hard to point out that the *news* is the subject of *had* from the constituent structure, while the dependency structure can clearly indicate this relation between the two words. In addition, dependency structures are more amenable to annotators who have good knowledge of the target domain but lack deep linguistic knowledge.

[2]From Joakim Nivre's tutorial at COLING-ACL, Sydney 2006.

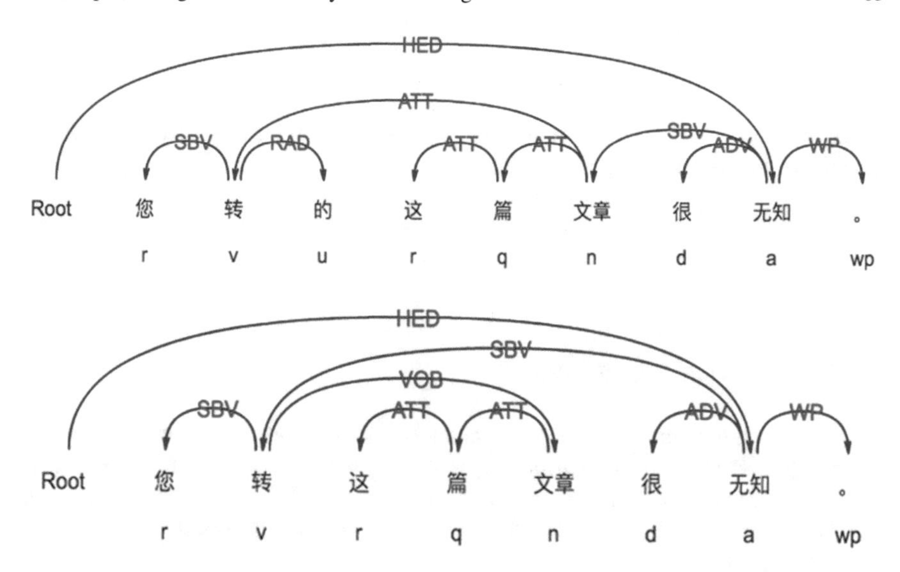

Fig. 4.2 The processing results of LTP

Syntactic parsing can provide useful structural information for applications. For example, the following two Chinese sentences "nin zhuan zhe pian wen zhang hen wu zhi" (*You are ignorant to retweet the article*) and "nin zhuan **de** zhe pian wen zhang hen wu zhi" (*The article you retweeted is ignorant*) have completely different meanings, although the second one only has an additional word "de", which is a possessive particle in Chinese. The main difference between the two sentences is that they have different subjects.

Dependency parsing can tell us this syntactic information directly. One example is LTP (Language Technology Platform)[3] developed by HIT (Harbin Institute of Technology), which provides a Chinese NLP preprocessing pipeline that includes Chinese word segmentation, POS tagging, dependency parsing, and so on. The LTP's processing results of the above two sentences are shown in Fig. 4.2. From these results, we can easily know that subjects of the two sentences are *wenzhang* (article) and *zhuan* (retweet) respectively. Many applications, such as sentiment analysis, can take advantage of these syntactic information. Although the sentiment of the two sentences can be easily determined by the polarity word *wuzhi* (ignorant), it is difficult to identify its targets or aspects if we do not know their syntactic structures.

[3]http://www.ltp.ai.

4.2.4 Structured Predication

These different natural language processing tasks can fall into three types of **structured prediction** problems (Smith 2011):

- Sequence segmentation
- Sequence labeling
- Parsing.

4.2.4.1 Sequence Segmentation

Sequence segmentation is the problem of breaking a sequence into contiguous parts called segments. More formally, if the input is $\mathbf{x} = x_1, \ldots, x_n$, then a segmentation can be written as $\langle x_1, \ldots, x_{y_1} \rangle, \langle x_{y_1+1}, \ldots, x_{y_2} \rangle, \ldots, \langle x_{y_m+1}, \ldots, x_n \rangle$, and the output is $\mathbf{y} = y_1, \ldots, y_m$ which corresponds to the segmental points, where $\forall i \in \{1, \ldots, m\}, 1 \leq y_i \leq n$.

Besides word segmentation, there exist other sequence segmentation problems such as **sentence segmentation** (breaking a piece of string into sentences which is an important postprocessing stage for speech transcription) and **chunking** (also known as **shallow parsing** to find important phrases from sentences, such as noun phrases).

4.2.4.2 Sequence Labeling

Sequence labeling (also named as **tagging**) is the problem of assigning a corresponding label or tag for each item of an input sequence. More formally, if the input sequence is $\mathbf{x} = x_1, \ldots, x_n$, then the output tag sequence is $\mathbf{y} = y_1, \ldots, y_n$, where each input x_i has a single output tag y_i.

POS tagging is perhaps the most classical, and most famous, example of this type of problem, where x_i is a word in a sentence, and y_i is its corresponding POS tag.

Besides POS tagging, many NLP tasks can be mapped to sequence labeling problems such as **named entity recognition** (locating and classifying named entities in text into predefined categories such as the names of persons, locations, and organizations). For this problem, the input is again a sentence. The output is the sentence with entity boundaries tags. We assume there are three possible entity types: PER, LOC, and ORG. Then for input sentence

- *Rachel Holt, Uber's regional general manager for U.S. and Canada, said in a statement provided to CNNTech.*[4]

the output of named entity recognition can be

[4]http://money.cnn.com/2017/04/14/technology/uber-financials/.

- *Rachel/B-PER Holt/I-PER,/O Uber/B-ORG's/O regional/O general/O manager/O for/O U.S./B-LOC and/O Canada/B-LOC , /O said/O in/O a/O statement/O provided/O to/O CNNTech/B-ORG. /O*

where each word in the sentence is either tagged as being the beginning of a particular entity type, B-XXX (e.g., the tag B-PER corresponds to words that are the first word in a person), as being the inside of a particular entity type, I-XXX (e.g., the tag I-PER corresponds to words that are part of a person name, but are not the first word), or otherwise (the tag O, i.e., not an entity).

Once this mapping has been performed on training examples, we can train a tagging model on these training examples. Given a new test sentence we can then predict a sequence of tags by the model, and then it is straightforward to identify the entities from the tagged sequence.

The above sequence segmentation problems can even be transformed into sequence tagging problems by designing proper tag sets. For Chinese word segmentation as an example, each character in a sentence can be annotated with either tag B (beginning of a word) or I (inside a word) (Xue 2003).

The purpose of transforming a sequence segmentation problem into a sequence labeling problem is that the latter is much easier to be modeled and decoded. For example, we will introduce a traditional popular sequence labeling model, conditional random field (CRF), in Sect. 4.3.1.1.

4.2.4.3 Parsing Algorithms

In general, we use **parsing** to denote all kinds of algorithms converting sentences to syntactic structures. As mentioned in Sect. 4.2.3, there are two popular syntactic paring representations, phrase-structure (or named as constituency) parsing and dependency parsing.

For constituent parsing, in general, a **grammar** is used to derive syntactic structures. In brief, a grammar consists of a set of rules, each corresponding to a derivation step that is possible to take under particular conditions. Context-free grammars (CFGs) are most frequently used in constituency parsing (Booth 1969). The parsing is viewed as choosing the maximum-scoring derivation from a grammar.

Graph-based and **transition-based** methods are currently two dominant dependency parsing algorithms (Kbler et al. 2009). Graph-based dependency parsing can be formalized as finding the maximum spanning tree (MST) from a directed graph with vertices (words) and edges (dependency arcs between two words). A transition-based dependency parsing algorithm can be formalized as a transition system consisting of a set of states and a set of transition actions. The transition system begins in start state and transitions are iteratively followed until a terminal state is reached. The common critical problem for graph-based and transition-based dependency parsing is how to calculate the score of a dependency arc or a transition action. We will introduce the two methods in detail at Sects. 4.3.1.2 and 4.3.2.1 respectively.

4.3 Structured Prediction Methods

In this section, we will introduce two types of state-of-the-art structured prediction methods: **graph-based** and **transition-based** respectively. Most deep learning algorithms for structured prediction problems are also derived from these methods.

4.3.1 Graph-Based Methods

The graph-based structured prediction methods differentiate output structures based on their characteristics directly. The **conditional random fields** (CRFs) are typical graph-based methods, which aim to maximize the **probability** of the correct output structure. The graph-based methods can also be applied to dependency parsing, where the aim changes to maximize the **score** of the correct output structure. Next, we will introduce these two methods in detail.

4.3.1.1 Conditional Random Fields

Conditional random fields, strictly speaking, are a variant of undirected graphical models (also called Markov random fields or Markov networks) in which some random variables are observed and others are modeled probabilistically. CRFs were introduced by Lafferty et al. (2001) for sequence labeling. They are also known as linear-chain CRFs. It has been the de facto method for sequence labeling problems before deep learning.

The CRFs define the distribution over label sequences $\mathbf{y} = y_1, \ldots, y_n$, given an observed sequence $\mathbf{x} = x_1, \ldots, x_n$, by a special case of log-linear models:

$$p(\mathbf{y}|\mathbf{x}) = \frac{\exp \sum_{i=1}^{n} \mathbf{w} \cdot \mathbf{f}(\mathbf{x}, y_{i-1}, y_i, i)}{\sum_{\mathbf{y}' \in \mathcal{Y}(\mathbf{x})} \exp \sum_{i=1}^{n} \mathbf{w} \cdot \mathbf{f}(\mathbf{x}, y'_{i-1}, y'_i, i)}, \tag{4.1}$$

where $\mathcal{Y}(\mathbf{x})$ is a set of all possible label sequences; $\mathbf{f}(\mathbf{x}, y_{i-1}, y_i, i)$ is the feature function that extracts a feature vector from position i of sequence $\mathbf{x}$, which can include the labels at the current position y_i and at the previous position y_{i-1}.

The attraction of the CRFs is that it permits the inclusion of any (local) features. For example, in POS tagging, the features can be word-tag pairs, pairs of adjacent tags, spelling features, such as whether the word starts with a capital letter or contains a digit, and prefix or suffix features. These features may be dependent, but the CRFs permit over-lapping features and learn to balance their effect on prediction against the other features. The reason why we name these features as local features is that we assume the label y_i only depends on y_{i-1}, but longer history. This is also named as (first order) Markov assumption.

The general Viterbi algorithm, a kind of dynamic programming algorithm, can be applied for decoding with CRFs. Then the first-order gradient-based (such as gradient descent) or second-order (such as L-BFGS) optimization methods can be used to learn proper parameters to maximize conditional probability in Eq. (4.1).

Besides sequence labeling problems, CRFs have been generalized in many ways for other structured prediction problems. For example, Sarawagi and Cohen (2004) proposed the semi-CRF model for sequence segmentation problems. In semi-CRF, the conditional probability of a semi-Markov chain on the input sequence is explicitly modeled, whose each state corresponds to a subsequence of input units. However, to achieve good segmentation performance, conventional semi-CRF models require carefully handcrafted features to represent the segment. Generally, these feature functions fall into two types: (1) the CRF style features which represent input unit-level information such as the specific words at a particular position; (2) the semi-CRF style features which represent segment-level information such as the length of the segment.

Hall et al. (2014) proposed a CRF-based constituency parsing model, where the features factor over anchored rules of a small backbone grammar, such as basic span features (first word, last word, and length of the span), span context features (the words immediately preceding or following the span), split point features (words at the split point inside the span), and span shape features (for each word in the span, indicating whether that word begins with a capital letter, lowercase letter, digit, or punctuation mark). The CKY algorithm[5] can be used to find the tree with maximum probabilities given learned parameters.

4.3.1.2 Graph-Based Dependency Parsing

Consider a directed graph with vertices V and edges E. Let $s(u, v)$ denote the score of an edge from vertex u to vertex v. A directed spanning tree is a subset of edges $E' \subset E$ such that all vertices have exactly one incoming arc in E, except the root vertex (which has none), and such that E' contains no cycles. Let $\mathscr{T}(E)$ denote the set of all possible directed spanning trees for E. The total score of a spanning tree E' is the sum of the scores of edges in E'. The maximum spanning tree (MST) is defined by

$$\max_{E' \in \mathscr{T}(E)} \sum_{s(u,v) \in E'} s(u, v). \tag{4.2}$$

Then the (unlabeled) dependency parsing decoding problem can be reduced to the maximum spanning tree problem if we view words in a sentence as vertices and edges as dependency arcs, where u is often named as a head (or parent) and v as a modifier (or child).

It is straightforward to extend this approach to labeled dependency parsing, if we have multiple edges from u to v, one associated with each label. The same algorithm

[5] https://en.wikipedia.org/wiki/CYK_algorithm.

applies. The most widely used decoding algorithm for the MST problem is the Eisner algorithm (Eisner 1996) for projective parsing and Chu-Liu-Edmonds algorithm (Chu and Liu 1965; Edmonds 1967) for non-projective parsing.

Here, we introduce the basic graph-based method, which is called the first-order model. The first-order graph-based model makes a strong independence assumption: the arcs in a tree are independent from each other. In other words, the score of an arc is not affected by other arcs. This method is also called the **arc-factorization** method.

So, the critical problem is, given an input sentence, how to determine the score $s(u, v)$ of each candidate arc. Traditionally, discriminative models were used which represent an arc with a feature vector extracted with feature function $\mathbf{f}(u, v)$. Then, the score of the arc is the dot product of a feature weight vector $\mathbf{w}$ and $\mathbf{f}$, i.e., $s(u, v) = \mathbf{w} \cdot \mathbf{f}(u, v)$.

Then how to define $\mathbf{f}(u, v)$ and how to learn optimizing parameters $\mathbf{w}$?

Feature Definition

The choice of features is central to the performance of a dependency parsing model. For each possible arc, the following features are readily considered:

- for each word involved, the surface form, its lemma, its POS, and any shape, spelling, or morphological features;
- words involved include the head, the modifier, context words on either side of the head and modifier, words in between the head and modifier;
- the length of the arc (number of words between the head and modifier), its direction, and (if the parse is to be labeled) the syntactic relation type.

Besides these atomic features, all kinds of combination features and back-off features can also be extracted.

Parameter Learning

Online structured learning algorithms such as the averaged perceptron (AP) (Freund and Schapire 1999; Collins 2002), online passive-aggressive algorithms (PA) (Crammer et al. 2006), or margin infused relaxed algorithm (MIRA) (Crammer and Singer 2003; McDonald 2006) are commonly used for learning parameters $\mathbf{w}$ in graph-based dependency parsing.

4.3.2 Transition-Based Methods

Different from graph-based methods, which differentiate structural outputs directly, a transition-based method can be formalized as a transition system consisting of a set of states S (possibly infinite), including a start state $s_0 \in S$ and a set of terminal states $S_t \in S$, and a set of transition actions T (Nivre 2008). The transition system begins in s_0 and transitions are iteratively followed until a terminal state is reached. Figure 4.3 shows a simple finite state transducer, where the start state is s_0, and the terminal states include s_6, s_7, s_8, s_{14}, s_{15}, s_{16}, s_{17} and s_{18}. The goal of a transition-based structured prediction model is to differentiate sequences of transition actions

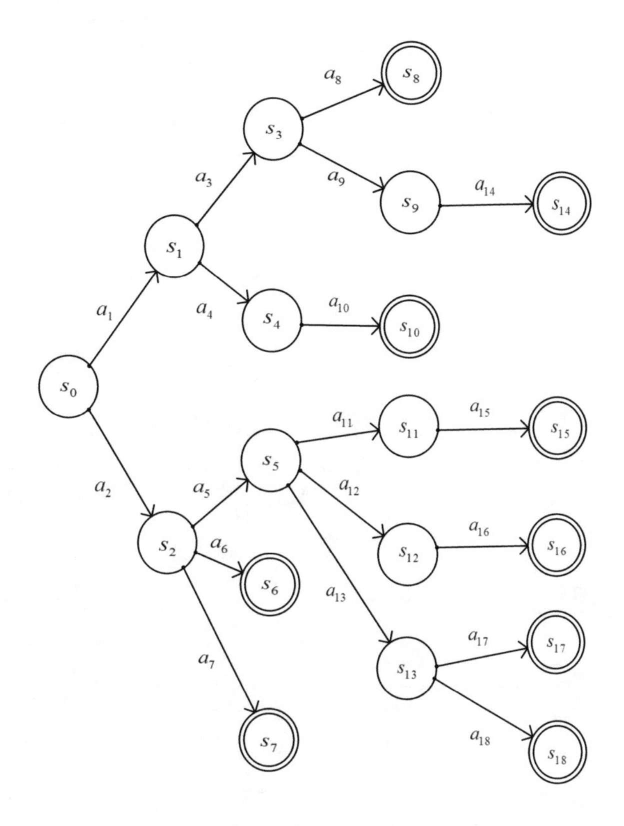

Fig. 4.3 Transition-based method for structured prediction

that lead to the terminal states, so that those that correspond to the correct output state are scored higher.

4.3.2.1 Transition-Based Dependency Parsing

The arc-standard transition system (Nivre 2008) is widely used for projective dependency parsing. In this system, each state corresponds to a stack σ containing partially built subtrees, a buffer β of as-yet-unprocessed words, and a set of dependency arcs A. The transition actions are shown as deductive rules in Fig. 4.4. A transition sequence for the sentence

- *Economic*$_1$ *news*$_2$ *had*$_3$ *little*$_4$ *effect*$_5$ *on*$_6$ *financial*$_7$ *markets*$_8$.$_9$

in Fig. 4.1 generated by the arc-standard algorithm is presented in Table 4.1.

In a greedy parser, the decision about what to do in state $s \in S$ is made by a classifier. Training the classifier is accomplished by considering gold-standard trees in the training section of a treebank, from which we can derive canonical gold-standard sequences (oracle sequences) of transition state and action pairs.

Information that can be obtained from a state $s = \langle \sigma, \beta, A \rangle$ includes:

Fig. 4.4 Transition actions in the deduction system (Nivre 2008)

Start state $([\text{ROOT}], [0...,n], \emptyset)$

$$\textsc{LeftArc}_l\ (\text{LA}_l)\quad \frac{([\sigma \mid s_1, s_0], \beta, A)}{([\sigma \mid s_0], \beta, A \cup \{s_1 \xleftarrow{l} s_0\})}$$

$$\textsc{RightArc}_l\ (\text{RA}_l)\quad \frac{([\sigma \mid s_1, s_0], \beta, A)}{([\sigma \mid s_1], \beta, A \cup \{s_1 \xrightarrow{l} s_0\})}$$

$$\textsc{Shift}\ (\text{SH})\quad \frac{(\sigma, [b \mid \beta], A)}{([\sigma \mid b], \beta, A)}$$

Terminal state $([\text{ROOT}], [\,], A)$

Table 4.1 Transitions by the arc-standard algorithm

State	Action	σ	β	A
0	Initialization	[0]	[1, ..., 9]	$\emptyset$
1	SH	[0, 1]	[2, ..., 9]	
2	SH	[0, 1, 2]	[3, ..., 9]	
3	LA_{nmod}	[0, 2]	[3, ..., 9]	$A \cup \{1 \xleftarrow{nmod} 2\}$
4	SH	[0, 2, 3]	[4, ..., 9]	
5	LA_{sbj}	[0, 3]	[4, ..., 9]	$A \cup \{2 \xleftarrow{sbj} 3\}$
6	SH	[0, 3, 4]	[5, ..., 9]	
7	SH	[0, 3, 4, 5]	[6, ..., 9]	
8	LA_{nmod}	[0, 3, 5]	[6, ..., 9]	$A \cup \{4 \xleftarrow{nmod} 5\}$
9	SH	[0, 3, 5, 6]	[7, ..., 9]	
10	SH	[0, 3, 5, 6, 7]	[8, 9]	
11	SH	[0, 3, 5, 6, 7, 8]	[9]	
12	LA_{nmod}	[0, 3, 5, 6, 8]	[9]	$A \cup \{7 \xleftarrow{nmod} 8\}$
13	RA_{pc}	[0, 3, 5, 6]	[9]	$A \cup \{6 \xrightarrow{pc} 8\}$
14	RA_{nmod}	[0, 3, 5]	[9]	$A \cup \{5 \xrightarrow{nmod} 6\}$
15	RA_{obj}	[0, 3]	[9]	$A \cup \{3 \xrightarrow{obj} 5\}$
16	SH	[0, 3, 9]	[ ]	
17	RA_{p}	[0, 3]	[ ]	$A \cup \{3 \xrightarrow{p} 9\}$
18	RA_{root}	[0]	[ ]	$A \cup \{0 \xrightarrow{root} 3\}$

- all the words and their corresponding POS tags;
- the head of a word and its label from partial parsed dependency arcs A;
- the position of a word on the stack σ and buffer β.

For example, Zhang and Nivre (2011) proposed 72 feature templates which include 26 baseline and 46 new feature templates. The baseline features mainly describe the words and POS tags at top of stack and buffer and their combination. The new features are: direction and distance between a pair of head and modifier; the number of modifiers to a given head; higher order partial parsed dependency arcs; the set of unique dependency labels from the modifiers of the top word in the stack and buffer. Finally, these new features boost about 1.5% UAS (unlabeled attachment score).

We usually use the term "feature engineering" to describe the need of the amount of linguistic expertise that has gone into designing features for various linguistic structured prediction tasks.

NLP researchers tend to adopt the strategy of incorporating as many features as they can think of into learning and allowing the parameter estimation method to determine which features are helpful and which should be ignored. Perhaps because of the heavy-tailed nature of linguistic phenomena and the continued growth in computational power available to researchers, the current consensus seems to be that more features are always welcome in an NLP model, especially in frameworks like log-linear models that can incorporate them.

To reduce error propagation in greedy transition-based algorithms, beam search decoding with global normalization is usually applied and large margin training with early update (Collins and Roark 2004) is used for learning from inexact search.

4.3.2.2 Transition-Based Sequence Labeling and Segmentation

Besides dependency parsing, the transition-based framework can be applied to most structured prediction tasks in NLP, to which a mapping can be found between structured outputs and state transition sequences. Take sequence labeling for example. The output can be constructed by incrementally assigning labels to each input from left to right. In this setting, the state is a pair (σ, β), where σ represents a partially labeled sequence and β represents a queue of unlabeled words. With the start state being ([ ], *input*) and the terminal states being (*output*, [ ]), each action advances a state by assigning a particular label on the front of β.

Sequence segmentation, such as word segmentation is a second example, for which a transition system can process input characters incrementally from left to right. A state takes the form (σ, β), where σ is a partially segmented word sequence and β is a queue of next incoming characters. In the start state, σ is empty and β consists of the full input sentence. In any terminal state, σ contains a full segmented sequence and β is empty. Each transition action advances the current state by processing the next incoming character, either separating (SEP) it at the beginning of a new word or appending (APP) it to the end of the last word in the partially segmented sequence. A gold-standard state transition sequence for the sentence "wo xi huan du shu (I like reading)" is shown in Table 4.2.

4.3.2.3 Advantages of Transition-Based Methods

Transition-based methods do not reduce structural ambiguity—the search space does not shrink in size for a given structured prediction task when the solution changes from a graph-based model to a transition-based model. The only difference is that structural ambiguities are transformed into ambiguities between different transition

Table 4.2 Gold state transition sequence for word segmentation

State	σ	β	Next action
0	[]	[wo, xi, huan, du, shu]	SEP
1	[wo (I)]	[xi, huan, du, shu]	SEP
2	[wo (I), xi]	[huan, du, shu]	APP
3	[wo (I), xihuan (like)]	[du, shu]	SEP
4	[wo (I), xihuan (like), dushu (reading)]	[]	APP

actions at each state. A question that naturally arises is why transition-based methods have attracted significant research attention.

The main answer lies in the features that can be utilized by transition-based models, or the information that is made available for ambiguity resolution. Traditional graph-based methods are typically constrained by efficiency of exact inference, which limits the range of features to use. For example, to train CRF models (Lafferty et al. 2001), it is necessary to efficiently estimate the marginal probabilities of small cliques, the sizes of which are decided by feature ranges. To allow efficient training, CRF models assume low-order Markov properties of their features. As a second example, CKY parsing (Collins 1997) requires that the features are constrained to local grammar rules, so that a tolerable polynomial dynamic program can be used to find the highest scored parse tree among an exponential of search candidates.

In contrast, early work on transition-based methods employ greedy local models (Yamada and Matsumoto 2003; Sagae and Lavie 2005; Nivre 2003), and are typically regarded as a very fast alternative to graph-based systems, running in linear time with regard to the input size. Thanks to the use of arbitrary nonlocal features, their accuracies are not far behind the state-of-the-art models. Since global training has been utilized for training sequences of actions (Zhang and Clark 2011b), fast and accurate transition-based models were made, which gives the state-of-the-art accuracies for tasks such as CCG parsing (Zhang and Clark 2011a; Xu et al. 2014), natural language synthesis (Liu et al. 2015; Liu and Zhang 2015; Puduppully et al. 2016), dependency parsing (Zhang and Clark 2008b; Zhang and Nivre 2011; Choi and Palmer 2011) and constituent parsing (Zhang and Clark 2009; Zhu et al. 2013). Take constituent parsing for example, ZPar (Zhu et al. 2013) gives competitive accuracies to Berkeley parser (Petrov et al. 2006), yet runs 15 times faster.

The efficiency advantage of transition-based systems further allows joint structured problems with highly complex search spaces to be exploited. Examples include joint word segmentation and POS tagging (Zhang and Clark 2010), joint segmentation, POS tagging and chunking (Lyu et al. 2016), joint POS tagging and parsing (Bohnet and Nivre 2012; Wang and Xue 2014), joint word segmentation, POS tagging and parsing (Hatori et al. 2012; Zhang et al. 2013, 2014), joint segmentation and normalization for microblog (Qian et al. 2015), joint morphological generation and text linearization (Song et al. 2014), and joint entity and relation extraction (Li and Ji 2014; Li et al. 2016).

4.4 Neural Graph-Based Methods

4.4.1 Neural Conditional Random Fields

Collobert and Weston (2008) was the first work to utilize deep learning for sequence labeling problems. This was almost the earliest work successfully using deep learning for addressing natural language processing tasks. They not only embedded words into a d-dimensional vector, but also embedded some additional features. Then words and corresponding features in a window were fed into an MLP (multiple layer perceptron) to predict a tag. Word-level log-likelihood, each word in a sentence being considered independently, was used as the training criterion. As mentioned above, there is often a correlation between the tag of a word in a sentence and its neighboring tags. Therefore, in their updated work (Collobert et al. 2011), tag transition scores were added in their sentence-level log-likelihood model. In fact, the model is the same with the CRF models except that the conventional CRF models use a linear model instead of a nonlinear neural network.

While, limited by Markov assumption, the CRF models can only make use of local features. It leads to the long-term dependency between tags cannot be modeled, which sometimes is important in many natural language processing tasks. Theoretically, recurrent neural networks (RNNs) can model arbitrarily sized sequence into fixed-size vectors without resorting to the Markov assumption. Then the output vector is used for further prediction. For example, it can be used to predict the conditional probability of a POS tag given an entire previous word sequence.

In more detail, RNNs are defined recursively, by means of a function taking as input a previous state vector and an input vector and returning a new state vector. So, intuitively, RNNs can be thought of as very deep feedforward networks, with shared parameters across different layers. The gradients then include repeated multiplication of the weight matrix, making it very likely for the values to vanish or explode. The gradient exploding problem has a simple but very effective solution: clipping the gradients if their norm exceeds a given threshold. While the gradient vanishing problem is much more complicated. The gating mechanism, such as the long short-term memory (LSTM) (Hochreiter and Schmidhuber 1997) and the gated recurrent unit (GRU) (Cho et al. 2014), can solve it more or less.

A natural extension of RNN is a bidirectional RNN (Graves 2008) (BiRNN, such as BiLSTM and BiGRU). In sequence labeling problems, predicting a tag not only depends on the previous words, but also depends on the successive words, which cannot be seen in a standard RNN. Therefore, BiRNN use two RNNs (forward and backward RNN) to represent the word sequences before and behind the current word. Then, the forward and backward states of the current word are concatenated together as input to predict the probability of a tag.

In addition, RNNs can be stacked in layers, where the inputs of an RNN are the outputs of the RNN below it. Such layered architectures are often called deep RNNs. Deep RNNs have shown power in many problems, such as semantic role labeling

(SRL) with sequence labeling method (Zhou and Xu 2015, https://www.aclweb.org/anthology/P/P17/P17-1044.bib).

Although RNNs have been successfully applied in many sequence labeling problems, they do not explicitly model the dependency between output tags like CRFs. Therefore, the transition score matrix between any tags can also be added to form a sentence-level log-likelihood model usually named as RNN-CRF model where RNN can also be LSTM, BiLSTM, GRU, BiGRU, and so on.

Like conventional CRFs, the neural CRFs can also be extend to handle the sequence segmentation problems. For example, Liu et al. (2016) proposed a neural semi-CRF, which used a segmental recurrent neural network (SRNN) to represent a segment by composing input units with an RNN. At the same time, additional segment-level representation using segment embedding is also regarded as inputs which encodes the entire segment explicitly. Finally, they achieve the state-of-the-art Chinese word segmentation performance.

Durrett and Klein (2015) extended their CRF phrase-structure parsing (Hall et al. 2014) to neural one. In their neural CRF parsing, instead of linear potential functions based on sparse features, they use nonlinear potentials computed via a feedforward neural network. The other components, such as decoding, are unchanged from the conventional CRF parsing. Finally, they achieve the state-of-the-art phrase-structure parsing performance.

4.4.2 Neural Graph-Based Dependency Parsing

Conventional graph-based models rely heavily on an enormous number of handcrafted features, which brings about serious problems. First, a mass of features could put the models in the risk of overfitting, especially in the combinational features capturing interactions between head and modifier could easily explode the feature space. In addition, feature design requires domain expertise, which means useful features are likely to be neglected due to a lack of domain knowledge.

To ease the problem of feature engineering, some recent works propose some general and effective neural network models for graph-based dependency parsing.

4.4.2.1 Multiple Layer Perceptron

Pei et al. (2015) used an MLP (multiple layer perceptron) model to score an edge. Instead of using millions of features as in conventional models, they only use atomic features such as word unigrams and POS tag unigrams, which are less likely to be sparse. Then these atomic features are transformed into their corresponding distributed representations (feature embeddings or feature vector) and push into MLP. Feature combinations are automatically learned with novel $tanh-cub$ activation function at the hidden layer, thus alleviating the heavy burden of feature engineering in conventional graph-based models.

The distributed representation can discover useful new features that have never been used in conventional parsers. For instance, context information of the dependency edge (h, m), such as words between h and m, has been widely believed to be useful in graph-based models. However, in conventional methods, the complete context cannot be used as features directly because of the data sparseness problem. Therefore, they are usually backed off to low-order representation such as bigrams and trigrams.

Pei et al. (2015) proposed to use distributed representation of the context. They simply average all word embeddings in a context to represent it. The method can not only effectively use every word in the context, but also can capture semantic information behind context, because similar words have similar embeddings.

At last, max-margin criterion is used to train the model. The training object is that the highest scoring tree is the correct one and its score will be larger up to a margin to other possible tree. The structured margin loss is defined as the number of word with an incorrect head and edge label in the predicted tree.

4.4.2.2 Convolutional Neural Networks

Pei et al. (2015) simply average embeddings in context to represent them, which ignore the word position information and cannot assign different weights for different words or phrases. Zhang et al. (2016b) introduce convolutional neural networks (CNN) to compute the representation of a sentence. Then use the representation to help scoring an edge. While the pooling regimes make CNN invariant to shifting, that is CNN ignore the position of words which is very important for dependency parsing. In order to overcome the problem, Zhang et al. (2016b) input the relative positions between a word and a head or modifier to CNN. Another difference from Pei et al. (2015) is that they utilize the probabilistic treatment for training: calculating the gradients according to probabilistic criteria. The probabilistic criteria can be viewed as a soft version of the max-margin criteria, and all the possible factors are considered when calculating gradients for the probabilistic way, while only wrongly predicted factors have nonzero subgradients for max-margin training.

4.4.2.3 Recurrent Neural Networks

Theoretically, recurrent neural networks (RNN) can model sequences with arbitrary length which is sensitive to the relative positions of words in the sequence. As an improvement of conventional RNN, LSTM can better represent a sequence. The BiLSTM (bidirectional LSTM) particularly excels at representing words in the sequence together with their contexts, capturing the word and an "infinite" window around it. Therefore, Kiperwasser and Goldberg (2016) represent each word by its BiLSTM hidden layer output, and use the concatenation of the head and modifier words' representation as the features, which is then passed to a nonlinear scoring function (MLP). To speed up parsing, Kiperwasser and Goldberg (2016) proposed a two-stage strategy.

First, they predict the unlabeled structure using the method given above, and then predict the label of each resulting edge. The labeling of an edge is performed using the same feature representation as above fed into a different MLP predictor. Finally, the max-margin criterion is used to train the model, i.e., let the correct tree is scored above incorrect ones with a margin.

Wang and Chang (2016) also use BiLSTM to represent the head and modifier words. Moreover, they introduce some additional features, such as distance between the two words and context like Pei et al. (2015). Different from Pei et al. (2015), they utilize LSTM-Minus to represent a context, in which distributed representation of a context is learned by using subtraction between LSTM hidden vectors. The similar idea was also been used by Cross and Huang (2016) for transition-based constituent parsing.

All above work contact the distributed representation of head and modifier words outputted by LSTM as input to MLP to calculate the score of a potential dependency edge. Borrowing the idea from Luong et al. (2015), Dozat and Manning (2016) used a bilinear transformation between representation of the head and modifier words to calculate the score. While, they also notice that there are two disadvantages of using the representation directly. The first is that they contain much more information than is necessary for calculating the score, because they are recurrent, they also contain information needed for calculating scores elsewhere in the sequence. Training on the entire vector then means training on superfluous information, which is likely to lead to overfitting. The second disadvantage is that the representation $\mathbf{r}_i$ consists of the concatenation of the left recurrent state $\overleftarrow{r_i}$ and the right recurrent state $\overrightarrow{r_i}$, meaning using by itself in the bilinear transformation keeps the features learned by the two LSTMs distinct; ideally we would like the model to learn features composed from both. Dozat and Manning (2016) address both of these issues simultaneously by first applying (distinct) MLP functions with a smaller hidden size to the two recurrent states $\mathbf{r}_i$ and $\mathbf{r}_j$ before the bilinear operation. This allows the model to combine the two recurrent states together while also reducing the dimensionality. Another change to the bilinear scoring mechanism is to add a linear transformation of the head word representation to scoring function, which captures the prior probability of a word taking any dependent. They name the new method as *biaffine* transformation. Their model is a two-stage one with additional dependency relation classification stage. The biaffine transformation scoring function again is used to predict a label for each dependency edge. Finally, they achieve the state-of-the-art performance on English Penn Treebank test set.

4.5 Neural Transition-Based Methods

4.5.1 Greedy Shift-Reduce Dependency Parsing

The outputs of dependency parsing are syntactic trees, which is a typical structure as sequences are. Graph-based dependency parsers score elements in dependency graphs, such as labels and sibling labels. In contrast, transition-based dependency parsers utilize shift-reduce actions to construct outputs incrementally. Seminal work use statistical models such as SVM to make greedy local decisions on the actions to take, as exemplified by MaltParser (Nivre 2003). Such greedy parsing processes can be illustrated in Table 4.1. At each step, the context, or parser configuration, I can be abstracted in Fig. 4.5, where the stack σ contains partially processed words s_0, s_1, from the top, and the buffer β contains the incoming words q_0, q_1, from the sentence. The task of a greedy local parser is to find the next parsing action given the current configuration, where an example set of actions is shown in Sect. 4.3.2.

MaltParser works by extracting features from the top nodes of σ and the front words of β. For example, the form and POS of s_0, s_1, q_0 and q_1 are all used as binary discrete features. In addition, the forms, POS, and dependency arc labels of dependents of s_0, s_1 and other nodes on σ can be used as additional features. Here, the dependency arc label of a word refers to the label of the arc between the word and the word it modifies. Given a parser configuration, all such features are extracted and fed to an SVM classifier, the output of which is a shift-reduce actions over a set of valid actions.

Chen and Manning (2014) built a neural network alternative of MaltParser, the structure of which is shown in Fig. 4.6a. Similar to MaltParser, features are extracted from the top of σ and the front of β given a parser configuration, and then used for predicting the next shift-reduce action to take. Chen and Manning (2014) follow Zhang and Nivre (2011) in defining the range of word, POS and label features. On the other hand, different from using discrete indicator features, embeddings are used to represent words, POS and arc labels. As shown in Fig. 4.6a, a neural network consisting of three layers is used to predict the next action given the input features. In the input layer, word, POS, and arc label embeddings from the context are concatenated. The hidden layer takes the resulting input vector, and apply a linear transformation before a cube activation function:

$$h = (Wx + b)^3.$$

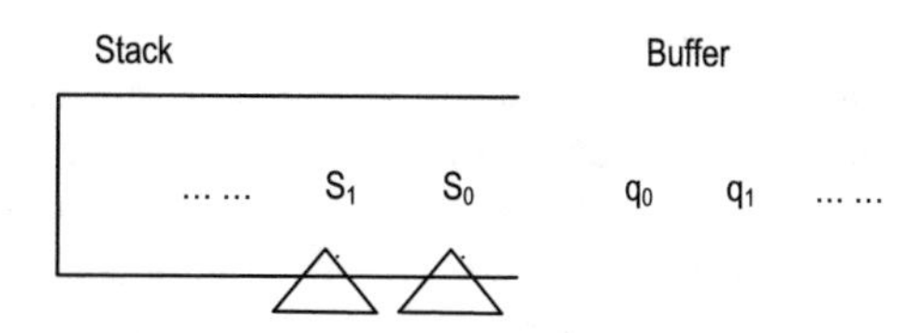

Fig. 4.5 Context of shift-reduce dependency parsing

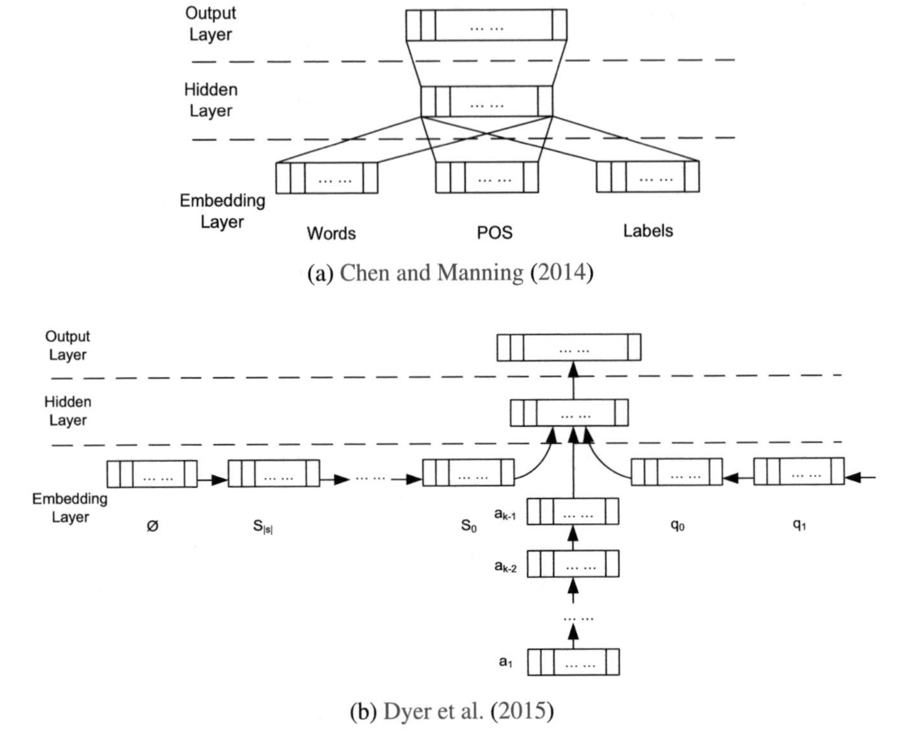

Fig. 4.6 Two greedy parsers

The motivation behind using a cube function as the nonlinear activation function instead of the standard *sigmoid* and *tanh* functions is that it can achieve arbitrary combination of three elements in the input layer, which has been traditionally defined manually in statistical parsing models. This method empirically worked better than alternative activation functions. Finally, the hidden layer is passed as input to a standard *softmax* layer to choose the action.

The parser of Chen and Manning (2014) outperformed MaltParser significantly on several benchmarks. The main reasons are twofold. First, the use of word embeddings allows syntactic and semantic information of words to be learned from large raw data via unsupervised pretaining, which increases the robustness of the model. Second, the hidden layer achieves the effect of complex feature combinations, which is done manually in statistical models. For example, a combined feature can be s_0wq_0p, which captures the form of s_0 and the POS of q_0 simultaneously. This can be a strong indicator of certain actions to take. However, such combinations can be exponentially many, which requires significant manual efforts in feature engineering. In addition, they can be highly sparse if more than two features are combined into one feature. Such sparsity can cause issues in both accuracies and speeds, since they can result in a statistical model with tens of millions of binary indicator features. In contrast, the neural model of Chen and Manning (2014) is compact and less sparse, making it strong in rendering contexts while less subject to overfitting.

The dense input feature representations of Chen and Manning (2014) are highly different from the manual feature templates of traditional statistical parsers, the former being real-valued and low-dimensional, while the latter being binary 0/1 valued and high-dimensional. Intuitively, they should capture different aspects of the same input sentences. Inspired by this observation, Zhang and Zhang (2015) built an extension of Chen and Manning (2014)'s parser, integrating traditional indicator features by concatenating a large sparse feature vector to the hidden vector of Chen and Manning (2014), before feeding it to the *softmax* classification layer. This combination can be regarded as an integration of decades of human labor in feature engineering, and the strong but relatively less interpretable power of automatic feature combination using neural network models. The results are much higher compared to the baseline Chen and Manning (2014) parser, showing that indicator features and neural features are indeed complimentary in this case.

Similar to the observation of Xu et al. (2015) over the super tagger of Lewis and Steedman (2014), Kiperwasser and Goldberg (2016) found the use of local context of Chen and Manning (2014) a potential limitation of their model. To address this issue, they extracted nonlocal features by using LSTMs over the input word and POS features of each word, resulting in a sequence of hidden vector representations for input words. Compared with the feature vectors of Chen and Manning (2014), these hidden feature vectors contain nonlocal sentential information. Kiperwasser and Goldberg (2016) utilized bidirectional LSTMs over the input word sequence, and stacked two LSTM layers to derive hidden vectors. Stack and buffer features are extracted from the corresponding hidden layer vectors, before being used for action classification. This method showed large accuracy improvements over Chen and Manning (2014), demonstrating the power of LSTM in collecting global information.

As shown in Fig. 4.6b, Dyer et al. (2015) took a different method to address the lack of nonlocal features in Chen and Manning (2014)'s model, using LSTMs to represent the stack σ, the buffer β and the sequence of actions that have already been take. In particular, words on the stack are modeled left to right, recurrently, while words on the buffer are modeled right-to-left. The action history is modeled recurrently in temporal order. Since the stack is dynamic, it is possible for words to be popped off the top of it. In this case, Dyer et al. (2015) use a "stack LSTM" structure to model the dynamics, recording the current top of stack with a pointer. When a word is pushed on top of s_0, the word and the hidden state of the stack LSTM for s_0 are used to advance the recurrent state, resulting in a new hidden vector for the new word, which becomes s_0, and s_0 becomes s_1 after the pushing step. In the reverse direction, if s_0 is popped off the stack, the top pointer is updated, moving from the hidden state of s_0 to that of s_1 of the stack LSTM, with s_1 becoming s_0 after the action. By using the hidden states of the top of σ, the front of β and the last action to represent the parser configuration, Dyer et al. (2015) obtained large improvements over the model of Chen and Manning (2014).

Dyer et al. (2015) represented input words with a retrained embedding, a randomly initialized but fine-tuned embedding and the embedding of their POS.

Ballesteros et al. (2015) extended the model of Dyer et al. (2015), by further using an LSTM to model the character sequence in each word. They experimented with multilingual data and observed consistently strong results. Further along this direction, Ballesteros et al. (2016) address the issue of inconsistence between action histories during training and during testing, by simulating testing scenarios during training, where the history of actions is predicted by the model rather than gold-standard action sequences, when a specific action is predicted. This idea is similar to the idea of scheduled sampling by Bengio et al. (2015).

4.5.2 Greedy Sequence Labeling

Given an input sentence, a greedy local sequence labeler works incrementally, assigning a label to each input word by making a local decision, and treating the assignment of labels as classification tasks. Strictly speaking, this form of sequence labeler can be regarded as either graph-based or transition-based, since each label assignment can be regarded as either disambiguating the graph structure ambiguities or transition action ambiguities. Here, we classify greedy local sequence labeling as transition-based due to the following reason. Graph-based sequence labeling models typically disambiguate whole sequences of labels as a single graph by making Markov assumptions on output labels, so that exact inference is feasible using the Viterbi algorithm. Such constraints imply that features can only be extracted over local label sequences, such as second-order and third-order transmission features. In contrast, transition-based sequence labeling models do not impose Markov properties on the outputs, and therefore typically extract highly nonlocal features. In consequence, they typically use greedy search or beam search algorithms for inference. All the examples below are greedy algorithms, and some use highly nonlocal features.

A strand of work has been done using neural models for CCG super tagging, which is a more challenging tasks compared to POS tagging. CCG is a lightly lexicalized grammar, where much syntactic information is conveyed in lexical categories, namely supertags in CCG parsing. Compared with shallow syntactic labels such as POS, super tags contain rich syntactic information, and also denote predicate-argument structures. There are over 1000 super tags that frequently occur in treebanks, which makes super tagging a challenging task.

Traditional statistical models for CCG super tagging employ CRF (Clark and Curran 2007) where features for each label are extracted over a word window context, and POS information is used as crucial features. This makes POS tagging a necessary preprocessing step before super tagging, thus making it possible for POS tagging errors to negatively affect super tagging quality.

Lewis and Steedman (2014) investigated a simple neural model for CCG super tagging, the structure of which is shown in Fig. 4.7a. In particular, given an input sentence, a three-layer neural network is used to assign super tags to each word. The first (bottom) layer is an embedding layer, which maps each word into its embedding form. In addition, a few binary-valued discrete features are concatenated to the

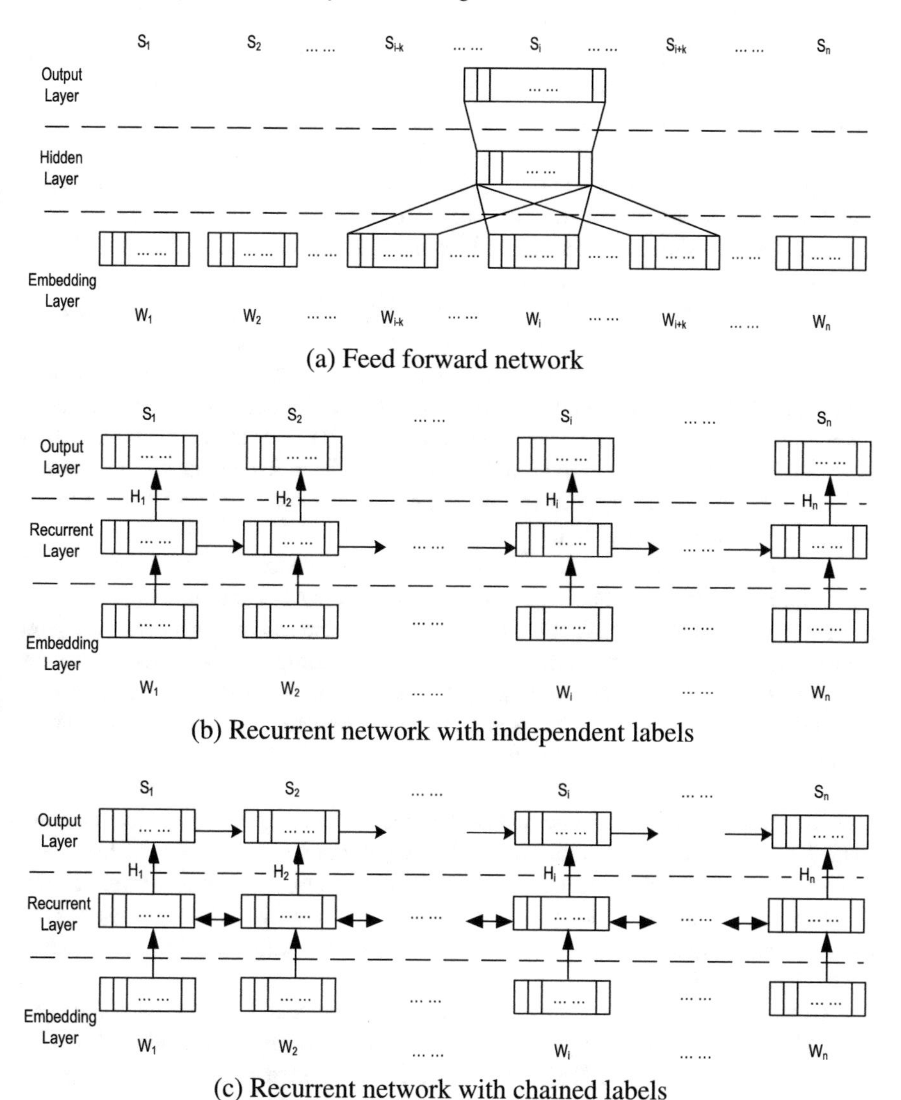

Fig. 4.7 Neural models for CCG supertagging

embedding vector, which include the two-letter suffix of the word, and a binary indicator whether the word is capitalized. The second layer is a hidden layer for feature integration. For a given word w_i, a context window of word w_{i-k}, w_i, w_{i+k} is used for feature extraction. Augmented input embeddings from each word in the context window are concatenated, and fed to the hidden layer, which uses a *tanh* activation function to achieve nonlinear feature combination. The final (top) layer is a *softmax* classification function, which assigns probabilities to all possible output labels.

This simple model worked surprisingly well, leading to better parsing accuracies for both in-domain data and cross-domain data compared to the CRF baseline tagger. Being a greedy model, it also runs significantly faster compared to a neural CRF alternative, while giving comparable accuracies. The success can be attributed to the power of neural network models in automatically deriving features, which makes POS tagging unnecessary. In addition, word embeddings can be retrained over large raw data, thereby alleviating the issue of feature sparsity in baseline discrete models, allowing better cross-domain tagging.

The context window of Lewis and Steedman (2014) follows the work of Collobert and Weston (2008), which is local and comparable to the context window of CRF (Clark and Curran 2007). On the other hand, recurrent neural networks have been used to extract nonlocal features from the whole sequence, achieving better accuracies for a range of NLP tasks. Motivated by this observation, Xu et al. (2015) extended the method of Lewis and Steedman (2014), by replacing the window-based hidden layer with a recurrent neural network layer (Elman 1990). The structure of this model is shown in Fig. 4.7b.

In particular, the input layer of Xu et al. (2015) is identical to the input layer of Lewis and Steedman (2014), where a word embedding is concatenated with two-character suffix and capitalization features. The hidden layers are defined by an Elman recurrent neural network, which recurrently computes the hidden state for w_i using the previous hidden state h_{i-1} and the current embedding layer of w_i. A *sigmoid* activation function is used to achieve nonlinearity. Finally, the same form of output layers is used to label each word locally.

Compared with the method of Lewis and Steedman (2014), the RNN method gives improved accuracies for both super tagging and subsequent CCG parsing using a standard parser model. In addition, the RNN super tagging also gives better 1-best super tagging accuracy compared to the CRF method of Clark and Curran (2007), while the NN method of Lewis and Steedman did not achieve. The main reason is the use of recurrent neural network structure, which models unbounded history context for the labeling of a word.

Lewis and Steedman (2014) made further improvements to the model of Xu et al. (2015) by using LSTMs to replace the Elman RNN structure in the hidden layer. In particular, a bidirectional LSTM is used to derive the hidden features h_1, h_2, h_n given the embedding layer. The input representations are also adjusted slightly, where the discrete components are discarded, and the 1- to 4-letter prefixes and suffixes of each word are represented with embedding vectors, and concatenated to the embeddings of words as input features. Thanks to these changes, the final model gives much improved accuracies for both super tagging and subsequent CCG parsing. In addition, by using tri-training techniques, the results are further raised, reaching 94.7% F1 on 1-best tagging.

The models of Xu et al. (2015) and Lewis and Steedman (2014) consider nonlocal dependencies between words in the input, yet does not capture nonlocal dependencies between output labels. In this respect, they are less expressive compared with the CRF model of Clark and Curran (2007), which considers the dependencies between three consecutive labels. To address this issue, Vaswani et al. (2016) leverage LSTM on the output label sequence also, by considering the label history s_1, s_2, s_{i-1} when the word w_i is labeled. The model structure is shown in Fig. 4.7c.

The input layer of this model uses the same representations as Lewis and Steedman (2014), and the hidden layer is similar to that of Lewis and Steedman (2014). In the output layer, the classification of each label s_i is based on both the corresponding hidden layer vector h_i and the previous label sequence, represented by the hidden states h^s_{i-1} of a label LSTM. The label LSTM is unidirectional, where each state h^s_i is derived from its previous state h^s_{i-1} and the previous label s_i. To further improve the accuracies, scheduled sampling (Bengio et al. 2015) is used to find training data that are more similar to test cases. During training, the history label sequence s_1, s_2, s_{i-1} for labeling s_i is sampled by choosing the predicted supertag at each position with a sampling probability p. This way, the model can learn better how to assign a correct label even if errors are made in the history during test time.

Vaswani et al. (2016) showed that by adding the output label LSTM, the accuracies can be slightly improved if scheduled sampling is applied, but decreases compared with the greedy local output model of Lewis and Steedman (2014) without scheduled sampling. This shows the usefulness of scheduled sampling, which avoids overfitting to gold label sequences and consequent tossing of test data robustness.

4.5.3 Globally Optimized Models

Greedy local neural models have demonstrated their advantage over their statistical counterparts by leveraging word embeddings to alleviate sparseness, and using deep neural networks to learn nonlocal features. Syntactic and semantic information over the whole sentence has been utilized for structured prediction, and nonlocal dependencies over labels are also modeled. On the other hand, the training of such models is local, and hence can potentially lead to label bias, since the optimal sequence of actions does not always contain locally topical actions. Globally optimized models, which have been the dominant approach for statistical NLP, have been applied to neural models also.

Such models typically apply beam search (in Algorithm 1), where an agenda is used to keep the B highest scored sequences of actions at each step. The beam search process for arc-eager dependency parsing is shown in Fig. 4.8. Here the blue circle illustrates the gold-standard sequence of actions. As shown in Fig. 4.8, at some steps, the gold-standard state may not be the highest scored in the agenda. In case of local search, such situation leads to search errors. For beam search, however, it is possible for the decoder to recover the gold-standard state in subsequent stages as the highest scored item in the agenda.

The beam search algorithm for transition-based structured prediction is formally shown in Algorithm 1. Initially, the agenda contains only the start state in the state transition system. At each step, all items in the agenda are expanded by applying all possible transition actions, leading to a set of new states. From these states, the highest scored B are selected, and used as agenda items for the next step. Such process

Algorithm 1 The generic beam search algorithm

1: **function** BEAM- SEARCH(*problem*, *agenda*, *candidates*, *B*)
2: *candidates* ← {`StartItem`(*problem*)}
3: *agenda* ← `Clear`(*agenda*)
4: **loop**
5: **for each** *candidate* ∈ *candidates* **do**
6: *agenda* ← `Insert`(`Expand`(*candidate*, *problem*), *agenda*)
7: **end for**
8: *best* ← `Top`(*agena*)
9: **if** `GoalTest`(*problem*, *best*) **then**
10: **return** *best*
11: **end if**
12: *candiates* ← `Top` − `B`(*agenda*, *B*)
13: *agenda* ← `Clear`(*agenda*)
14: **end loop**
15: **end function**

repeats until terminal states have been reached, and the highest scored state in the agenda is taken as the output. Similar to greedy search, the beam search algorithm has a linear time complexity with respect to the action sequence length.

The items in the agenda are ranked using their global scores, which are the total scores of all transition actions in the sequence. Different from greedy local models, the training objective of globally optimized models is to different full sequences of actions based on their global scores. There are two general training approaches, with one being to maximize the likelihood of gold-standard sequences of actions, other being to maximize the score margin between the gold-standard sequence of

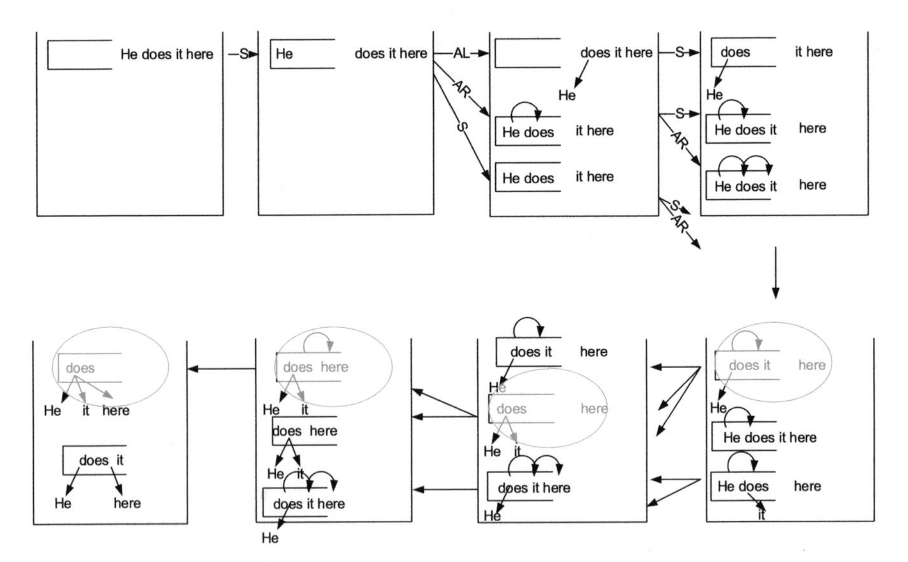

Fig. 4.8 Parsing process given state transition system with beam search

action and non-gold-standard sequences of actions. Other training objectives are occasionally used, as will be shown later.

Following Zhang and Clark (2011b), most globally optimized models regard training as beam search optimization, where negative training examples are sampled by the beam search process itself, and used together with the gold-standard positive example to update the model. Here we use Zhang and Clark (2011b) as one example to illustrate the training method. Online learning is used, where an initial model is applied to decode the training examples. During the decoding of each sample, the gold-standard sequence of actions is available. The same beam search algorithm above is used, as in test cases. At any step, if the gold-standard sequence of actions falls out of the agenda, a search error is unavoidable. At this situation, search is stopped, and the model is updated by using the gold-standard sequence of actions till this step as the positive example, and the current highest scored sequence of actions in the beam as a negative example. Zhang and Clark (2011b) used a statistical model, where model parameters are updated using the perceptron algorithm of Collins (2002). The early stopping of beam search is known as *early update* (Collins and Roark 2004). In the case where the gold-standard sequence of action remains in the agenda until decoding finishes, the training algorithm checks if it is the highest scoring in the last step. If so, the current training sample is finished without parameter update; otherwise the current highest scored sequence of actions in the beam is taken as a negative example to update parameters. The same process can repeat over the training examples for multiple iterations, and the final model is used for testing.

We discuss a strand of work using global training for neural transition-based structured prediction below, categorized by their training objectives.

4.5.3.1 Large Margin Methods

The large margin objective maximizes the score difference between gold-standard output structures and incorrect output structures; it has been used by discrete structured prediction methods such as the structured perceptron (Collins 2002) and MIRA (Crammer and Singer 2003). The ideal large margin training objective should ensure that the gold-standard structure is scored higher than all incorrect structures by a certain margin. However, for structured prediction tasks, the number of incorrect structures can be exponentially many, hence making the exact objective intractable in most cases. The perceptron approximates this objective by making model adjustments for the most violated margin, and has theoretical guarantee of convergence in training. In particular, given the gold-standard structure as a positive example, and the max-violation incorrect structure as a negative example, the perceptron algorithm adjusts model parameters by adding the feature vector of the positive example to the model, and subtracting the feature vector of the negative example from the model parameter vector. By repeating this procedure for all training examples, the model converges to scoring gold-standard structures higher than incorrect structures. The perceptron algorithm finds a negative example for each gold-standard training example, such that the violation of the ideal score margin is the largest. This typically implies the searching for a highest scored incorrect output, or one that ranks the highest by considering both its current model score and it deviation from the gold-standard. In the latter case, the structured dilation is the cost of the incorrect output, where outputs with similar structures to the gold-standard have less cost. By considering not only the model score but also the cost, this training objective allows model scores to differentiate not only gold-standard and incorrect structures, but also between different incorrect structures by their similarity to the correct structure.

With neural networks, the training objective translates to maximizing the score difference between a given positive example and a corresponding negative example. This objective is typically achieved by taking the derivative of the score difference with respect to all model parameters, updating model parameters using gradient-based methods such as AdaGrad (Duchi et al. 2011).

Zhang et al. (2016a) used such a large margin objective for transition-based word segmentation. As shown in Sect. 1.1, a state for this task can be encoded in a a pair $s = (\sigma, \beta)$, where σ contains a list of recognized words, and β contains the list of next incoming characters. Zhang et al. (2016a) use a word LSTM to represent σ, and a bidirectional character LSTM to represent β. In addition, following Dyer et al. (2015), they also use an LSTM to represent the sequence of actions that have been taken. Given a state s, the three LSTM context representations are integrated and used to score SEP and APP actions. Formally, given a state s, the score of action a can be denoted as $f(s, a)$, where f is the network model. As a global model, Zhang et al. (2016a) calculate the score of a sequence of actions for ranking the state they lead to, where

$$score(s_k) = \sum_{i=1}^{k} f(s_{i-1}, a_i).$$

Following Zhang and Clark (2011b), online learning with early update is used. Each training example is decoded using beam search, until the gold-standard sequence of transition actions fall out of beam, or does not rank highest by a score margin after decoding finishes. Here the margin between the gold-standard structure and an incorrect structure is defined by the number of incorrect actions Δ, weighted by a factor η. Therefore, given a state after k actions, the corresponding loss function for training the network is defined as follows:

$$L(s_k) = \max\left(score(s_k) - score(s_k^g) + \eta\Delta(s_k, s_k^g), 0\right),$$

where s_k^g is the corresponding gold-standard structure after k transitions.

During training, Zhang et al. (2016a) use the current model score $score(s_k)$ plus $\Delta(s_k, s_k^g)$ to rank states in the agenda, so that structural differences are considered for finding the maximum violation. Given this ranking, a negative example can be chosen in the early update and final update cases. Model parameters are updated according to the less function between s_k and s_k^g above. Since $score(s_k)$ is the sum of all action scores, the loss is evenly distributed to each action. In practice, back-propagation is used to train the network, where the derivative of the lost function is taken with respect to model parameters via the network $f(s_{i-1}, a_i)$ for $i \in [1..k]$. Since each action a_i shares the same representation layers as described earlier, their losses accumulate for model parameter updates. AdaGrad is used to change the model.

Cai and Zhao (2016) adopted a very similar neural model for word segmentation. Both the models of Zhang et al. (2016a) and Cai and Zhao (2016) can be regarded as extensions of the method of Zhang and Clark (2007) using neural network. On the other hand, the scoring function of Cai and Zhao (2016) is different from that of Zhang et al. (2016a), Cai and Zhao (2016) also uses beam search, segmenting a sentence incrementally. But their incremental steps are based on words, rather than characters. They used multiple beams to store partial segmentation outputs containing the same numbers of characters, which is similar to Zhang and Clark (2008a). As a result, constraints to the word size must be used to ensure linear time complexity. For training, exactly the same large margin objective is taken.

A slightly different large margin objective is used by Watanabe and Sumita (2015) for constituent parsing. They adopt the transition system of Sagae et al. (2005) and Zhang and Clark (2009), where a state can be defined as a pair (σ, β), similar to the dependency parsing case in Sect. 1.1. Here σ contains partially constructed constituent trees, and β contains next incoming words. A set of transition actions including SHIFT, REDUCE and UNARY are used to consume input words and construct output structures. Interested readers can refer to (Sagae and Lavie 2005) and (Zhang and Clark 2009) for more details on the state transition system.

Watanabe and Sumita (2015) represent σ using a stack LSTM structure, which dynamically change, and is similar to that of Dyer et al. (2015). β is represented using a standard LSTM. Given this context representation, the score of a next action a can be denoted as $f(s, a)$, where s represents the current state and f is the network structure. Similar to the case of Zhang et al. (2016a), the score of a state s_k is the sum of all actions that lead to the state, as shown in Fig. 4.9:

$$score(s_k) = \sum_{i=1}^{k} f(s_{i-1}, a_i)$$

Similar to Zhang et al. (2016a), beam search is used to find the highest scored state over all structures. For training, however, max-violation update is used instead of early update (Huang et al. 2012), where the negative example is chosen by running beam search until the terminal state is reached, and then finding the intermediate state that gives the largest violation of the score margin between gold-standard and incorrect structures. Update is executed at the max-violation step. In addition, rather than using the maximum violation state as the negative example, all incorrect states in the beam are used as negative examples to enlarge the sample space, and the training objective is defined to minimize the loss:

$$L = \max\left(\mathbf{E}_{s_k \in A} score(s_k) - score(s_k^g + 1)\right).$$

Here A represents the agenda, and the expectation $\mathbf{E}_{s_k \in A} score(s_k)$ is calculated based on probabilities of each s_k in the agenda using model scores:

$$p(s_k) = \frac{\exp(score(s_k))}{\sum_{s_k \in A} \exp(score(s_k))}.$$

4.5.3.2 Maximum Likelihood Methods

Maximum likelihood objectives for neural structured prediction are inspired by log-linear models. In particular, given the score of an output y $score(y)$, a log-linear model calculates its probability as

$$p(y) = \frac{\exp(score(y))}{\sum_{y \in Y} \exp(score(y))},$$

where Y represents the set of all outputs. When y is a structure, this log-linear model becomes CRF under certain constraints.

A line of work investigate a similar objective by assuming the structured score calculation in Fig. 4.9 for transition-based models, where the score for a state s_k is calculated as

$$score(s_k) = \sum_{i=1}^{k} f(s_{i-1}, a_i).$$

The definition of f and a are the same as in the previous section. Given this score calculation, the probability of the state s_k is

$$p(s_k) = \frac{\exp(score(s_k))}{\sum_{s_k \in S} \exp(score(s_k))},$$

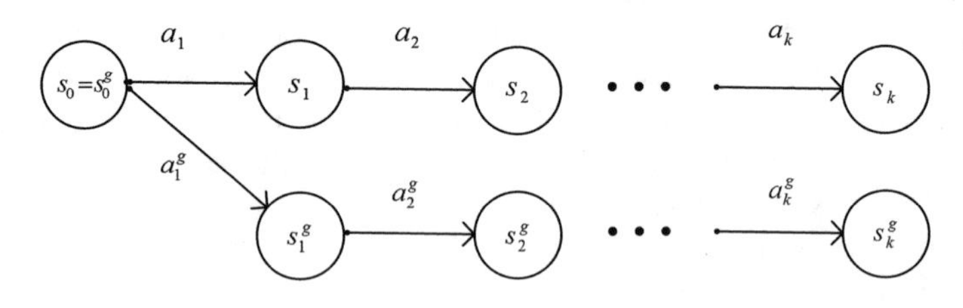

Fig. 4.9 Structured score calculation

where S denotes all possible states after k transition actions. Apparently, the number of states in S grows exponentially with k, as the number of structures they contain. As a result, it is difficult to estimate the denominator for maximum likelihood training, as in the case of CRF. For CRF, the issue is solved by imposing constraints on feature locality, so that marginal probabilities of features can be used to estimate the partition function. For transition-based models, however, such feature locality is nonexistent.

Zhou et al. (2015) first addressed this issue by using all states in the agenda to approximate S during beam search. They perform beam search and online learning, using early update in the same way as Zhang et al. (2016a). On the other hand, during each update, rather than calculating a score margin between positive and negative examples, Zhou et al. (2015) maximize the approximated likelihood of the gold-standard stage s_g, where

$$p(s_g) = \frac{\exp(score(s_g))}{\sum_{s_k \in A} \exp(score(s_k))},$$

where A represents the agenda, as in the last section.

This method uses the probability mass of states in the agenda to approximate the partition function, and hence is referred to as beam contrastive learning by Zhou et al. (2015). Zhou et al. (2015) applied the training objective to the task of transition-based dependency parsing, achieving better results compared to Zhang and Nivre (2011).

Andor et al. (2016) applied this method to more structured prediction tasks, including part-speech-tagging. They also obtained significantly better results than Zhou et al. (2015) by using a better baseline method and doing more thorough hyper-parameter search. In addition, Andor et al. (2016) gave a theoretical justification that the globally normalized model outperforms locally trained baselines.

4.5.3.3 Maximum Expected F1

Another training objective that has been tried is maximum F1, which Xu et al. (2016) used for transition-based CCG parsing (Zhang and Clark 2011a). In particular, Xu et al. (2016) use beam search to find the highest scored state, where the score of each state is given by the calculation method of Fig. 4.9. Given state s_k, the score is calculated as

$$score(s_k) = \sum_{i=1}^{k} g(s_{i-1}, a_i).$$

Here the function g represents a network model, and a represents a transition action. The difference between the network function g of Xu et al. (2016) and the network function of all aforementioned methods is that g uses a *softmax* layer to normalize the output actions, while f does not use nonlinear activation functions over scores of different actions given a state.

The training objective of Xu et al. (2016) is

$$E_{s_k \in A} F1(s_k) = \sum_{s_k \in A} p(s_k) F1(s_k),$$

where A denotes the beam after parsing finishes, and $F1(s_k)$ denotes the F1 score of s_k as evaluated by standard metrics against the gold-standard structure.

Xu et al. (2016) calculates $p(s_k)$ using

$$p(s_k) = \frac{\exp(score(s_k))}{\sum_{s_k \in A} \exp(score(s_k))}$$

which is consistent to all aforementioned methods.

4.6 Summary

In this chapter, we provided an overview on the application of deep learning to lexical analysis and parsing, two standard tasks in NLP, and compare the deep learning approach with traditional statistical methods.

First, we introduced the definitions of lexical analysis and parsing. They model structured properties of words and their relationships to each other. The commonly used techniques in these tasks include word segmentation, part-of-speech tagging, and parsing. The most important characteristic of lexical analysis and parsing is that the outputs are structured.

Then, we introduced two types of traditional methods usually used to solve these structured prediction tasks: graph-based methods and transition-based methods. Graph-based methods exploit output structures based on their characteristics directly, while transition-based methods transform the output construction processes into state transition processes, and subsequently process sequences of transition actions.

Finally, we in this chapter introduced methods using neural network and deep learning models in both graph-based and transition-based structured prediction.

While recent advances have shown that neural network models can be used effectively to augment or replace statistical models in the traditional graph-based and transition-based frameworks for lexical analysis and parsing, they have begun to illustrate the strong representation power of neural networks which can go beyond the function of mere modeling. For example, in the traditional statistical modeling approach, it has been commonly understood that local training leads to weaknesses such as label bias (Lafferty et al. 2001). However, the model and method described in (Dozat and Manning 2016) achieve state-of-the-art accuracy results using a neural model that factors out single dependency arcs as training objectives, without globally training the probabilities of a dependency tree. This suggests that structural correlations between output labels can be obtained by the strong representation of word sequences using LSTMs. The future direction for lexical analysis and parsing in NLP will likely be a unification between well-established research on structured learning and the emerging power of deep learning.

References

Andor, D., Alberti, C., Weiss, D., Severyn, A., Presta, A., Ganchev, K., et al. (2016). Globally normalized transition-based neural networks. In *Proceedings of the 54th Annual Meeting of the Association for Computational Linguistics* (Vol. 1: Long Papers, pp. 2442–2452). Berlin, Germany: Association for Computational Linguistics.

Ballesteros, M., Dyer, C., & Smith, N. A. (2015). Improved transition-based parsing by modeling characters instead of words with LSTMs. In *Proceedings of the 2015 Conference on Empirical Methods in Natural Language Processing* (pp. 349–359). Lisbon, Portugal: Association for Computational Linguistics.

Ballesteros, M., Goldberg, Y., Dyer, C., & Smith, N. A. (2016). Training with exploration improves a greedy stack LSTM parser. In *Proceedings of the 2016 Conference on Empirical Methods in Natural Language Processing* (pp. 2005–2010). Austin, Texas: Association for Computational Linguistics.

Bengio, S., Vinyals, O., Jaitly, N., & Shazeer, N. (2015). Scheduled sampling for sequence prediction with recurrent neural networks. In *Proceedings of the 28th International Conference on Neural Information Processing Systems, NIPS'15* (pp. 1171–1179). Cambridge, MA, USA: MIT Press.

Bohnet, B. & Nivre, J. (2012). A transition-based system for joint part-of-speech tagging and labeled non-projective dependency parsing. In *Proceedings of the 2012 Joint Conference on Empirical Methods in Natural Language Processing and Computational Natural Language Learning* (pp. 1455–1465). Jeju Island, Korea: Association for Computational Linguistics.

Booth, T. L. (1969). Probabilistic representation of formal languages. *2013 IEEE 54th Annual Symposium on Foundations of Computer Science, 00*, 74–81.

Cai, D., & Zhao, H. (2016). Neural word segmentation learning for Chinese. In *Proceedings of the 54th Annual Meeting of the Association for Computational Linguistics* (Vol. 1: Long Papers, pp. 409–420). Berlin, Germany: Association for Computational Linguistics.

Carnie, A. (2012). *Syntax: A Generative Introduction* (3rd ed.). New York: Wiley-Blackwell.

Chen, D., & Manning, C. (2014). A fast and accurate dependency parser using neural networks. In *Proceedings of EMNLP-2014.*

Cho, K., van Merrienboer, B., Gulcehre, C., Bahdanau, D., Bougares, F., Schwenk, H., & Bengio, Y. (2014). Learning phrase representations using RNN encoder–decoder for statistical machine translation. In *Proceedings of the 2014 Conference on Empirical Methods in Natural Language Processing (EMNLP)* (pp. 1724–1734). Doha, Qatar: Association for Computational Linguistics.

Choi, J. D., & Palmer, M. (2011). Getting the most out of transition-based dependency parsing. In *Proceedings of the 49th Annual Meeting of the Association for Computational Linguistics: Human Language Technologies* (pp. 687–692). Portland, Oregon, USA: Association for Computational Linguistics.

Chu, Y., & Liu, T. (1965). On the shortest arborescence of a directed graph. *Scientia Sinica, 14*, 1396–1400.

Clark, S., & Curran, J. R. (2007). Wide-coverage efficient statistical parsing with ccg and log-linear models. *Computational Linguistics, 33*(4), 493–552.

Collins, M. (1997). Three generative, lexicalised models for statistical parsing. In *Proceedings of the 35th Annual Meeting of the Association for Computational Linguistics* (pp. 16–23). Madrid, Spain: Association for Computational Linguistics.

Collins, M. (2002). Discriminative training methods for hidden Markov models: Theory and experiments with perceptron algorithms. In *Proceedings of the 2002 Conference on Empirical Methods in Natural Language Processing* (pp. 1–8). Association for Computational Linguistics.

Collins, M., & Roark, B. (2004). Incremental parsing with the perceptron algorithm. In *Proceedings of the 42nd Meeting of the Association for Computational Linguistics (ACL'04), Main Volume* (pp. 111–118). Barcelona, Spain.

Collobert, R., & Weston, J. (2008). A unified architecture for natural language processing: Deep neural networks with multitask learning. In *Proceedings of the 25th International Conference on Machine Learning, ICML '08* (pp. 160–167). New York, NY, USA: ACM.

Collobert, R., Weston, J., Bottou, L., Karlen, M., Kavukcuoglu, K., & Kuksa, P. (2011). Natural language processing (almost) from scratch. *Journal of Machine Learning Research, 12*, 2493–2537.

Crammer, K., Dekel, O., Keshet, J., Shalev-Shwartz, S., & Singer, Y. (2006). Online passive-aggressive algorithms. *Journal of Machine Learning Research, 7*, 551–585.

Crammer, K., & Singer, Y. (2003). Ultraconservative online algorithms for multiclass problems. *Journal of Machine Learning Research, 3*, 951–991.

Cross, J., & Huang, L. (2016). Span-based constituency parsing with a structure-label system and provably optimal dynamic oracles. In *Proceedings of the 2016 Conference on Empirical Methods in Natural Language Processing* (pp. 1–11). Austin, Texas: Association for Computational Linguistics.

Dozat, T., & Manning, C. D. (2016). Deep biaffine attention for neural dependency parsing. *CoRR*, abs/1611.01734.

Duchi, J., Hazan, E., & Singer, Y. (2011). Adaptive subgradient methods for online learning and stochastic optimization. *Journal of Machine Learning Research, 12*, 2121–2159.

Durrett, G., & Klein, D. (2015). Neural CRF parsing. In *Proceedings of the 53rd Annual Meeting of the Association for Computational Linguistics and the 7th International Joint Conference on Natural Language Processing* (Vol. 1: Long Papers, pp. 302–312). Beijing, China: Association for Computational Linguistics.

Dyer, C., Ballesteros, M., Ling, W., Matthews, A., & Smith, N. A. (2015). Transition-based dependency parsing with stack long short-term memory. In *Proceedings of the 53rd Annual Meeting of the Association for Computational Linguistics and the 7th International Joint Conference on Natural Language Processing* (Vol. 1: Long Papers, pp. 334–343). Beijing, China: Association for Computational Linguistics.

Edmonds, J. (1967). Optimum branchings. *Journal of Research of the National Bureau of Standards, 71B*, 233–240.

Eisner, J. (1996). Efficient normal-form parsing for combinatory categorial grammar. In *Proceedings of the 34th Annual Meeting of the Association for Computational Linguistics* (pp. 79–86). Santa Cruz, California, USA: Association for Computational Linguistics.

Elman, J. L. (1990). Finding structure in time. *Cognitive Science, 14*(2), 179–211.

Freund, Y., & Schapire, R. E. (1999). Large margin classification using the perceptron algorithm. *Machine Learning, 37*(3), 277–296.

Graves, A. (2008). *Supervised sequence labelling with recurrent neural networks*. Ph.D. thesis, Technical University Munich.

Hall, D., Durrett, G., & Klein, D. (2014). Less grammar, more features. In *Proceedings of the 52nd Annual Meeting of the Association for Computational Linguistics* (Vol. 1: Long Papers, pp. 228–237). Baltimore, MD: Association for Computational Linguistics.

Hatori, J., Matsuzaki, T., Miyao, Y., & Tsujii, J. (2012). Incremental joint approach to word segmentation, pos tagging, and dependency parsing in Chinese. In *Proceedings of the 50th Annual Meeting of the Association for Computational Linguistics* (Vol. 1: Long Papers, pp. 1045–1053), Jeju Island, Korea: Association for Computational Linguistics.

Hochreiter, S., & Schmidhuber, J. (1997). Long short-term memory. *Neural Computation, 9*(8), 1735–1780.

Huang, L., Fayong, S., & Guo, Y. (2012). Structured perceptron with inexact search. In *Proceedings of the 2012 Conference of the North American Chapter of the Association for Computational Linguistics: Human Language Technologies* (pp. 142–151). Montréal, Canada: Association for Computational Linguistics.

Jurafsky, D., & Martin, J. H. (2009). *Speech and language processing* (2nd ed.). Upper Saddle River, NJ, USA: Prentice-Hall Inc.

Kbler, S., McDonald, R., & Nivre, J. (2009). Dependency parsing. *Synthesis Lectures on Human Language Technologies*, *2*(1), 1–127.

Kiperwasser, E., & Goldberg, Y. (2016). Simple and accurate dependency parsing using bidirectional lstm feature representations. *Transactions of the Association for Computational Linguistics*, *4*, 313–327.

Lafferty, J. D., McCallum, A., & Pereira, F. C. N. (2001). Conditional random fields: Probabilistic models for segmenting and labeling sequence data. In *Proceedings of the Eighteenth International Conference on Machine Learning, ICML '01* (pp. 282–289), San Francisco, CA, USA: Morgan Kaufmann Publishers Inc.

Lewis, M., & Steedman, M. (2014). A* CCG parsing with a supertag-factored model. In *Proceedings of the 2014 Conference on Empirical Methods in Natural Language Processing (EMNLP)* (pp. 990–1000). Doha, Qatar: Association for Computational Linguistics.

Li, F., Zhang, Y., Zhang, M., & Ji, D. (2016). Joint models for extracting adverse drug events from biomedical text. In *Proceedings of the Twenty-Fifth International Joint Conference on Artificial Intelligence, IJCAI 2016* (pp. 2838–2844). New York, NY, USA, 9–15 July 2016.

Li, Q., & Ji, H. (2014). Incremental joint extraction of entity mentions and relations. In *Proceedings of the 52nd Annual Meeting of the Association for Computational Linguistics* (Vol. 1: Long Papers, pp. 402–412). Baltimore, MD: Association for Computational Linguistics.

Liu, J., & Zhang, Y. (2015). An empirical comparison between n-gram and syntactic language models for word ordering. In *Proceedings of the 2015 Conference on Empirical Methods in Natural Language Processing* (pp. 369–378). Lisbon, Portugal: Association for Computational Linguistics.

Liu, Y., Che, W., Guo, J., Qin, B., & Liu, T. (2016). Exploring segment representations for neural segmentation models. In *Proceedings of the Twenty-Fifth International Joint Conference on Artificial Intelligence, IJCAI 2016* (pp. 2880–2886). New York, NY, USA, 9–15 July 2016.

Liu, Y., Zhang, Y., Che, W., & Qin, B. (2015). Transition-based syntactic linearization. In *Proceedings of the 2015 Conference of the North American Chapter of the Association for Computational Linguistics: Human Language Technologies* (pp. 113–122). Denver, Colorado: Association for Computational Linguistics.

Luong, T., Pham, H., & Manning, C. D. (2015). Effective approaches to attention-based neural machine translation. In *Proceedings of the 2015 Conference on Empirical Methods in Natural Language Processing* (pp. 1412–1421). Lisbon, Portugal: Association for Computational Linguistics.

Lyu, C., Zhang, Y., & Ji, D. (2016). Joint word segmentation, pos-tagging and syntactic chunking. In *Proceedings of the Thirtieth AAAI Conference on Artificial Intelligence, AAAI'16* (pp. 3007–3014). AAAI Press.

Manning, C. D., & Schütze, H. (1999). *Foundations of Statistical Natural Language Processing*. Cambridge, MA, USA: MIT Press.

McDonald, R. (2006). *Discriminative learning spanning tree algorithm for dependency parsing*. PhD thesis, University of Pennsylvania.

Nivre, J. (2003). An efficient algorithm for projective dependency parsing. In *Proceedings of the 8th International Workshop on Parsing Technologies (IWPT)* (pp. 149–160).

Nivre, J. (2008). Algorithms for deterministic incremental dependency parsing. *Computational Linguistics*, *34*(4), 513–554.

Pei, W., Ge, T., & Chang, B. (2015). An effective neural network model for graph-based dependency parsing. In *Proceedings of the 53rd Annual Meeting of the Association for Computational Linguistics and the 7th International Joint Conference on Natural Language Processing* (Vol. 1: Long Papers, pp. 313–322), Beijing, China: Association for Computational Linguistics.

Petrov, S., Barrett, L., Thibaux, R., & Klein, D. (2006). Learning accurate, compact, and interpretable tree annotation. In *Proceedings of the 21st International Conference on Computational Linguistics and 44th Annual Meeting of the Association for Computational Linguistics* (pp. 433–440), Sydney, Australia: Association for Computational Linguistics.

Puduppully, R., Zhang, Y., & Shrivastava, M. (2016). Transition-based syntactic linearization with lookahead features. In *Proceedings of the 2016 Conference of the North American Chapter of the Association for Computational Linguistics: Human Language Technologies* (pp. 488–493). San Diego, CA: Association for Computational Linguistics.
Qian, T., Zhang, Y., Zhang, M., Ren, Y., & Ji, D. (2015). A transition-based model for joint segmentation, pos-tagging and normalization. In *Proceedings of the 2015 Conference on Empirical Methods in Natural Language Processing* (pp. 1837–1846), Lisbon, Portugal: Association for Computational Linguistics.
Sagae, K., & Lavie, A. (2005). A classifier-based parser with linear run-time complexity. In *Proceedings of the Ninth International Workshop on Parsing Technology, Parsing '05* (pp. 125–132). Stroudsburg, PA, USA: Association for Computational Linguistics.
Sagae, K., Lavie, A., & MacWhinney, B. (2005). Automatic measurement of syntactic development in child language. In *Proceedings of the 43rd Annual Meeting of the Association for Computational Linguistics (ACL'05)* (pp. 197–204). Ann Arbor, MI: Association for Computational Linguistics.
Sarawagi, S., & Cohen, W. W. (2004). Semi-Markov conditional random fields for information extraction. In L. K. Saul, Y. Weiss, & L. Bottou (Eds.), *Advances in neural information processing systems 17* (pp. 1185–1192). Cambridge: MIT Press.
Shaalan, K. (2014). A survey of arabic named entity recognition and classification. *Computational Linguistics*, *40*(2), 469–510.
Smith, N. A. (2011). *Linguistic structure prediction*. Morgan and Claypool: Synthesis Lectures on Human Language Technologies.
Song, L., Zhang, Y., Song, K., & Liu, Q. (2014). Joint morphological generation and syntactic linearization. In *Proceedings of the Twenty-Eighth AAAI Conference on Artificial Intelligence, AAAI'14* (pp. 1522–1528). AAAI Press.
Vaswani, A., Bisk, Y., Sagae, K., & Musa, R. (2016). Supertagging with LSTMs. In *Proceedings of the 2016 Conference of the North American Chapter of the Association for Computational Linguistics: Human Language Technologies* (pp. 232–237). San Diego, CA: Association for Computational Linguistics.
Wang, W., & Chang, B. (2016). Graph-based dependency parsing with bidirectional LSTM. In *Proceedings of the 54th Annual Meeting of the Association for Computational Linguistics* (Vol. 1: Long Papers, pp. 2306–2315). Berlin, Germany: Association for Computational Linguistics.
Wang, Z., & Xue, N. (2014). Joint pos tagging and transition-based constituent parsing in Chinese with non-local features. In *Proceedings of the 52nd Annual Meeting of the Association for Computational Linguistics* (Vol. 1: Long Papers, pp. 733–742). Baltimore, MD: Association for Computational Linguistics.
Watanabe, T., & Sumita, E. (2015). Transition-based neural constituent parsing. In *Proceedings of the 53rd Annual Meeting of the Association for Computational Linguistics and the 7th International Joint Conference on Natural Language Processing* (Vol. 1: Long Papers, pp. 1169–1179). Beijing, China: Association for Computational Linguistics.
Wong, K.-F., Li, W., Xu, R., & Zhang, Z.-s., (2009). Introduction to Chinese natural language processing. *Synthesis Lectures on Human Language Technologies, 2*(1), 1–148.
Xu, W., Auli, M., & Clark, S. (2015). CCG supertagging with a recurrent neural network. In *Proceedings of the 53rd Annual Meeting of the Association for Computational Linguistics and the 7th International Joint Conference on Natural Language Processing* (Vol. 2: Short Papers, pp. 250–255). Beijing, China: Association for Computational Linguistics.
Xu, W., Auli, M., & Clark, S. (2016). Expected f-measure training for shift-reduce parsing with recurrent neural networks. In *Proceedings of the 2016 Conference of the North American Chapter of the Association for Computational Linguistics: Human Language Technologies* (pp. 210–220). San Diego, CA: Association for Computational Linguistics.
Xu, W., Clark, S., & Zhang, Y. (2014). Shift-reduce CCG parsing with a dependency model. In *Proceedings of the 52nd Annual Meeting of the Association for Computational Linguistics* (Vol. 1: Long Papers).

Xue, N. (2003). Chinese word segmentation as character tagging. *International Journal of Computational Linguistics and Chinese Language Processing*, *8*, 29–48.
Yamada, H., & Matsumoto, Y. (2003). Statistical dependency analysis with support vector machines. In *In Proceedings of IWPT* (pp. 195–206).
Zhang, M., & Zhang, Y. (2015). Combining discrete and continuous features for deterministic transition-based dependency parsing. In *Proceedings of the 2015 Conference on Empirical Methods in Natural Language Processing* (pp. 1316–1321). Lisbon, Portugal: Association for Computational Linguistics.
Zhang, M., Zhang, Y., Che, W., & Liu, T. (2013). Chinese parsing exploiting characters. In *Proceedings of the 51st Annual Meeting of the Association for Computational Linguistics* (Vol. 1: Long Papers, pp. 125–134). Sofia, Bulgaria: Association for Computational Linguistics.
Zhang, M., Zhang, Y., Che, W., & Liu, T. (2014). Character-level Chinese dependency parsing. In *Proceedings of the 52nd Annual Meeting of the Association for Computational Linguistics* (Vol. 1: Long Papers, pp. 1326–1336). Baltimore, MD: Association for Computational Linguistics.
Zhang, M., Zhang, Y., & Fu, G. (2016a). Transition-based neural word segmentation. In *Proceedings of the 54th Annual Meeting of the Association for Computational Linguistics* (Vol. 1: Long Papers, pp. 421–431), Berlin, Germany: Association for Computational Linguistics.
Zhang, Y., & Clark, S. (2007). Chinese segmentation with a word-based perceptron algorithm. In *Proceedings of the 45th Annual Meeting of the Association of Computational Linguistics* (pp. 840–847), Prague, Czech Republic: Association for Computational Linguistics.
Zhang, Y., & Clark, S. (2008a). Joint word segmentation and POS tagging using a single perceptron. In *Proceedings of ACL-08: HLT* (pp. 888–896). Columbus, OH: Association for Computational Linguistics.
Zhang, Y., & Clark, S. (2008b). A tale of two parsers: Investigating and combining graph-based and transition-based dependency parsing. In *Proceedings of the 2008 Conference on Empirical Methods in Natural Language Processing* (pp. 562–571), Honolulu, HI: Association for Computational Linguistics.
Zhang, Y., & Clark, S. (2009). Transition-based parsing of the Chinese Treebank using a global discriminative model. In *Proceedings of the 11th International Conference on Parsing Technologies, IWPT '09* (pp. 162–171). Stroudsburg, PA, USA: Association for Computational Linguistics.
Zhang, Y., & Clark, S. (2010). A fast decoder for joint word segmentation and POS-tagging using a single discriminative model. In *Proceedings of the 2010 Conference on Empirical Methods in Natural Language Processing* (pp. 843–852). Cambridge, MA: Association for Computational Linguistics.
Zhang, Y., & Clark, S. (2011a). Shift-reduce CCG parsing. In *Proceedings of the 49th Annual Meeting of the Association for Computational Linguistics: Human Language Technologies* (pp. 683–692). Portland, OR, USA: Association for Computational Linguistics.
Zhang, Y., & Clark, S. (2011b). Syntactic processing using the generalized perceptron and beam search. *Computational Linguistics, 37*(1).
Zhang, Y., & Nivre, J. (2011). Transition-based dependency parsing with rich non-local features. In *Proceedings of the 49th Annual Meeting of the Association for Computational Linguistics: Human Language Technologies* (pp. 188–193). Portland, OR, USA: Association for Computational Linguistics.
Zhang, Z., Zhao, H., & Qin, L. (2016b). Probabilistic graph-based dependency parsing with convolutional neural network. In *Proceedings of the 54th Annual Meeting of the Association for Computational Linguistics* (Vol. 1: Long Papers, pp. 1382–1392), Berlin, Germany: Association for Computational Linguistics.
Zhou, H., Zhang, Y., Huang, S., & Chen, J. (2015). A neural probabilistic structured-prediction model for transition-based dependency parsing. In *Proceedings of the 53rd Annual Meeting of the Association for Computational Linguistics and the 7th International Joint Conference on Natural Language Processing* (Vol. 1: Long Papers, pp. 1213–1222), Beijing, China: Association for Computational Linguistics.

Zhou, J., & Xu, W. (2015). End-to-end learning of semantic role labeling using recurrent neural networks. In *Proceedings of the 53rd Annual Meeting of the Association for Computational Linguistics and the 7th International Joint Conference on Natural Language Processing* (Vol. 1: Long Papers, pp. 1127–1137), Beijing, China: Association for Computational Linguistics.

Zhu, M., Zhang, Y., Chen, W., Zhang, M., & Zhu, J. (2013). Fast and accurate shift-reduce constituent parsing. In *Proceedings of the 51st Annual Meeting of the Association for Computational Linguistics* (Vol. 1: Long Papers, pp. 434–443), Sofia, Bulgaria: Association for Computational Linguistics.

Chapter 5
Deep Learning in Knowledge Graph

Zhiyuan Liu and Xianpei Han

Abstract Knowledge Graph (KG) is a fundamental resource for human-like commonsense reasoning and natural language understanding, which contains rich knowledge about the world's entities, entities' attributes, and semantic relations between different entities. Recent years have witnessed the remarkable success of deep learning techniques in KG. In this chapter, we introduce three broad categories of deep learning-based KG techniques: (1) *knowledge representation learning* techniques which embed entities and relations in a KG into a dense, low-dimensional, and real-valued semantic space; (2) *neural relation extraction* techniques which extract facts/relations from text, which can then be used to construct/complete KG; (3) *deep learning-based entity linking* techniques which bridge Knowledge Graph with textual data, which can facilitate many different tasks.

5.1 Introduction

With the thriving development of Internet in twenty-first century, the amount of web information shows an explosive trend, during which people find it is getting harder and less efficient to extract valuable information, or more precisely, knowledge, from the huge noisy plaintexts. And then, people start to realize that the world is made up of entities instead of strings, just as Dr. Singhal said, "things, not strings". As a result, the concept of Knowledge Graph comes into the public view.

Knowledge Graph (KG), also known as Knowledge Base, is a significant dataset organizing human knowledge about the world in a structured form, where the knowledge is represented as concrete entities and the multi-relational abstract concepts among them. There are mainly two methods when it comes to the construction of Knowledge Graph. One is using the existing semantics web datasets in Resource

Z. Liu
Tsinghua University, Beijing, China
e-mail: liuzy@tsinghua.edu.cn

X. Han (✉)
Institute of Software, Chinese Academy of Sciences, Beijing, China
e-mail: xianpei@iscas.ac.cn

L. Deng and Y. Liu (eds.), *Deep Learning in Natural Language Processing*, https://doi.org/10.1007/978-981-10-5209-5_5

Description Framework (RDF) with the help of manually annotation. The other is using machine learning or deep learning method to automatically extract knowledge from enormous plaintexts in Internet.

Due to such well-structured united knowledge representation, KG can provide effective structured information about the complicated real world. Hence, it starts to play an important role in many applications of artificial intelligence, especially the field of natural language processing and information retrieval such as web search, question answering, speech recognition, etc., in recent years, attracting wide attention from both academia and industry.

In this chapter, we will first introduce the basic concepts and typical Knowledge Graphs in Sect. 5.1, and then, we introduce recent advances of knowledge representation learning in Sect. 5.2, relation extraction in Sect. 5.3, and entity linking in Sect. 5.4. Finally, we give a brief conclusion in Sect. 5.5.

5.1.1 Basic Concepts

A typical KG is usually composed of two elements, entities (i.e., concrete entities and abstract concepts in real world) and relations between entities. Thus, it arranges all kinds of knowledge into large quantities of triple facts in the form of (e_1, `relation`, e_2) where e_1 indicates the head entity and e_2 indicates the tail entity. For instance, we know that *Donald Trump* is the president of *United States*. This knowledge could be represented as (*Donald Trump*, `president_of`, *United States*). Furthermore, it should be noted that in real world, the same head entity and relation may have several different tail entities. For example, *Kaká* was a soccer player in *Real Madrid* and *A.C. Milan* football club. We can get such two triples from this common knowledge: (*Kaká*, `player_of_team`, *Real Madrid FC*), (*Kaká*, `player_of_team`, *A.C. Milan*). Reversely this situation could also happen when tail entity and relation are fixed. It is also possible when head and tail entity are both multiple (e.g., the relation `author_of_paper`). From this aspect, we can see that KG has great flexibility as well as the ability to represent knowledge. Through all these triples, knowledge is thus represented as a huge directed graph, in which entities are considered as nodes and relations as edges.

5.1.2 Typical Knowledge Graphs

The current Knowledge Graphs can be divided into two categories from the aspect of capacity and knowledge domain. The graphs in the first category contain great quantities of triples and well-known common relation, such as Freebase. The graphs in the second category are comparatively smaller but focus on specific knowledge domain and usually fine-grained.

There are several Knowledge Graphs widely used in applications and having great influence. In the following sections, we will introduce some well-known Knowledge Graphs.

5.1.2.1 Freebase

Freebase is one of the most popular Knowledge Graphs in the world. It is a large collaborative database consisting of data composed mainly of its community members. It is an online collection of structured data harvested from many sources, including Wikipedia, Fashion Model Directory, NNDB, MusicBrainz, and other individual, user-submitted wiki contributions. It also announced an open API, RDF endpoint and a dataset dump for its users for both commercial and noncommercial use.

Freebase was developed by the American software company Metaweb and ran publicly since March 2007. In July 2010, Metaweb was acquired by Google, and Google Knowledge Graph was powered in part by Freebase. In December 2014, the Freebase team officially announced that the Freebase website would be shut down together with its API by June 2015. Up to March 24, 2016, Freebase has 58,726,427 topics and 3,197,653,841 facts.

For instance, Fig. 5.1 is the example page of former American president John F. Kennedy in Freebase. It is easy to notice that the information such as date of birth, gender, and career are listed in structured form just like a resume.

5.1.2.2 DBpedia

DBpedia ("DB" stands for "dataset") is a crowdsourced community effort to extract structured information from Wikipedia and make this information available on the web. DBpedia allows users to ask sophisticated queries against Wikipedia, and to link the different datasets on the web to Wikipedia resources, which will make it easier for the huge amount of information in Wikipedia to be used in some new interesting ways. The project was started by people at the Free University of Berlin and Leipzig University, in collaboration with OpenLink Software, and the first publicly available dataset was published in 2007. The whole DBpedia dataset describes 4.58 million entities, out of which 4.22 million are classified in a consistent ontology, including 1,445,000 persons, 735,000 places, 123,000 music albums, 87,000 films, 19,000 video games, 241,000 organizations, 251,000 species and 6,000 diseases. The dataset also features labels and abstracts for these entities in up to 125 different languages. What is more, due to the reason that DBpedia is linked to Wikipedia's infobox, it can make dynamic updates as the information changes.

5.1.2.3 Wikidata

Wikidata is a collaboratively edited Knowledge Base operated by the Wikimedia Foundation. It is intended to provide a common source of data which can be used by

John F. Kennedy

Discuss "John F. Kennedy" Show Empty Fields

image 1 of 1

Types: Film Actor (Film), Person (People), US Senator (Government), US President (Government), US Politician (Government), Deceased Person (People)

Also known as: JFK, John F Kennedy, President John F. Kennedy

Gender: Male

Date of Birth: May 29, 1917

Place of Birth: Brookline, Massachusetts

Country Of Nationality: United States

Profession: President of the United States

Spouse(s): Jacqueline Kennedy Onassis - 1968 - 1975

Children: John F. Kennedy, Jr.

IMDB Entry: http://www.imdb.com...

Presidency Number: 35

Vice President: Lyndon B. Johnson

Party: Democratic Party

Date of Death: Nov 22, 1963

Place of Death: Dallas

Cause Of Death: Assassination

Description

John Fitzgerald Kennedy (May 29, 1917 – November 22, 1963), also referred to as **John F. Kennedy**, **JFK**, **John Kennedy** or **Jack Kennedy**, was the 35th President of the United States. He served from 1961 until his assassination in 1963. Major events during his presidency include the Bay of Pigs Invasion, the Cuban Missile Crisis, the building of the Berlin Wall, the Space Race, the American Civil Rights Movement and early events of the Vietnam War.

Fig. 5.1 The Freebase page of John F. Kennedy

Wikimedia projects such as Wikipedia, and by anyone else. The creation of the project was funded by donations from the Allen Institute for Artificial Intelligence, the Gordon and Betty Moore Foundation, and Google, Inc., totaling *euro* 1.3 million. As for the inside detailed structure, Wikidata is a document-oriented database, focused on items. Each item represents a topic (or an administrative page used to maintain Wikipedia) and is identified by a unique number. Information is added to items by creating statements. Statements take the form of key-value pairs, with each statement consisting of a property (the key) and a value linked to the property. Up to May 2017, the Knowledge Base contains 25,887,362 data items that anyone can edit.

Fig. 5.2 The Wikidata page of John F. Kennedy

For instance, Fig. 5.2 is the example page of John F. Kennedy in Wikidata. It could be noticed that each relation is also attached to references which anyone can add or edit.

5.1.2.4 YAGO

YAGO, which stands for Yet Another Great Ontology, is a huge high-quality Knowledge Base developed by Max Planck Institute for Informatics and the Telecom ParisTech University. The knowledge inside is derived from Wikipedia, WordNet, and GeoNames. Currently, it has knowledge of more than 10 million entities (like persons, organizations, cities, etc.) and contains more than 120 million facts about these entities. The highlight spots in YAGO could be concluded as follows: First, the accuracy of YAGO has been manually evaluated, proving a confirmed accuracy of 95% and every relation is annotated with its confidence value. Second, YAGO combines the clean taxonomy of WordNet with the richness of the Wikipedia category system, assigning the entities to more than 350,000 classes. Third, YAGO is an ontology that is anchored in both time and space which means it attaches a temporal dimension and a spacial dimension to many of its facts and entities.

5.1.2.5 HowNet

HowNet (Dong and Dong 2003) is an online commonsense Knowledge Base unveiling inter-conceptual relations and inter-attribute relations of concepts as connoting in lexicons of the Chinese and their English equivalents. The main philosophy behind HowNet is its understanding and interpretation of the objective world. HowNet states that all matters (physical and metaphysical) are in constant motion and are ever changing in a given time and space. Things evolve from one state to another as recorded in the corresponding change in their attributes. Take the instance "human", for example, it is described by the following state of living: birth, aging, sickness, and death. As a person grows, his age (the attribute-value) also adds up. At the meantime, his hair color (an attribute) turns white (the attribute-value). It could be concluded that every object carries a set of attributes and the similarities and the differences between the objects are determined by the attributes they each carries. Besides `attribute`, `part` is also a significant key philosophy concept in HowNet. It could be understood that all objects are probably parts of something else while at the same time, all objects are also the whole of something else. For example, doors and windows are parts of buildings while meantime buildings are also part of a community. In total, HowNet contains 271 information structure patterns, 58 semantic structure patterns, 11,000 word instances, and 60,000 Chinese words in all (Fig. 5.3).

In addition, HowNet also lays emphasis on sememes in the construction process, which are defined as the minimum semantic units of word meanings, and there exists a limited close set of sememes to compose the semantic meanings of an open set of concepts. HowNet annotates precise senses to each word, and for each sense, HowNet annotates the significance of parts and attributes represented by sememes. For example, the word "apple" actually has two main senses: one is a sort of fruit

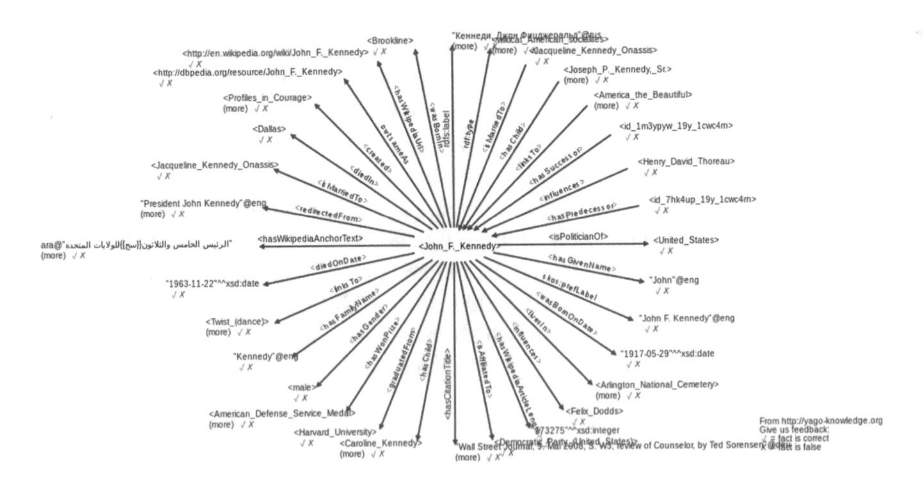

Fig. 5.3 The YAGO page of John F. Kennedy

and the other is a computer brand. Thus, in the first sense, it has the sememe `fruit` and in the second it has sememes `computer`, `bring`, and `SpeBrand`.

5.2 Knowledge Representation Learning

In the past years, various specific algorithms have been designed to store and utilize the information in KG according to its traditional representation (i.e., network representation), which is usually very time-consuming and suffers from data sparsity. Recently, representation learning, which is a subarea of deep learning, has attracted lots of attentions in different areas including natural language processing and artificial intelligence. Representation learning aims at embedding the objects into a dense, low-dimensional, and real-valued semantic space. And knowledge representation learning is a subarea of representation learning, which focuses on embedding the entities and relations in KG.

Recent studies reveal that translation-based representation learning methods are efficient and effective to encode relational facts in KG with low-dimensional representations of both entities and relations, which can alleviate the issue of data sparsity and be further employed to knowledge acquisition, fusion, and inference. TransE (Bordes et al. 2013) is one of typical translation-base knowledge representation learning methods, which learns low-dimensional vectors for both entities and relations and is very simple and effective. TransE regards the relation in a relational triple as a translation between the embeddings of the head and tail entities, that is, $\mathbf{h} + \mathbf{r} \approx \mathbf{t}$ when the triple (h, r, t) holds. And it achieves amazing performance in the task of Knowledge Graph completion.

Although TransE has achieved great success, it still has issues when modeling 1-to-N, N-to-1, and N-to-N relations. The entity embeddings learnt by TransE are lacking in discrimination due to these complex relations. Therefore, how to deal with complex relations is one of the key challenges in knowledge representation learning. Recently, there are lots of extensions of TransE which focus on this challenge. TransH (Wang et al. 2014b) and TransR (Lin et al. 2015b) are proposed to represent an entity with different representations when involved in different relations. TransH models the relation as a translation vector on a hyperplane and projects the entity embeddings into the hyperplane with a normal vector. TransR represents entities in the entity semantic space and uses a relation-specific transform matrix to project it into the different relation spaces when involved in different relations. Further, researchers propose two extension of TransR including TransD (Ji et al. 2015) which considers the information of entities in the projecting matrices and TranSparse (Ji et al. 2016) which considers the heterogeneity and imbalance of relations via sparse matrices. In addition, there are many other extensions of TransE which focus on different characteristics of relations including TransG (Xiao et al. 2015) and KG2E (He et al. 2015) adopt Gaussian embeddings to model both entities and relations; ManifoldE (Xiao et al. 2016) employs a manifold-based embedding principle in knowledge representation learning; and so on.

Besides, TransE still has a problem that only considering direct relations between entities. To address this issue, Lin et al. (2015a) propose Path-based TransE which extends TransE to model relational paths by selecting reasonable relational paths and representing them with low-dimensional vectors. Almost at the same time, there are others researchers considering relational paths in KG successfully (Garcıa-Durán et al. 2015) using neural network. Besides, relational path learning has also been used in the KG-based QA (Gu et al. 2015).

Most existing knowledge representation learning methods discussed above only focus on the structure information in KG, regardless of the rich multisource information such as textual information, type information, and visual information. These cross-modal information can provide supplementary knowledge of the entities specially for those entities with less relational facts and is significant when learning knowledge representations. For textural information, Wang et al. (2014a) and Zhong et al. (2015) propose to jointly embed both entities and words into a unified semantic space by aligning them with entity names, descriptions, and Wikipedia anchors. Further, Xie et al. (2016b) propose to learn entity representations based on their descriptions with CBOW or CNN encoders. For type information, Krompaß et al. (2015) take type information as constraints of head and tail entity set for each relation to distinguish entities which belong to the same types. Instead of merely considering type information as type constraints, Xie et al. (2016c) utilize hierarchical type structures to enhance TransR via guiding the construction of projection matrices. For visual information, Xie et al. (2016a) propose image-embodied knowledge representation learning to take visual information into consideration via learning entity representations using their corresponding figures. It is natural that we learn things in real world with all kinds of multisource information. Multisource information such as plaintexts, hierarchical types, or even images and videos is of great importance when modeling the complicated world and constructing cross-modal representations. Moreover, other types of information could also be encoded into knowledge representation learning to enhance the performance.

5.3 Neural Relation Extraction

To enrich existing KGs, researchers have invested in automatically finding unknown relational facts, i.e., relation extraction (RE). Relation extraction aims at extracting relational data from plaintexts. In recent years, as the development of deep learning (Bengio 2009) techniques, neural relation extraction adopts an end-to-end neural network to model the relation extraction task. The framework of neural relation extraction includes a sentence encoder to capture the semantic meaning of the input sentence and represents it as a sentence vector, and a relation extractor to generate the probability distribution of extracted relations according to sentence vectors. We will give an in-depth review of recent works on neural relation extraction.

Neural relation extraction (NRE) has two main tasks including sentence-level NRE and document-level NRE. In this section, we will introduce these two tasks in detail, respectively.

5.3.1 Sentence-Level NRE

Sentence-level NRE aims at predicting the semantic relations between the entity (or nominal) pair in a sentence. Formally, given the input sentence x which consists of m words $x = (w_1, w_2, \ldots, w_m)$ and its corresponding entity pair e_1 and e_2 as inputs, sentence-level NRE wants to obtain the conditional probability $p(r|x, e_1, e_2)$ of relation r ($r \in \mathbb{R}$) via a neural network, which can be formalized as

$$p(r|x, e_1, e_2) = p(r|x, e_1, e_2, \theta), \tag{5.1}$$

where θ is parameter of the neural network, and r is a relation in the relation set $\mathbf{R}$.

A basic form of sentence-level NRE consists of three components: (a) an input encoder which gives a representation for the input words, (b) a sentence encoder which computes either a single vector or a sequence of vectors representing the original sentence, and (c) a relation classifier which calculates the conditional probability distribution of all relations.

5.3.1.1 Input Encoder

First, a sentence-level NRE system projects discrete source sentence words into continuous vector space, and obtain the input representation $\mathbf{w} = \{\mathbf{w}_1; \mathbf{w}_2; \cdots; \mathbf{w}_m$ of the source sentence.

Word embeddings learn low-dimensional real-valued representation of words, which can reflect syntactic and semantic relationships between words. Formally, each word w_i is encoded by the corresponding column vector in an embedding matrix $\mathbf{V} \in \mathbb{R}^{d^a \times |V|}$, where V indicates a fix-sized vocabulary.

Position embeddings aim to specify the position information of the word with respect to two corresponding entities in the sentence. Formally, each word w_i is encoded by two position vectors with respect to the relative distances from the word to two target entities, respectively. For example, in the sentence *New York* is a city of *United States*, the relative distance from the word city to *New York* is 3 and *United States* is -2.

Part-of-speech tag embeddings represent the lexical information of target word in the sentence. Due to the fact that word embeddings are obtained from a generic corpus on a large scale, the information they contain may not be in accordance with the meaning in a specific sentence, it is necessary to align each word with its linguistic information, e.g., noun, verb, etc. Formally, each word w_i is encoded by the corresponding column vector in an embedding matrix $\mathbf{V}^p \in \mathbb{R}^{d^p \times |V^p|}$, where d^p

is the dimension of embedding vector and V^p indicates a fix-sized part-of-speech tag vocabulary.

WordNet hypernym embeddings aim to take advantages of the prior knowledge of hypernym to contribute to relation extraction. It is easier to build the link between different but conceptual similar words when given each word's hypernym information in WordNet, e.g., noun.food, verb.motion, etc. Formally, each word w_i is encoded by the corresponding column vector in an embedding matrix $\mathbf{V}^h \in \mathbb{R}^{d^h \times |V^h|}$, where d^h is the dimension of embedding vector and V^h indicates a fix-sized hypernym vocabulary.

5.3.1.2 Sentence Encoder

Next, the sentence encoder encodes input representations into either a single vector or a sequence of vectors $\mathbf{x}$. We will introduce the different sentence encoders in the following.

Convolution neural network encoder (Zeng et al. 2014) is proposed to embed input sentence using a convolutional neural network (CNN) which extracts local feature by a convolution layer and combines all local features via a max-pooling operation to obtain a fixed-sized vector for the input sentence. Formally, as illustrated in Fig. 5.4, convolution operation is defined as a matrix multiplication between a sequence of vectors and a convolution matrix $\mathbf{W}$ and a bias vector $\mathbf{b}$ with a sliding window. Let us define the vector $\mathbf{q}_i$ as the concatenation of a sequence of input representations in the i-th window, we have

$$[\mathbf{x}]_j = \max_i [f(\mathbf{W}\mathbf{q}_i + \mathbf{b})]_j, \tag{5.2}$$

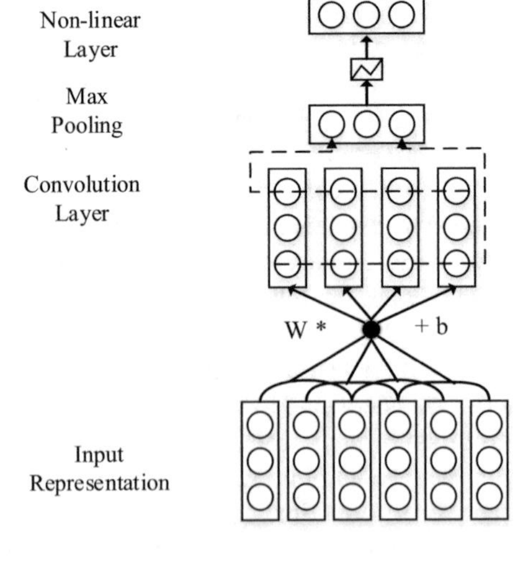

Fig. 5.4 The architecture of CNN encoder

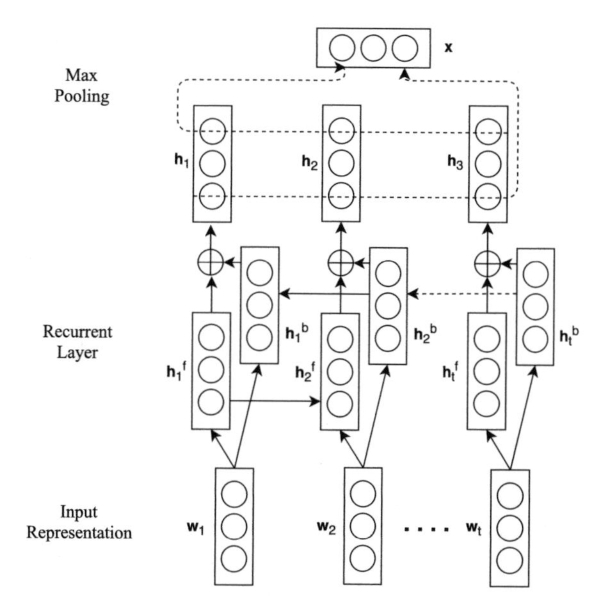

Fig. 5.5 The architecture of recurrent encoder

where f indicates a nonlinear function such as sigmoid or tangent function.

Further, to better capture the structural information between two entities, piecewise max-pooling operation (Zeng et al. 2015) is proposed instead of traditional max-pooling operation. Piecewise max-pooling operation returns the maximum value in three segments of the input sentence which are divided into two target entities.

Recurrent neural network encoder (Zhang and Wang 2015) is proposed to embed input sentence using a recurrent neural network (RNN) which has the capability to learn the temporal features. As illustrated in Fig. 5.5, each word representation vectors are put into recurrent layer step-by-step. For each step i, the network takes the word representation vector $\mathbf{w}_i$ and the previous step $i-1$'s output $\mathbf{h}_{i-1}$ as inputs, and then we have

$$\mathbf{h}_i = f(\mathbf{w}_t, \mathbf{h}_{i-1}), \tag{5.3}$$

where f indicates the transform function inside the RNN cell, which can be the LSTM units (Hochreiter and Schmidhuber 1997) (LSTM-RNNs) or the GRU units (Cho et al. 2014) (GRU-RNNs). In addition, a bidirectional RNN network is employed to fully utilize the information of future words when predicting the semantic meaning in the middle of a sentence.

Next, RNN combines the information from forward and backward network as a local feature and uses a max-pooling operation to extract the global feature, which forms the representation of the whole input sentence. The max-pooling layer could be formulated as

$$[\mathbf{x}]_j = \max_i [\mathbf{h}_i]_j. \tag{5.4}$$

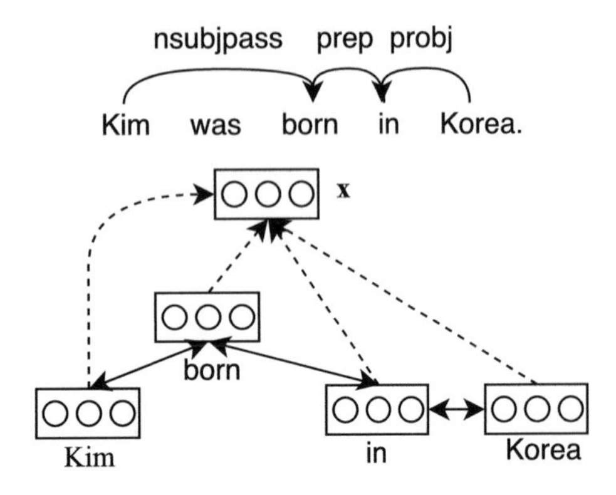

Fig. 5.6 The architecture of dependency tree-structured LSTM

Besides max-pooling, word attention can also combine all local feature vectors together. It uses attention mechanism (Bahdanau et al. 2014) to learn attention weights on each step. Suppose $\mathbf{H} = [\mathbf{h}_1, \mathbf{h}_2, \ldots, \mathbf{h}_m]$ is the matrix consisting of all output vectors that produced by the recurrent layer, the whole sentence's feature vector $\mathbf{x}$ is formed by a weighted sum of each step's output:

$$\alpha = \text{softmax}(\mathbf{s}^T \tanh(\mathbf{H})) \tag{5.5}$$

$$\mathbf{x} = \mathbf{H}\alpha^T, \tag{5.6}$$

where $\mathbf{s}$ is a trainable query vector and s^T indicates its transposition.

Besides, Miwa and Bansal (2016) proposed a model that captures both word sequence and dependency tree substructure information by stacking bidirectional path-based LSTM-RNNs (i.e., bottom-up and top-down) on bidirectional sequential LSTM-RNNs. As illustrated in Fig. 5.6, it focuses on the shortest path between the target entities in the dependency tree because experimental result in (Xu et al. 2015) shows that these paths are effective in relation classification.

Recursive neural network encoder aims to extract features from the information of syntactic parsing tree structure because the syntactic information is important for extracting relations from sentences. Generally, these encoders treat the tree structure inside the syntactic parsing tree as a strategy of composition as well as direction for recursive neural network to combine each word's embedding vector.

Socher et al. (2012) proposed a recursive matrix-vector model (MV-RNN) which captures constituent parsing tree structure information by assigning a matrix-vector representation for each constituent. The vector captures the meaning of constituent itself and the matrix represents how it modifies the meaning of the word it combines with. Suppose we have two children components l, r and their father component p, the composition can be formulated as follows:

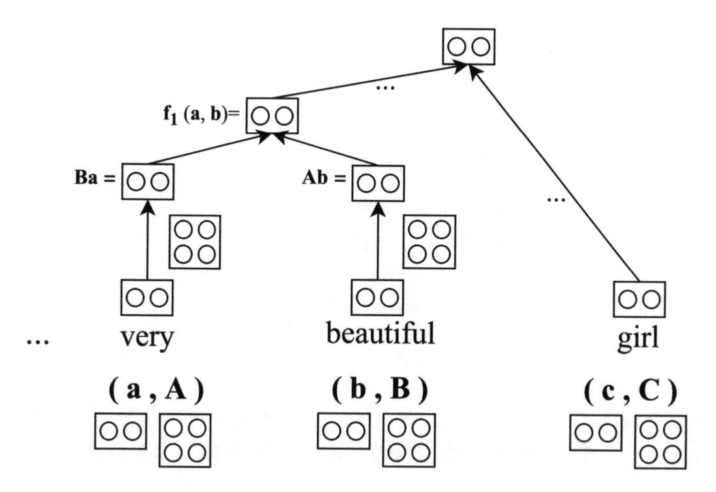

Fig. 5.7 The architecture of matrix-vector recursive encoder

$$\mathbf{p} = f_1(l, r) = g\left(\mathbf{W}_1 \begin{bmatrix} \mathbf{Ba} \\ \mathbf{Ab} \end{bmatrix}\right) \tag{5.7}$$

$$\mathbf{P} = f_2(l, r) = \mathbf{W}_2 \begin{bmatrix} \mathbf{A} \\ \mathbf{B} \end{bmatrix}, \tag{5.8}$$

where $\mathbf{a}, \mathbf{b}, \mathbf{p}$ are embedding vectors for each components and $\mathbf{A}, \mathbf{B}, \mathbf{P}$ are matrices, $\mathbf{W}_1$ is a matrix that maps transformed words into another semantic space, the element-wise function g is an activation function, and $\mathbf{W}_2$ is a matrix that maps two matrices into one combined matrix P with the same dimension. The whole process is illustrated in Fig. 5.7. And then, MV-RNN selects the highest node of the path in the parse tree between the two target entities to represent the input sentence.

In fact, the RNN unit here can be replaced by LSTM units or GRU units. Tai et al. (2015) propose two types of tree-structured LSTMs including the Child-Sum Tree-LSTM and the N-ary Tree-LSTM to capture constituent or dependency parsing tree structure information. For the Child-Sum Tree-LSTM, given a tree, let $C(t)$ denote the set of children of node t. Its transition equations are defined as follows:

$$\hat{\mathbf{h}}_t = \sum_{k \in C(t)} \mathbf{h}_k, \tag{5.9}$$

$$\mathbf{i}_t = \sigma(\mathbf{W}^{(i)}\mathbf{w}_t + \mathbf{U}^i\hat{\mathbf{h}}_t + \mathbf{b}^{(i)}), \tag{5.10}$$

$$\mathbf{f}_{tk} = \sigma(\mathbf{W}^{(f)}\mathbf{w}_t + \mathbf{U}^f\mathbf{h}_k + \mathbf{b}^{(f)}) \ \ (k \in C(t)), \tag{5.11}$$

$$\mathbf{o}_t = \sigma(\mathbf{W}^{(o)}\mathbf{w}_t + \mathbf{U}^o\hat{\mathbf{h}}_t + \mathbf{b}^{(o)}), \tag{5.12}$$

$$\mathbf{u}_t = \tanh(\mathbf{W}^{(u)}\mathbf{w}_t + \mathbf{U}^u\hat{\mathbf{h}}_t + \mathbf{b}^{(u)}), \tag{5.13}$$

$$\mathbf{c}_t = \mathbf{i}_t \odot \mathbf{u}_t + \sum_{k \in C(t)} \mathbf{f}_{tk} \odot \mathbf{c}_{t-1}, \tag{5.14}$$

$$\mathbf{h}_t = \mathbf{o}_t \odot \tanh(\mathbf{c}_t). \tag{5.15}$$

The N-ary Tree-LSTM has similar transition equations with the Child-Sum Tree-LSTM. The only difference is that it limits the tree structures has at most N branches.

5.3.1.3 Relation Classifier

Finally, when obtaining the representation $\mathbf{x}$ of the input sentence, relation classifier calculates the conditional probability $p(r|x, e_1, e_2)$ via a softmax layer as follows:

$$p(r|x, e_1, e_2) = \text{softmax}(\mathbf{M}\mathbf{x} + \mathbf{b}), \tag{5.16}$$

where $\mathbf{M}$ indicates the relation matrix and $\mathbf{b}$ is a bias vector.

5.3.2 Document-Level NRE

Although existing neural models have achieved great success for extracting novel relational facts, it always suffers from the insufficiency of training data. To address the issue, researchers proposed distant supervision assumption to automatically generate training data via aligning KGs and plaintexts. The intuition of distant supervision assumption is that all sentences that contain two entities will express their relations in KGs. For example, (*New York*, `city of`, *United States*) is a relational fact in KGs. Distant supervision assumption will regard all sentences that contain these two entities as valid instances for relation `city of`. It offers a natural way to utilize information from multiple sentences (document-level) rather than single sentence (sentence-level) to decide if a relation holds between two entities.

Therefore, document-level NRE aims to predict the semantic relations between an entity pair using all involved sentences. Given the input sentence set S which consists of n sentences $S = (x_1, x_2, \ldots, x_n)$ and its corresponding entity pair e_1 and e_2 as inputs, document-level NRE wants to obtain the conditional probability $p(r|S, e_1, e_2)$ of relation r ($r \in \mathbb{R}$) via a neural network, which can be formalized as

$$p(r|S, e_1, e_2) = p(r|S, e_1, e_2, \theta). \quad (5.17)$$

A basic form of document-level NRE consists of four components: (a) an input encoder similar to sentence-level NRE, (b) a sentence encoder similar to sentence-level NRE, (c) a document encoder which computes either vector representing all related sentences, and (d) a relation classifier similar to sentence-level NRE which takes document vector as input instead of sentence vector. In this next, we will introduce the document encoder and in detail.

5.3.2.1 Document Encoder

The document encodes all sentence vectors into either single vector $\mathbf{S}$. We will introduce the different document encoders in the following.

Random Encoder. It simply assumes that each sentence can express the relation between two target entities and randomly select one sentence to represent the document. Formally, the document representation is defined as

$$\mathbf{S} = \mathbf{x}_i \ (i = 1, 2, \ldots, n), \quad (5.18)$$

where $\mathbf{x}_i$ indicates the sentence representation of x_i and i is a random index.

Max Encoder. In fact, as introduced above, not all sentences containing two target entities can express their relations. For example, the sentence "New York City is the premier gateway for legal immigration to the United States" does not express the relation `city_of`. Hence, in (Zeng et al. 2015), they follow the at-least-one assumption which assumes that at least one sentence that contains these two target entities can express their relations, and select the sentence with highest probability for the relation to represent the document. Formally, the document representation is defined as

$$\mathbf{S} = \mathbf{x}_i \ (i = \text{argmax}_i \, p(r|x_i, e_1, e_2)). \quad (5.19)$$

Average Encoder. Both random encoder or max encoder use only one sentence to represent the document, which ignores the rich information of different sentences. To exploit the information of all sentences, Lin et al. (2016) believe that the representation $\mathbf{S}$ of the document depends on all sentences' representations $\mathbf{x}_1, \mathbf{x}_2, \ldots, \mathbf{x}_n$. Each sentence representation $\mathbf{x}_i$ can give the relation information about two entities for input sentence x_i. The average encoder assumes that all sentences contribute equally to the representation of the document. It means the embedding $\mathbf{S}$ of the document is the average of all the sentence vectors:

$$\mathbf{S} = \sum_i \frac{1}{n} \mathbf{x}_i. \quad (5.20)$$

Attentive Encoder. Due to the wrong label issue brought by distant supervision assumption inevitably, the performance of average encoder will be influenced by those sentences that contain no related information. To address this issue, Lin et al. (2016) further propose to employ a selective attention to de-emphasize those noisy sentence. Formally, the document representation is defined as a weighted sum of sentence vectors:

$$\mathbf{S} = \sum_i \alpha_i \mathbf{x}_i, \tag{5.21}$$

where α_i is defined as

$$\alpha_i = \frac{\exp(\mathbf{x}_i \mathbf{A} \mathbf{r})}{\sum_j \exp(\mathbf{x}_j \mathbf{A} \mathbf{r})}, \tag{5.22}$$

where $\mathbf{A}$ is a diagonal matrix and $\mathbf{r}$ is the representation vector of relation r.

5.3.2.2 Relation Classifier

Similar to sentence-level NRE, when obtaining the document representation $\mathbf{S}$, relation classifier calculates the conditional probability $p(r|S, e_1, e_2)$ via a softmax layer as follows:

$$p(r|S, e_1, e_2) = \text{softmax}(\mathbf{M}'\mathbf{S} + \mathbf{b}'), \tag{5.23}$$

where $\mathbf{M}'$ indicates the relation matrix and $\mathbf{b}'$ is a bias vector.

5.4 Bridging Knowledge with Text: Entity Linking

Knowledge Graph contains rich knowledge about the world's entities, their attributes, and semantic relations between different entities. Bridging Knowledge Graph with textual data can facilitate many different tasks, such as information extraction, text classification, and question answering. For example, it is helpful for understanding "*Jobs leaves Apple*" if we knew "*Steve Jobs* is CEO of *Apple Inc.*".

Currently, the main research issue in bridging Knowledge Graph with textual data is *entity linking (EL)* (Ji et al. 2010). Given a set of name mentions $M = \{m_1, m_2, \ldots, m_k\}$ in a document d, and a Knowledge Graph KB containing a set of entities $E = \{e_1, e_2, \ldots, e_n\}$, an entity linking system is a function $\delta : M \rightarrow E$ which maps name mentions to their referent entities in KB. Fig. 5.8 shows an example, where an EL system will identify the referent entities of the three entity mentions *WWDC, Apple, and Lion* correspondingly are *Apple Worldwide Developers Conference, Apple Inc.* and, *Mac OS X Lion*. Based on the entity linking results, all knowledge about these entities in KB can be used to understand the text, for example, we can classify the given document into *IT* category, rather than into *Animal* category based on the knowledge "*Lion is an Operation System*".

The main challenges for entity linking are the *name ambiguity problem* and the *name variation problem*. The name ambiguity problem is related to the fact that a name may refer to different entities in different contexts. For example, the name *Apple* can refer to more than 20 entities in Wikipedia, such as *fruit Apple, the IT company Apple Inc.*, and *the Apple Bank*. The name variation problem means that an entity can be mentioned in different ways, such as its full name, aliases, acronyms, and misspellings. For example, the IBM company can be mentioned using more than 10 names, such as *IBM, International Business Machine*, and its nickname *Big Blue*.

To solve the name ambiguity problem and the name variation problem, many approaches have been proposed for entity linking (Milne and Witten 2008; Kulkarni et al. 2009; Ratinov et al. 2011; Han and Sun 2011; Han et al. 2011; Han and Sun 2012). In the following, we first describe a general framework for entity linking, and then we introduce how deep learning techniques can be used to enhance EL performance.

5.4.1 *The Entity Linking Framework*

Given a document d and a Knowledge Graph KB, an entity linking system links the name mentions in the document as follows.

Name Mention Identification. In this step, all name mentions in a document will be identified for entity linking. For example, an EL system should identify three mentions {*WWDC, Apple, Lion*} from the document in Fig. 5.8. Currently, most EL systems employ two techniques for this task. One is the classical named entity recognition (NER) technique (Nadeau and Sekine 2007), which can recognize names of *Person, Location*, and *Organization* in a document, and then these entity names will be used as name mentions for entity linking. The main drawback of NER technique is that it can only identify limited types of entities, while ignores many

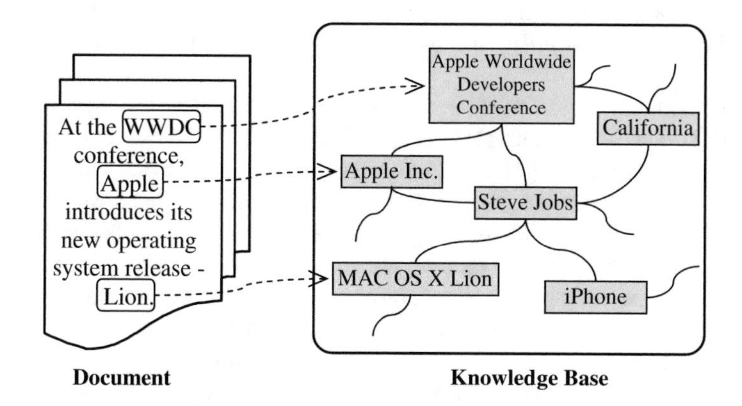

Fig. 5.8 A demo of entity linking

commonly used entities such as *Music, Film*, and *Book*. The other technique for name mention detection is dictionary-based matching, which first constructs a name dictionary for all entities in a Knowledge Graph (e.g., collected from anchor texts in Wikipedia Mihalcea and Csomai 2007), and then all names matched in a document will be used as name mentions. The main drawback of dictionary-based matching is that it may match many noisy name mentions, e.g., even the stop words *is* and *an* are used as entity names in Wikipedia. To resolve this problem, many techniques (Mihalcea and Csomai 2007; Milne and Witten 2008) have been proposed to filter out noisy name mentions.

Candidate Entity Selection. In this step, an EL system selects candidate entities for each name mention detected in Step 1. For example, a system may identify {*Apple(fruit), Apple Inc., Apple Bank*} as the possible referents for name *Apple*. Due to the name variation problem, most EL systems rely on a reference table for candidate entity selection. Specifically, a reference table records all possible referents of a name using (*name, entity*) pairs, and reference tables can be collected from Wikipedia anchor texts (Milne and Witten 2008), web (Bollegala et al. 2008), or query log (Silvestri et al. 2009).

Local Compatibility Computation. Given a name mention m in document d and its candidate referent entities $E = \{e_1, e_2, \ldots, e_n\}$, a critical step of EL systems is to compute the local compatibility $sim(m, e)$ between mention m and entity e, i.e., estimate how likely the mention m will be linked to the entity e. Based on the local compatibility scores, a name mention m will be linked to the entity which has the largest compatibility score with it:

$$e^* = \text{argmax}_e \quad \text{sim}(m, e). \tag{5.24}$$

For example, to determine the referent entity of the name *apple* in the following sentence:

The ***apple*** *tree is a deciduous tree in the rose family*

we need to compute its compatibility with entities *Apple(fruit)* and *Apple Inc.*, and finally link *apple* with *Apple(fruit)* based on the contextual words "*tree*", "*rose family*", etc.

Currently, many approaches have been proposed for local compatibility computation (Milne and Witten 2008; Mihalcea and Csomai 2007; Han and Sun 2011). The essential idea is to extract discriminative features (e.g., important words, frequent co-occur entities, attribute values) from the mention's context and the description of a specific entity (e.g., the Wikipedia page of the entity), and then the compatibility is determined by their shared common features.

Global Inference. It has long been proven that global inference can significantly increase the performance of entity linking. The underlying assumption of global inference is the topic coherence assumption, i.e., *all entities in a document should semantically related to the document's main topics*. Based on this assumption, a referent entity should not only compatible with its local context but also should coherent with other referent entities in the same document. For example, if we know the referent entity of the name mention *Lion* is *Mac OSX(Lion)* in Fig. 5.8, we can easily determine the referent entity of *Apple* is *Apple Inc.* using the semantic relation Product-of(*Apple Inc., Mac OSX(Lion)*). These examples strongly suggest that the entity linking performance could be improved by resolving the entity linking problems in the same document jointly, rather than independently.

Formally, given all mentions $M = \{m_1, m_2, \ldots, m_k\}$ in a document d, a global inference algorithm aims to find the optimal referent entities which will maximize the global coherence score:

$$[e_1^*, \ldots, e_k^*] = \text{argmax} \left(\sum_i \text{sim}(m_i, e_i) + \text{Coherence}(e_1, e_2, \ldots, e_k) \right). \quad (5.25)$$

In recent years, many global inference algorithms have been proposed for entity linking, including graph-based algorithms (Han et al. 2011; Chen and Ji 2011), topic model-based methods (Ganea et al. 2016; Han and Sun 2012), and optimization-based algorithms (Ratinov et al. 2011; Kulkarni et al. 2009). These methods differ with each other in how their model the document coherence, and how they infer the global optimal EL decisions. For example, Han et al. (2011) model the coherence as the sum of semantic relatedness between all referent entities:

$$\text{Coherence}(e_1, e_2, \ldots, e_k) = \sum_{(i,j)} \text{SemanticRelatedness}(e_i, e_j) \quad (5.26)$$

then the global optimal decisions are obtained through a graph random walk algorithm. By contrast, Han and Sun (2012) propose an entity-topic model, where the coherence is modeled as the probability of generating all referent entities from a document's main topics, and the global optimal decisions are obtained through a Gibbs sampling algorithm.

5.4.2 Deep Learning for Entity Linking

In this section, we introduce how to employ deep learning techniques for entity linking. As introduced above, one main problem of EL is the name ambiguity problem; thus, the key challenge is how to compute the compatibility between a name mention and an entity by effectively using contextual evidences.

It has been observed that the performance of entity linking heavily depend on the local compatibility model. Existing studies typically use handcrafted features to represent different types of contextual evidences (e.g., mention, context, and entity description), and measure the local compatibility using heuristic similarity measures (Milne and Witten 2008; Mihalcea and Csomai 2007; Han and Sun 2011). These feature-engineering-based approaches, however, have the following drawbacks:

- Feature engineering is labor-intensive, and it is difficult to manually design discriminative features. For example, it is challenging to design features which can capture the semantic similarity between the words *cat* and *dog*.
- The contextual evidences for entity linking are usually heterogeneous and may be at different granularities. The modeling and exploitation of heterogeneous evidences are not straightforward using handcrafted features. Till now, many different kinds of contextual evidences have been used for entity linking, including entity name, entity category, entity description, entity popularity, semantic relations between entities, mention name, mention context, mention document, etc. It is hard to design features which can project all these evidences into the same feature space, or to summarize all these evidences into a uniform framework for EL decisions.
- Finally, traditional entity linking methods usually define the compatibility between a mention and an entity heuristically, which is weak in discovering and capturing all useful factors for entity linking decisions.

To resolve the above drawbacks of feature-engineering-based approaches, in recent years many deep learning techniques have been employed for entity linking (He et al. 2013; Sun et al. 2015; Francis-Landau et al. 2016; Tsai and Roth 2016). In following, we first describe how to represent heterogeneous evidences via neural networks, then we introduce how to model the semantic interactions between different types of contextual evidences, and finally, we describe how to optimize local compatibility measures for entity linking using deep learning techniques.

5.4.2.1 Representing Heterogeneous Evidences via Neural Networks

One main advantage of neural network is it can learn good representations automatically from different types of raw inputs, such as text, image, and video (Bengio 2009). In entity linking, neural networks have been exploited to represent heterogeneous contextual evidences, such as mention name, mention context and entity description. By encoding all contextual evidences in the continuous vector space which are suitable for entity linking, neural networks avoid the need of designing handcrafted features. In following, we introduce how to represent different types of contextual evidences in detail.

Name Mention Representation. A mention $m = [m_1, m_2, ...]$ is typically composed of one to three words, such as *Apple Inc., President Obama*. Previous methods mostly represent a mention as the average of embeddings of the words it contains

$$\mathbf{v}_m = \text{average}(\mathbf{e}_{m_1}, \mathbf{e}_{m_2}, \ldots), \tag{5.27}$$

where $\mathbf{e}_{m_i}$ is the embeddings of word m_i, which can be learned using CBOW or Skip-Gram models (Mikolov et al. 2013).

The above embedding average representation fails to take the importance and the position of a word into consideration. To resolve this problem, some methods employ

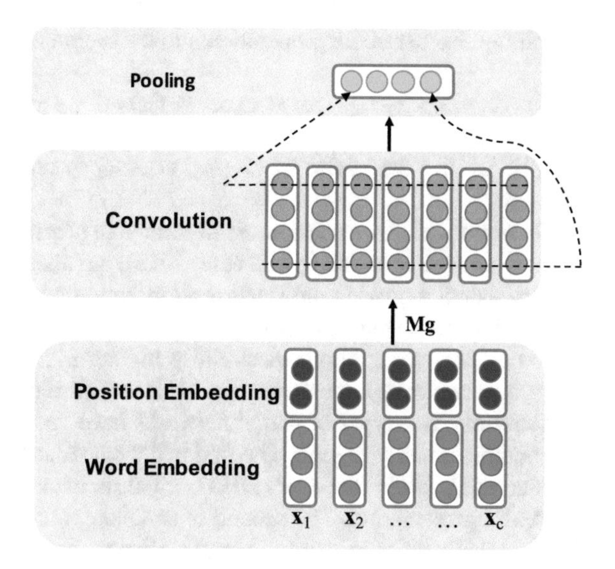

Fig. 5.9 Representing local context via convolutional neural network

convolutional neural networks (CNN) (Francis-Landau et al. 2016) to represent a mention, which provides more flexible ability to represent name mentions.

Local Context Representation. The local context around a mention provides critical information for entity linking decisions. For example, the context words {*tree, deciduous, rose family*} in "*The* ***apple*** *tree is a deciduous tree in the rose family*" provide critical information for linking the name mention *apple*. Sun et al. (2015) propose to represent local context using CNN, where the representation of a context is composed of the words it contains, by taking both the semantics of words and their relative positions to the mention into consideration.

Figure 5.9 demonstrates how to represent local context using CNN. Formally, given the words in a context $c = [w_1, w_2, \ldots, w_{|c|}]$, we represent each word w as $\mathbf{x} = [\mathbf{e}_w, \mathbf{e}_p]$, where $\mathbf{e}_w$ is the embeddings of word w and $\mathbf{e}_p$ is the position embeddings of word w, with d_w and d_p are the dimensions of word vector and position vector. A word w_i's position is its distance to the mention in the local context.

To represent the context c, we first concatenate all vectors of its words as

$$\mathbf{X} = [\mathbf{x}_1, \mathbf{x}_2, \ldots, \mathbf{x}_{|c|}] \tag{5.28}$$

then a convolution operation is applied to $\mathbf{X}$, and the output of convolution layer is

$$\mathbf{Z} = [\mathbf{M_g}\mathbf{X}_{[1,K+1]}, \mathbf{M_g}\mathbf{X}_{[2,K+2]}, \ldots, \mathbf{M_g}\mathbf{X}_{[|c|-K,|c|]}], \tag{5.29}$$

where $\mathbf{M_g} \in \mathbb{R}^{n_1 \times n_2}$ is the linear transformation matrix, and K is the context size of convolution layer.

Since the local context is of variable length, and in order to determine the most useful feature in each dimension of the feature vector, we perform a max-pooling operation(or other pooling operations) to the output of the convolution layer as

$$m_i = \max \quad \mathbf{Z}(i, .) \quad 0 \leq i \leq |c|. \tag{5.30}$$

Finally, we use the vector $\mathbf{m}_c = [m_1, m_2, \ldots]$ to represent the local context c of mention m.

Document Representation. As described in previous researches (He et al. 2013; Francis-Landau et al. 2016; Sun et al. 2015), the document and the local context of a name mention provide information at different granularities for entity linking. For example, a document usually captures larger topic information than local context. Based on this observation, most entity linking systems treat document and local context as two different evidences, and learn their representations individually.

Currently, two types of neural networks have been exploited for document representation in entity linking. The first is the convolutional neural network (Francis-Landau et al. 2016; Sun et al. 2015), which is the same as we introduced in local context representation. The second is denoising autoencoder (DA) (Vincent et al. 2008), which seeks to learn a compact document representation which can retain maximum information in original document d. Specifically, a document is first represented as a binary bag-of-words vector $\mathbf{x}_d$ (He et al. 2013), where each dimension of $\mathbf{x}$ indicates whether word w_i is appeared. Given the document representation $\mathbf{x}$, a denoising autoencoder seeks to learn a model which can reconstruct $\mathbf{x}$ given a random corruption $\mathbf{x}'$ of $\mathbf{x}$ through the following process: (1) randomly corrupt $\mathbf{x}$ by applying masking noise(randomly mask 1 or 0) to the original $\mathbf{x}$; (2) encode $\mathbf{x}$ into a compact representation $h(\mathbf{x})$ through an encoding process; (3) reconstruct $\mathbf{x}$ from $h(\mathbf{x})$ through a decoding process $g(h(\mathbf{x}))$. The learning goal of DA is to minimize the reconstruction error $L(x, g(h(\mathbf{x})))$. Figure 5.10 demonstrates the encoding and decoding process of DA.

DA has several advantages for document representation (He et al. 2013). First, the autoencoder tries to learn a compact representation of a document, and therefore can group similar words into clusters. Second, by randomly corrupting original inputs, DA can capture general topics and ignore meaningless words, such as function words

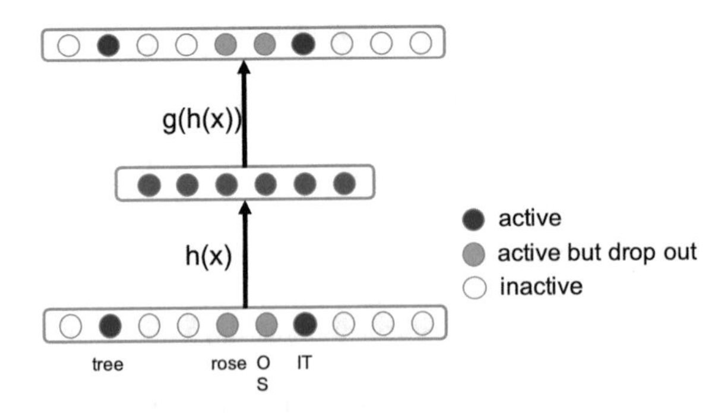

Fig. 5.10 DA and reconstruction sampling

is, and, or, etc. Third, autoencoder can be repeatedly stacked on top of previous learned $h(\mathbf{x})$; therefore, DA can learn multiple levels of representation of a document.

Entity Knowledge Representation. Currently, most entity linking systems use Wikipedia (or Knowledge Bases derived from Wikipedia, such as Yago, DBPedia, etc.) as its target Knowledge Base. Wikipedia contains rich knowledge about entities, such as title, description, infobox containing its important attributes, semantic categories, and sometimes its relations with other entities. For example, Fig. 5.11 shows *Apple Inc.*'s knowledge contained in Wikipedia. In following, we describe how to represent the evidence from entity knowledge using neural networks.

- **Entity Title Representation**. As the same with name mention, an entity title is typically composed of one to three words; therefore, most entity linking systems employ the same neural networks as in name mention representation to represent entity titles, i.e., average of word embeddings or CNN.
- **Entity Description**. Currently, most entity linking systems model entity description as a plain document, and learn its representation as the same with document representation, i.e., via CNN or DA.

From the above introduction, deep learning techniques propose a family of neural networks for representing contextual evidences, from word embeddings, denoising auto-encoder, to convolutional neural networks. These neural networks can effectively learn the representations of contextual evidences, without the need of handcrafted features.

In recent years, many other types of evidences have also been exploited for entity linking. For instance, entity popularity which tells the likelihood of an entity appearing in a document, semantic relations which capture the semantic association/relation between different entities (e.g., CEO-of(*Steve Jobs, Apple Inc.*) and Employee-of(*Michael I. Jordan, UC Berkeley*)), categories which provide key generalization information for an entity(e.g., *apple ISA fruit, Steve Jobs is a Businessman, Michael Jeffery Jordan ISA NBA player*). The representation of these contextual evidences using neural networks is still not straightforward. For future work, it may be helpful to design other neural networks which can effectively represent these contextual evidences.

5.4.2.2 Modeling Semantic Interactions Between Contextual Evidences

As shown in above, there exist many types of contextual evidences for entity linking. To make accurate EL decision, an EL system needs to take all different types of contextual evidences into consideration. Furthermore, in recent years, the task of cross-lingual entity linking makes it essential to compare contextual evidences in different languages. For example, an EL system needs to compare the Chinese name mention "***pingguo***(*Apple*) *fabu*(*released*) *xin*(*new*) *iPhone*" with the English description of "*Apple Inc.*" in Wikipedia for Chinese-to-English entity linking.

To take all contextual evidences into consideration, recent studies have employed neural networks to model the semantic interactions between different context evi-

Title:
Apple Inc.
Description:
Apple is an American multinational technology company headquartered in...
InfoBox:
{
Type: *Public*
Founders: [*Steve Jobs*, *Steve Wozniak*, *Ronald Wayne*]
...
}
Categories:
1976 establishments, IT Companies

Fig. 5.11 The information of *Apple Inc.* in Wikipedia

dences. Generally, two strategies have been used to model the semantic interactions between different contextual evidences:

- The first is to map different types of contextual evidences to the same continuous feature space via neural networks, and then the semantic interactions between contextual evidences can be captured using the similarities (mostly the cosine similarity) between their representations.
- The second is to learn a new representation which can summarize information from different contextual evidences, and then to make entity linking decisions based on the new representation.

In following, we describe how these two strategies are used in entity linking systems.

In Francis-Landau et al. (2016), it learns convolutional neural networks to project name mention, mention's local context, source document, entity title, and entity description into the same continuous feature space; then, the semantic interactions between different evidences are modeled as the similarities between their representations. Specifically, given the continuous vector representations learned by CNN, Francis-Landau et al. (2016) capture the semantic interactions between a mention and an entity as

$$\mathbf{f}(c, e) = [\cos(s_d, e_n), \cos(s_c, e_n), \cos(s_m, e_n), \cos(s_d, e_d), \cos(s_c, e_d), \cos(s_m, e_d)], \tag{5.31}$$

where s_d, s_m, and s_c correspondingly are the learned vectors of mention's document, context, and name, and e_n and e_d are correspondingly the learned vectors of entity's name and description. Finally, the above semantic similarities are combined with other signals such as link counts to predict the local compatibility.

In Sun et al. (2015), it learns a new representation for every mention, which consists of evidences from mention's name and local context based on their representations. Specifically, the new representation uses neural tensor network to compose mention vector($\mathbf{v}_m$) and context vector($\mathbf{v}_c$):

$$\mathbf{v}_{mc} = [\mathbf{v}_m, \mathbf{v}_c]^T [M_i^{appr}]^{[1,L]}[\mathbf{v}_m, \mathbf{v}_c]. \tag{5.32}$$

In this way, the semantic interactions between different contextual evidences are summarized into the new feature vector $\mathbf{v}_{mc}$. Sun et al. (2015) also learn a new representation for each entity by composing its entity name representation and entity category representation. Finally, the local compatibility between a mention and an entity is calculated as the cosine similarity between their new representations.

In Tsai and Roth (2016), it proposes a multilingual embedding method for cross-lingual entity linking. Cross-lingual entity linking aims to ground mentions written in non-English documents to entries in the English Wikipedia. Tsai and Roth (2016) project words and entity names in both the foreign language and in English into a new continuous vector space, and then the similarity between a foreign language mention and English Wikipedia entries can be effectively calculated for entity linking. Specifically, given the embeddings of the aligned English and foreign language titles $\mathbf{A}_{en} \in \mathbb{R}^{a \times k_1}$ and $\mathbf{A}_f \in \mathbb{R}^{a \times k_2}$, where a is the aligned title number, k_1 and k_2 correspondingly are the embedding dimensions of English and foreign language, Tsai and Roth (2016) apply a canonical correlation analysis (CCA) to these two matrices:

$$[\mathbf{P}_{en}, \mathbf{P}_f] = \text{CCA}(\mathbf{A}_{en}, \mathbf{A}_f). \tag{5.33}$$

Then, the English embeddings and the foreign language embeddings are projected into a new feature space as

$$\mathbf{E}'_{en} = \mathbf{E}_{en}\mathbf{P}_{en}, \tag{5.34}$$

$$\mathbf{E}'_f = \mathbf{E}_f\mathbf{P}_f, \tag{5.35}$$

where $\mathbf{E}_{en}$ and $\mathbf{E}_f$ is the original embeddings of all words in English and foreign language, and $\mathbf{E}'_{en}$ and $\mathbf{E}'_f$ is the new embeddings of all words in English and foreign language.

5.4.2.3 Learning Local Compatibility Measures

Both the contextual evidence representation learning and the semantic interaction modeling rely on a large set of parameters for good performance. Deep learning techniques provide an end-to-end framework, which can effectively optimize all parameters using back-propagation algorithm and gradient-based optimization algorithms. In Fig. 5.12, we show a commonly used architecture for local compatibility learning. We can see that mention's evidence and entity's evidence will be first encoded into a continuous feature space using contextual evidence representation neural networks, then compatibility signals between mention and entity will be computed using semantic interaction modeling neural networks, and finally, all these signals will be summarized into the local compatibility score.

To learn the above neural network for local compatibility, we need to collect entity linking annotations (d, e, m) from different resources, e.g., from Wikipedia hyperlinks. Then, the training objective is to minimize the ranking loss:

$$L = \sum_{(m,e)} L(m, e), \tag{5.36}$$

where $L(m, e) = \max\{0, 1 - sim(m, e) + sim(m, e')\}$ is the pairwise ranking criterion for each training instance (m, e), which gives a penalize if the top 1 ranked entity e' is not the true referent entity e.

We can see that, in the above learning process, deep learning techniques can optimize the similarity measure by fine-tuning the mention representation and entity representation, and learning the weights for different compatibility signals. In this way, it usually can achieve better performance than heuristically designed similarity measures.

5.5 Summary

Knowledge Graph is a fundamental knowledge repository for natural language understanding and commonsense reasoning, which contains rich knowledge about the world's entities, their attributes, and semantic relations between entities.

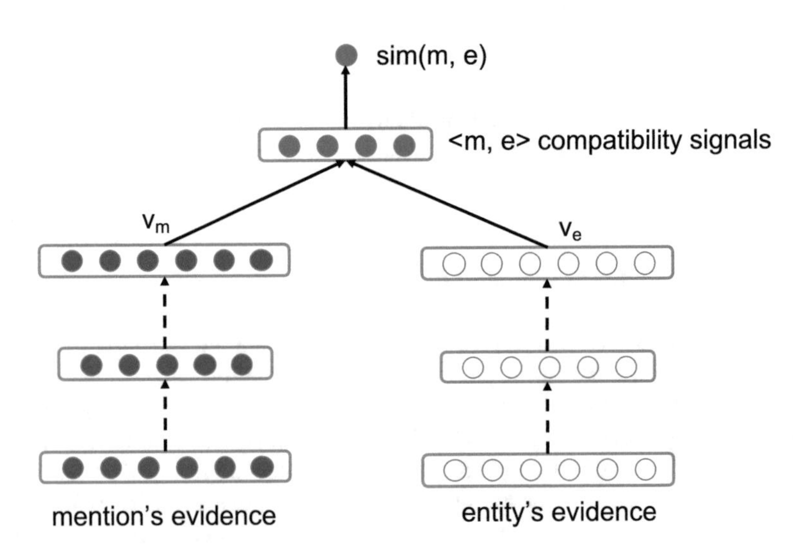

Fig. 5.12 A general framework for local compatibility learning

In this chapter, we introduce several important Knowledge Graphs, including DBPedia, Freebase, Wikidata, Yago, and HowNet. Afterwards, we introduce three important tasks for Knowledge Graph and describe how deep learning techniques can be applied to these issues: the first is representation learning, which can be used to embed entities, relations into a continuous feature space; the second is neural relation extraction, which shows how to construct Knowledge Graph by extracting knowledge from web pages and texts; the third is entity linking, which can be used to bridge knowledge with text. The deep learning techniques are used to embed entities and relations for Knowledge Graph representation, and to represent relation instances in relation extraction for Knowledge Graph construction, and to represent heterogeneous evidences for entity linking. The above techniques will provide a solid foundation for understanding, representing, constructing, and utilizing KGs in different tasks, e.g., question answering, text understanding and commonsense reasoning.

Besides benefiting KG construction, knowledge representation learning provides us an exciting approach for the application of KGs. In future, it will be important to explore how to better take KGs into consideration of deep learning models for natural language understanding and generation, and develop knowledgeable neural models for natural language processing.

References

Bahdanau, D., Cho, K., & Bengio, Y. (2014). Neural machine translation by jointly learning to align and translate. arXiv:1409.0473.

Bengio, Y. (2009). Learning deep architectures for AI. Foundations and trends®. *Machine Learning*, *2*(1), 1–127.

Bollegala, D., Honma, T., Matsuo, Y., & Ishizuka, M. (2008). Mining for personal name aliases on the web. In *Proceedings of the 17th International Conference on World Wide Web* (pp. 1107–1108). New York: ACM.

Bordes, A., Usunier, N., Garcia-Duran, A., Weston, J., & Yakhnenko, O. (2013). Translating embeddings for modeling multi-relational data. In *Proceedings of NIPS* (pp. 2787–2795).

Chen, Z., & Ji, H. (2011). Collaborative ranking: A case study on entity linking. In *Proceedings of the Conference on Empirical Methods in Natural Language Processing* (pp. 771–781). Association for Computational Linguistics.

Cho, K., Van Merriënboer, B., Gulcehre, C., Bahdanau, D., Bougares, F., Schwenk, H., & Bengio, Y. (2014). Learning phrase representations using RNN encoder-decoder for statistical machine translation. arXiv:1406.1078.

Dong, Z. & Dong, Q. (2003). Hownet-a hybrid language and knowledge resource. In *2003 International Conference on Natural Language Processing and Knowledge Engineering, 2003. Proceedings* (pp. 820–824). IEEE.

Francis-Landau, M., Durrett, G., & Klein, D. (2016). Capturing semantic similarity for entity linking with convolutional neural networks. In *Proceedings of NAACL-HLT* (pp. 1256–1261).

Ganea, O.-E., Ganea, M., Lucchi, A., Eickhoff, C., & Hofmann, T. (2016). Probabilistic bag-of-hyperlinks model for entity linking. In *Proceedings of the 25th International Conference on World Wide Web* (pp. 927–938). International World Wide Web Conferences Steering Committee.

Garcıa-Durán, A., Bordes, A., & Usunier, N. (2015). Composing relationships with translations. *Proceedings of EMNLP*.

Gu, K., Miller, J., & Liang, P. (2015). Traversing knowledge graphs in vector space. *Proceedings of EMNLP*.
Han, X., & Sun, L. (2011). A generative entity-mention model for linking entities with knowledge base. In *Proceedings of the 49th Annual Meeting of the Association for Computational Linguistics: Human Language Technologies* (Vol. 1, pp. 945–954). Association for Computational Linguistics.
Han, X., & Sun, L. (2012). An entity-topic model for entity linking. In *Proceedings of the 2012 Joint Conference on Empirical Methods in Natural Language Processing and Computational Natural Language Learning* (pp. 105–115). Association for Computational Linguistics.
Han, X., Sun, L., & Zhao, J. (2011). Collective entity linking in web text: A graph-based method. In *Proceedings of the 34th international ACM SIGIR conference on Research and development in Information Retrieval* (pp. 765–774). New York: ACM.
He, S., Liu, K., Ji, G., & Zhao, J. (2015). Learning to represent knowledge graphs with Gaussian embedding. In *Proceedings of the 24th ACM International on Conference on Information and Knowledge Management* (pp. 623–632). New York: ACM.
He, Z., Liu, S., Li, M., Zhou, M., Zhang, L., & Wang, H. (2013). Learning entity representation for entity disambiguation. *ACL*, *2*, 30–34.
Hochreiter, S., & Schmidhuber, J. (1997). Long short-term memory. *Neural Computation*, *9*(8), 1735–1780.
Ji, H., Grishman, R., Dang, H. T., Griffitt, K., & Ellis, J. (2010). Overview of the TAC 2010 knowledge base population track. In *Third Text Analysis Conference (TAC 2010)* (p. 3).
Ji, G., He, S., Xu, L., Liu, K., & Zhao, J. (2015). Knowledge graph embedding via dynamic mapping matrix. In *Proceedings of ACL* (pp. 687–696).
Ji, G., Liu, K., He, S., & Zhao, J. (2016). Knowledge graph completion with adaptive sparse transfer matrix.
Krompaß, D., Baier, S., & Tresp, V. (2015). Type-constrained representation learning in knowledge graphs. In *Proceedings of the 13th International Semantic Web Conference (ISWC)*.
Kulkarni, S., Singh, A., Ramakrishnan, G., & Chakrabarti, S. (2009). Collective annotation of Wikipedia entities in web text. In *Proceedings of the 15th ACM SIGKDD international conference on Knowledge discovery and data mining* (pp. 457–466). New York: ACM.
Lin, Y., Liu, Z., & Sun, M. (2015a). Modeling relation paths for representation learning of knowledge bases. *Proceedings of EMNLP*.
Lin, Y., Liu, Z., Sun, M., Liu, Y., & Zhu, X. (2015b). Learning entity and relation embeddings for knowledge graph completion. In *Proceedings of AAAI* (pp. 2181–2187).
Lin, Y., Shen, S., Liu, Z., Luan, H., & Sun, M. (2016). Neural relation extraction with selective attention over instances. *Proceedings of ACL*, *1*, 2124–2133.
Mihalcea, R., & Csomai, A. (2007). Wikify!: linking documents to encyclopedic knowledge. In *Proceedings of the sixteenth ACM conference on Conference on information and knowledge management* (pp. 233–242). New York: ACM.
Mikolov, T., Chen, K., Corrado, G., & Dean, J. (2013). Efficient estimation of word representations in vector space. arXiv:1301.3781.
Milne, D., & Witten, I. H. (2008). Learning to link with Wikipedia. In *Proceedings of the 17th ACM Conference on Information and Knowledge Management* (pp. 509–518). New York: ACM.
Miwa, M., & Bansal, M. (2016). End-to-end relation extraction using LSTMs on sequences and tree structures. arXiv:1601.00770.
Nadeau, D., & Sekine, S. (2007). A survey of named entity recognition and classification. *Lingvisticae Investigationes*, *30*(1), 3–26.
Ratinov, L., Roth, D., Downey, D., & Anderson, M. (2011). Local and global algorithms for disambiguation to Wikipedia. In *Proceedings of the 49th Annual Meeting of the Association for Computational Linguistics: Human Language Technologies* (Vol. 1, pp. 1375–1384). Association for Computational Linguistics.
Silvestri, F., et al. (2009). Mining query logs: Turning search usage data into knowledge. *Foundations and Trends in Information Retrieval*, *4*(1–2), 1–174.

Socher, R., Huval, B., Manning, C. D., & Ng, A. Y. (2012). Semantic compositionality through recursive matrix-vector spaces. In *Proceedings of EMNLP* (pp. 1201–1211).
Sun, Y., Lin, L., Tang, D., Yang, N., Ji, Z., & Wang, X. (2015). Modeling mention, context and entity with neural networks for entity disambiguation. In *IJCAI* (pp. 1333–1339).
Tai, K. S., Socher, R., & Manning, C. D. (2015). Improved semantic representations from tree-structured long short-term memory networks. In *Proceedings of ACL* (pp. 1556–1566).
Tsai, C.-T., & Roth, D. (2016). Cross-lingual wikification using multilingual embeddings. In *Proceedings of NAACL-HLT* (pp. 589–598).
Vincent, P., Larochelle, H., Bengio, Y., & Manzagol, P.-A. (2008). Extracting and composing robust features with denoising autoencoders. In *Proceedings of the 25th International Conference on Machine Learning* (pp. 1096–1103). New York: ACM.
Wang, Z., Zhang, J., Feng, J., & Chen, Z. (2014a). Knowledge graph and text jointly embedding. In *Proceedings of EMNLP* (pp. 1591–1601).
Wang, Z., Zhang, J., Feng, J., & Chen, Z. (2014b). Knowledge graph embedding by translating on hyperplanes. In *Proceedings of AAAI* (pp. 1112–1119).
Xiao, H., Huang, M., & Zhu, X. (2016). From one point to a manifold: Orbit models for knowledge graph embedding. In *Proceedings of IJCAI* (pp. 1315–1321).
Xiao, H., Huang, M., Hao, Y., & Zhu, X. (2015). Transg: A generative mixture model for knowledge graph embedding. arXiv:1509.05488.
Xie, R., Liu, Z., & Sun, M. (2016c). Representation learning of knowledge graphs with hierarchical. In *Proceedings of IJCAI*.
Xie, R., Liu, Z., Chua, T.-s., Luan, H., & Sun, M. (2016a). Image-embodied knowledge representation learning. arXiv:1609.07028.
Xie, R., Liu, Z., Jia, J., Luan, H., & Sun, M. (2016b). Representation learning of knowledge graphs with entity descriptions. In *Proceedings of AAAI*.
Xu, K., Feng, Y., Huang, S., & Zhao, D. (2015). Semantic relation classification via convolutional neural networks with simple negative sampling. arXiv:1506.07650.
Zeng, D., Liu, K., Chen, Y., & Zhao, J. (2015). Distant supervision for relation extraction via piecewise convolutional neural networks. In *Proceedings of EMNLP*.
Zeng, D., Liu, K., Lai, S., Zhou, G., & Zhao, J. (2014). Relation classification via convolutional deep neural network. In *Proceedings of COLING* (pp. 2335–2344).
Zhang, D., & Wang, D. (2015). Relation classification via recurrent neural network. arXiv:1508.01006.
Zhong, H., Zhang, J., Wang, Z., Wan, H., & Chen, Z. (2015). Aligning knowledge and text embeddings by entity descriptions. In *Proceedings of EMNLP* (pp. 267–272).

Chapter 6
Deep Learning in Machine Translation

Yang Liu and Jiajun Zhang

Abstract Machine translation (MT) is an important natural language processing task that investigates the use of computers to translate human languages automatically. Deep learning-based methods have made significant progress in recent years and quickly become the new *de facto* paradigm of MT in both academia and industry. This chapter introduces two broad categories of deep learning-based MT methods: (1) *component-wise deep learning for machine translation* that leverages deep learning to improve the capacity of the main components of SMT such as translation models, reordering models, and language models; and (2) *end-to-end deep learning for machine translation* that uses neural networks to directly map between source and target languages based on the encoder–decoder framework. The chapter closes with a discussion on challenges and future directions of deep learning-based MT.

6.1 Introduction

Machine translation, which aims at translating natural languages automatically using machines, is an important task in natural language processing. Due to the increasing availability of parallel corpora, data-driven machine translation has become the dominant method in the MT community since 1990s. Given sentence-aligned bilingual training data, the goal of data-driven MT is to acquire translation knowledge from data automatically, which is then used to translate unseen source language sentences.

Statistical machine translation (SMT) is a representative data-driven approach that advocates the use of probabilistic models to describe the translation process. While early SMT focused on generative models treating words as the basic unit (Brown et al. 1993), discriminative models (Och and Ney 2002) that use features defined on phrases and parses (Koehn et al. 2003; Chiang 2007) have been widely used since

Y. Liu
Tsinghua University, Beijing, China
e-mail: liuyang2011@tsinghua.edu.cn

J. Zhang (✉)
Institute of Automation, Chinese Academy of Sciences, Beijing, China
e-mail: jjzhang@nlpr.ia.ac.cn

L. Deng and Y. Liu (eds.), *Deep Learning in Natural Language Processing*, https://doi.org/10.1007/978-981-10-5209-5_6

2002. However, discriminative SMT models face a severe challenge: data sparsity. Using discrete symbolic representations, SMT is prone to learn poor estimates of model parameters on low-count events. In addition, it is hard to design features manually to capture all translation regularities due to the diversity and complexity of natural languages.

Recent years have witnessed the remarkable success of deep learning applications in MT. Surpassing SMT in leading international MT evaluation campaigns, deep learning-based MT has quickly become the new *de facto* paradigm for commercial online MT services. This chapter introduces two broad categories of deep learning-based MT methods: (1) *component-wise deep learning for machine translation* (Devlin et al. 2014) that leverages deep learning to improve the capacity of the main components of SMT such as translation models, reordering models, and language models; and (2) *end-to-end deep learning for machine translation* (Sutskever et al. 2014; Bahdanau et al. 2015) that uses neural networks to directly map between source and target languages based on an encoder–decoder framework.

This chapter is organized as follows. We will first introduce the basic concepts of SMT (Sect. 6.2.1) and discuss existing problems of string matching-based SMT (Sect. 6.2.2). Then, we will review the applications of deep learning in SMT in detail (Sects. 6.3.1–6.3.5). Section 6.4 is devoted to end-to-end neural machine translation, covering the standard encoder–decoder framework (Sect. 6.4.1), the attention mechanism (Sect. 6.4.2), and recent advances (Sects. 6.4.3–6.4.6). The chapter closes with a summary (Sect. 6.5).

6.2 Statistical Machine Translation and Its Challenges

6.2.1 Basics

Let $\mathbf{x}$ be a source language sentence, $\mathbf{y}$ be a target language sentence, $\boldsymbol{\theta}$ be a set of model parameters, and $P(\mathbf{y}|\mathbf{x};\boldsymbol{\theta})$ be the translation probability of $\mathbf{y}$ given $\mathbf{x}$. The goal of machine translation is to find the translation with the highest probability $\hat{\mathbf{y}}$:

$$\hat{\mathbf{y}} = \underset{\mathbf{y}}{\mathrm{argmax}} \Big\{ P(\mathbf{y}|\mathbf{x};\boldsymbol{\theta}) \Big\}. \tag{6.1}$$

Brown et al. (1993) use the Bayes' theorem to rewrite the decision rule in Eq. (6.1) equivalently as

$$\hat{\mathbf{y}} = \underset{\mathbf{y}}{\mathrm{argmax}} \left\{ \frac{P(\mathbf{y};\boldsymbol{\theta}_{lm})P(\mathbf{x}|\mathbf{y};\boldsymbol{\theta}_{tm})}{P(\mathbf{x})} \right\}, \tag{6.2}$$

$$= \underset{\mathbf{y}}{\mathrm{argmax}} \Big\{ P(\mathbf{y};\boldsymbol{\theta}_{lm})P(\mathbf{x}|\mathbf{y};\boldsymbol{\theta}_{tm}) \Big\}. \tag{6.3}$$

where $P(\mathbf{x}|\mathbf{y};\boldsymbol{\theta}_{tm})$ is referred to as a *translation model* and $P(\mathbf{y};\boldsymbol{\theta}_{lm})$ as a *language model*. $\boldsymbol{\theta}_{tm}$ and $\boldsymbol{\theta}_{lm}$ are translation and language model parameters, respectively.

The translation model $P(\mathbf{x}|\mathbf{y};\boldsymbol{\theta}_{tm})$ is usually defined as a generative model, which is further decomposed via latent structures (Brown et al. 1993):

$$P(\mathbf{x}|\mathbf{y};\boldsymbol{\theta}_{tm}) = \sum_{\mathbf{z}} P(\mathbf{x},\mathbf{z}|\mathbf{y};\boldsymbol{\theta}_{tm}), \tag{6.4}$$

where $\mathbf{z}$ denotes a latent structure such as word alignment that indicates the correspondence between words in source and target languages.

However, a key limitation of latent-variable generative translation models is that they are hard to extend due to the intricate dependencies between sub-models. As a result, Och and Ney (2002) advocate the use of log-linear models for statistical machine translation to incorporate arbitrary knowledge sources:

$$P(\mathbf{y}|\mathbf{x};\boldsymbol{\theta}) = \frac{\sum_{\mathbf{z}} \exp(\boldsymbol{\theta}\cdot\boldsymbol{\phi}(\mathbf{x},\mathbf{y},\mathbf{z}))}{\sum_{\mathbf{y}'}\sum_{\mathbf{z}'} \exp(\boldsymbol{\theta}\cdot\boldsymbol{\phi}(\mathbf{x}',\mathbf{y},\mathbf{z}'))}, \tag{6.5}$$

where $\boldsymbol{\phi}(\mathbf{x},\mathbf{y},\mathbf{z})$ is a set of *features* that characterize the translation process and $\boldsymbol{\theta}$ is a set of corresponding feature weights. Note that the latent-variable generative model in Eq. (6.4) is a special case of the log-linear model because both translation and language models can be treated as features.

The phrase-based translation model (Koehn et al. 2003) is the most widely used SMT method in both academia and industry due to its simplicity and effectiveness. The basic idea of phrase-based translation is to use phrases to memorize word selection and reordering sensitive to local context, making it very effective in handling word insertion and omission, short idioms, and free translation.

As shown in Fig. 6.1, the translation process of phrase-based SMT can be divided into three steps: (1) segmenting the source sentence into a sequence of phrases, (2) transforming each source phrase to a target phrase, and (3) rearranging target phrases in an order of target language. The concatenation of target phrases forms

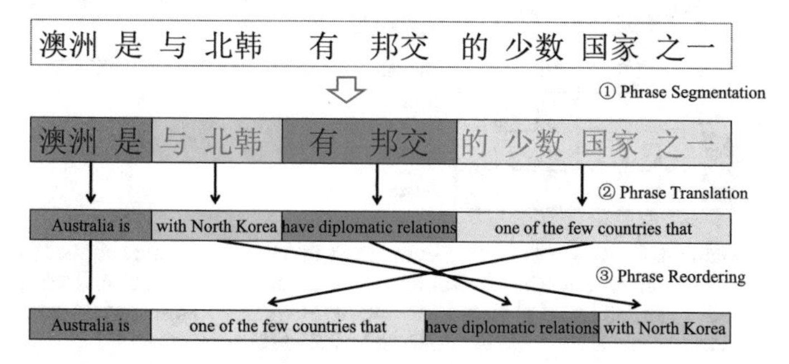

Fig. 6.1 The translation process of phrase-based SMT. It involves three steps: phrase segmentation, phrase translation, and phrase reordering

a target sentence. Therefore, phrase-based translation models often consist of three sub-models: phrase segmentation, phrase reordering, and phrase translation. These sub-models serve as main features in the log-linear model framework.

The central feature in discriminative phrase-based translation model is translation rule table or bilingual phrase table. Figure 6.2 illustrates translation rule extraction for phrase-based SMT. Given a parallel sentence pair, word alignment first runs to find the correspondence between words in the source and target sentences. Then, bilingual phrases (i.e., translation rules) satisfying a heuristic constraint defined on word alignment (Och and Ney 2002) are extracted from the word-aligned sentence pair. Then, the probabilities and lexical weights of bilingual phrases can be estimated from the training data. Note that the phrase reordering model can also be trained on the word-aligned parallel corpus.

In a latent-variable log-linear translation model, a latent structure $\mathbf{z}$ is often referred to as a derivation, which describes how a translation is generated. During decoding, searching for the translation with the highest probability needs to take all possible derivations into consideration:

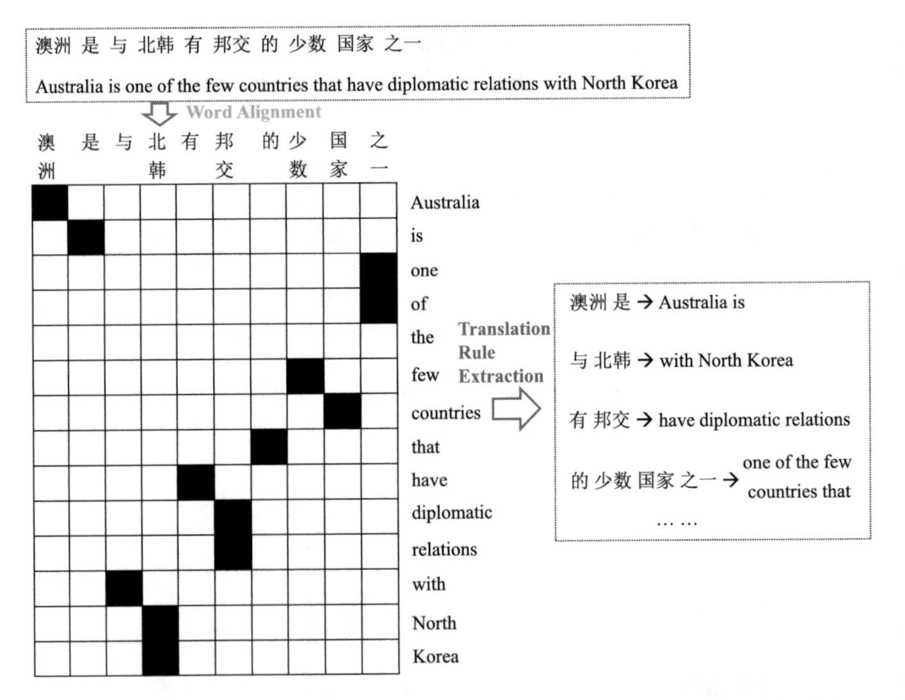

Fig. 6.2 Translation rule extraction for phrase-based SMT. Given a sentence-aligned parallel corpus, word alignment that indicates the correspondence between words in source and target sentences is first calculated. Then, bilingual phrases that capture semantically equivalent source and target word sequences are extracted from the word-aligned parallel corpus

$$\hat{\mathbf{y}} = \underset{\mathbf{y}}{\operatorname{argmax}} \left\{ \sum_{\mathbf{z}} \exp\left(\boldsymbol{\theta} \cdot \boldsymbol{\phi}(\mathbf{x}, \mathbf{y}, \mathbf{z})\right) \right\}. \tag{6.6}$$

Unfortunately, it is intractable to calculate the summation because there are exponentially many latent derivations. As a result, standard SMT systems usually approximate Eq. (6.6) with the derivation with the highest probability:

$$\hat{\mathbf{y}} \approx \underset{\mathbf{y}}{\operatorname{argmax}} \left\{ \max_{\mathbf{z}} \left\{ \boldsymbol{\theta} \cdot \boldsymbol{\phi}(\mathbf{x}, \mathbf{y}, \mathbf{z}) \right\} \right\}. \tag{6.7}$$

Then, polynomial-time dynamic programming algorithms can be designed to generate translations efficiently.

6.2.2 *Challenges in Statistical Machine Translation*

From the SMT training procedure, we can easily see that word alignment is the core basis and directly influences the quality of the translation rules and the reordering model. The SMT decoding shows that the probability estimation of translation rules, the reordering model, and the language model are three key factors which are combined within a log-linear framework to produce the final translation results.

For word alignment, the popular solution in SMT is to use unsupervised generative models (Brown et al. 1993). The generative approaches use symbolic representations of words, calculate the statistics of word co-occurrences, and learn word-to-word mapping probabilities to maximize the likelihood of training data. Then, translation rule probabilities are calculated using maximum likelihood estimation according to their co-occurrence statistics in the word-aligned sentence pairs (Koehn et al. 2003). The phrase reordering instances are extracted from the word-aligned bitexts, and the reordering model is then formalized as a classification problem using the discrete words as features (Galley and Manning 2008). The language model is often built with an n-gram model and the conditional probability of the current word given the $n-1$ history words is estimated based on the relative frequency of the word sequence (Chen and Goodman 1999).

According to the above analysis, two crucial challenges hinder the improvement of conventional SMT. The first challenge is *data sparsity*. Using discrete symbolic representations, conventional SMT is prone to learn poor estimates of model parameters on low-count events. This is undesirable because complex features, which can capture more contextual information, tend to be observed infrequently on the training data. As a result, conventional SMT has to use simple features. For example, the maximum phrase length is usually set to 7 and the language model only uses 4-grams (Koehn et al. 2003).

The second challenge is *feature engineering*. Although log-linear models are capable of incorporating a large number of features (Chiang et al. 2009), it is still hard to find features expressive enough to cover all translation phenomena. Standard practice in feature design for SMT usually begins with designing feature templates manually, which capture local lexical and syntactic information. Then, millions of features can be generated by applying the templates to training data. Most of these features are highly sparse, making it very challenging to estimate feature weights.

In recent years, deep learning techniques have been exploited to address the above two challenges for SMT. Deep learning is not only capable of alleviating the data sparsity problem by introducing distributed representations instead of discrete symbolic representations, but also circumventing the feature engineering problem by learning representations from data. In the following, we will introduce how deep learning is used to improve a variety of key components of SMT: word alignment (Sect. 6.3.1), translation rule probability estimation (Sect. 6.3.2), phrase reordering model (Sect. 6.3.3), language model (Sect. 6.3.4), and model feature combination (Sect. 6.3.5).

6.3 Component-Wise Deep Learning for Machine Translation

6.3.1 Deep Learning for Word Alignment

6.3.1.1 Word Alignment

Word alignment aims to identify the correspondence between words in parallel sentences (Brown et al. 1993; Vogel et al. 1996). Given a source sentence $\mathbf{x} = x_1, \ldots, x_j, \ldots, x_J$ and its target translation $\mathbf{y} = y_1, \ldots, y_i, \ldots, y_I$, the word alignment between $\mathbf{x}$ and $\mathbf{y}$ is defined as $\mathbf{z} = z_1, \ldots, z_j, \ldots, ..., z_J$ in which $z_j \in [0, I]$ and $z_j = i$ indicates that x_j and y_i are aligned. Figure 6.2 shows an alignment matrix.

In SMT, word alignment often serves as a latent variable in generative translation models [see Eq. (6.4)]. As a result, a word alignment model is usually represented as $P(\mathbf{x}, \mathbf{z}|\mathbf{y}; \boldsymbol{\theta})$. The HMM model (Vogel et al. 1996) is one of the most widely used alignment models, which is defined as

$$P(\mathbf{x}, \mathbf{z}|\mathbf{y}; \boldsymbol{\theta}) = \prod_{j=1}^{J} p(z_j|z_{j-1}, I) \times p(x_j|y_{z_j}), \tag{6.8}$$

where alignment probabilities $p(z_j|z_{j-1}, I)$ and translation probabilities $p(x_j|y_{z_j})$ are model parameters.

Let $\{\langle \mathbf{x}^{(s)}, \mathbf{y}^{(s)} \rangle\}_{s=1}^{S}$ be a set of sentence pairs. The standard training objective is to maximize the log-likelihood of the training data:

$$\hat{\boldsymbol{\theta}} = \operatorname*{argmax}_{\boldsymbol{\theta}} \left\{ \sum_{s=1}^{S} \log P(\mathbf{x}^{(s)}|\mathbf{y}^{(s)}; \boldsymbol{\theta}) \right\}. \tag{6.9}$$

Given learned model parameters $\hat{\boldsymbol{\theta}}$, the best alignment of a sentence pair $\langle \mathbf{x}, \mathbf{y} \rangle$ can be obtained by

$$\hat{\mathbf{z}} = \underset{\mathbf{z}}{\text{argmax}} \left\{ P(\mathbf{x}, \mathbf{z} | \mathbf{y}; \hat{\boldsymbol{\theta}}) \right\}. \tag{6.10}$$

6.3.1.2 Feed-Forward Neural Networks for Word Alignment

Although simple and tractable, classical alignment models that use discrete symbolic representations suffer from a major limitation: they fail to capture more contextual information due to data sparsity. For example, both the alignment probabilities $p(z_j | z_{j-1}, I)$ and translation probabilities $p(x_j | y_{z_j})$ fail to include the surrounding context in $\mathbf{x}$ and $\mathbf{y}$ to better capture alignment regularities.

To address this problem, Yang et al. (2013) propose a context-dependent deep neural network for word alignment. The basic idea is to enable the alignment model to capture more context information by exploiting continuous representations. This can be done by using feed-forward neural networks.

Given a source sentence $\mathbf{x} = x_1, \ldots, x_j, \ldots, x_J$, we use $\boldsymbol{x}_j$ to denote the vector representation of the j-th source word x_j. Similarly, $\boldsymbol{y}_i$ denotes the vector representation of the i-th target word y_i. Yang et al. (2013) propose to model $p(x_j | y_i, C(\mathbf{x}, j, w), C(\mathbf{y}, i, w))$ instead of $p(x_j | y_i)$ to include more contextual information, where w is a window size and the source and target contexts are defined as

$$C(\mathbf{x}, j, w) = x_{j-w}, \ldots, x_{j-1}, x_{j+1}, \ldots, x_{j+w} \tag{6.11}$$

$$C(\mathbf{y}, i, w) = y_{i-w}, \ldots, y_{i-1}, y_{i+1}, \ldots, y_{i+w}. \tag{6.12}$$

Therefore, the feed-forward neural network takes the concatenation of word embeddings of the source and target sub-strings as input:

$$\mathbf{h}^{(0)} = [\boldsymbol{x}_{j-w}; \ldots; \boldsymbol{x}_{j+w}; \boldsymbol{y}_{i-w}; \ldots; \boldsymbol{y}_{i+w}]. \tag{6.13}$$

Then, the first hidden layer is calculated as

$$\mathbf{h}^{(1)} = f(\mathbf{W}^{(1)} \mathbf{h}^{(0)} + \mathbf{b}^{(1)}), \tag{6.14}$$

where $f(\cdot)$ is a nonlinear activation function,[1] $\mathbf{W}^{(1)}$ is the weight matrix at the first layer, and $\mathbf{b}^{(1)}$ is the bias term at the first layer.

Generally, the l-th hidden layer can be recursively computed by

$$\mathbf{h}^{(l)} = f(\mathbf{W}^{(l)} \mathbf{h}^{(l-1)} + \mathbf{b}^{(l)}). \tag{6.15}$$

[1] Yang et al. (2013) employ $f(\cdot) = h\text{tanh}(\cdot)$ in their work.

Yang et al. (2013) define the final layer as a linear transformation without activation function:

$$t_{lex}(x_j, y_i, C(\mathbf{x}, j, w), C(\mathbf{y}, i, w), \boldsymbol{\theta}) = \mathbf{W}^{(L)}\mathbf{h}^{(L-1)} + \mathbf{b}^{(L)}. \tag{6.16}$$

Note that $t_{lex}(x_j, y_i, C(\mathbf{x}, j, w), C(\mathbf{y}, i, w), \boldsymbol{\theta}) \in \mathbb{R}$ is a real-valued score that indicates how likely x_j is a translation of y_i.

Therefore, the context-dependent translation probability can be obtained by normalizing the scores:

$$p(x_j|y_i, C(\mathbf{x}, j, w), C(\mathbf{y}, i, w) = \frac{\exp\big(t_{lex}(x_j, y_i, C(\mathbf{x}, j, w), C(\mathbf{y}, i, w), \boldsymbol{\theta})\big)}{\sum_{x \in V_x} \exp\big(t_{lex}(x, y_i, C(\mathbf{x}, j, w), C(\mathbf{y}, i, w), \boldsymbol{\theta})\big)}, \tag{6.17}$$

where V_x is the source language vocabulary.

In practice, as it is computationally expensive to enumerate all source words to compute translation probabilities, Yang et al. (2013) only use the translation score $t_{lex}(x_j, y_i, C(\mathbf{x}, j, w), C(\mathbf{y}, i, w), \boldsymbol{\theta})$ instead. Figure 6.3a illustrates the network structure for the translation score calculation.

As for the alignment probability $p(z_j|z_{j-1}, I)$, Yang et al. (2013) employ the unnormalized alignment score $t_{aign}(z_j|z_{j-1}, \mathbf{x}, \mathbf{y})$ and simplify the calculation as follows:

$$t_{align}(z_j|z_{j-1}, \mathbf{x}, \mathbf{y}) = t_{align}(z_j - z_{j-1}), \tag{6.18}$$

where $t_{align}(z_j - z_{j-1})$ is modeled by 17 parameters, each of which is associated with a specific alignment distance $d = z_j - z_{j-1}$ (from $d = -7$ to $d = 7$ and $d \leq -8$, $d \geq 8$).

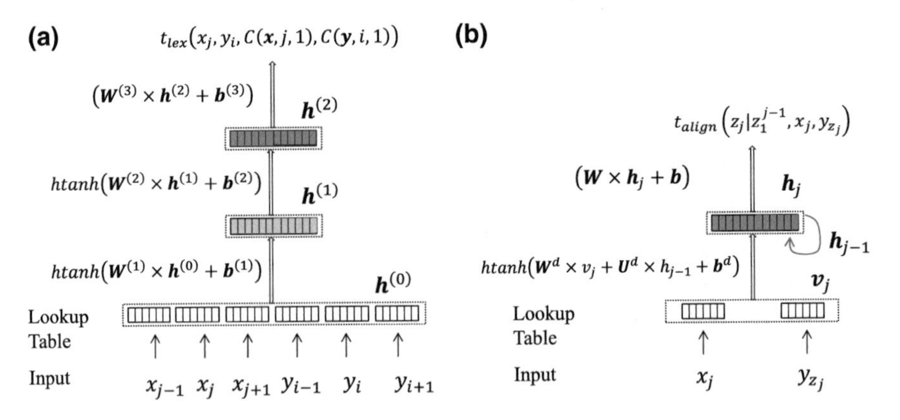

Fig. 6.3 Deep learning-based word alignment model: **a** feed-forward neural network for the lexical translation score prediction; **b** recurrent neural network for distortion score calculation

6.3.1.3 Recurrent Neural Networks for Word Alignment

The feed-forward neural network considers only the previous alignment z_{j-1} when computing the alignment score $t_{align}(z_j|z_{j-1}, \mathbf{x}, \mathbf{y}))$ and neglects the history information before z_{j-1}. Instead of applying the generative model (see Eq. 6.8) to search the best word alignment, Tamura et al. (2014) resort to using a recurrent neural network (RNN) to directly calculate the alignment score of $\mathbf{z} = z_1^J$:

$$s_{RNN}(z_1^J|\mathbf{x}, \mathbf{y}) = \prod_{j=1}^{J} t_{align}(z_j|z_1^{j-1}, x_j, y_{z_j}). \tag{6.19}$$

It is easy to see that RNN predicts the alignment score of z_j by conditioning it on all of the history alignments z_1^{j-1}. Figure 6.3b gives an illustration of the RNN structure to calculate the score of z_j ($t_{align}(z_j|z_1^{j-1}, x_j, y_{z_j})$).

First, the source word x_j and the target word y_{z_j} are projected into the vector representations which are further concatenated to form an input $\mathbf{v}_j$. The previous RNN hidden state $\mathbf{h}_{j-1}$ is another input and the new hidden state $\mathbf{h}_j$ is calculated as follows:

$$h_j = f(\mathbf{W}^d \mathbf{v}_j + \mathbf{U}^d \mathbf{h}_{j-1} + \mathbf{b}^d) \tag{6.20}$$

in which $f(\cdot) = h\text{tanh}(\cdot)$, $\mathbf{W}^d$ and $\mathbf{U}^d$ are weight matrices, and $\mathbf{b}^d$ is the bias term. Note that, in contrast to the classic RNN in which the same weight matrix is used at different time steps, $\mathbf{W}^d$, $\mathbf{U}^d$ and $\mathbf{b}^d$ are dynamically determined according to the alignment distance $d = z_j - z_{j-1}$. Following Yang et al. (2013), Tamura et al. (2014) also choose 17 values for d, and there are 17 different matrices for $\mathbf{W}^d$ ($\mathbf{W}^{\leq -8}, \mathbf{W}^{-7}, \cdots, \mathbf{W}^{7}, \mathbf{W}^{\geq 8}$). $\mathbf{U}^d$ and $\mathbf{b}^d$ are similar.

Then, the alignment score of z_j is obtained with a linear transformation of the current RNN hidden state:

$$t_{align}(z_j|z_1^{j-1}, x_j, y_{z_j}) = \mathbf{W}\mathbf{h}_j + \mathbf{b}. \tag{6.21}$$

Through extensive experiments, Tamura et al. (2014) report that recurrent neural networks outperform feed-forward neural networks in word alignment quality on the same test set and suggest that recurrent neural networks are able to capture long dependency by trying to memorize all the history information.

6.3.2 Deep Learning for Translation Rule Probability Estimation

Given the word-aligned training sentence pairs, all the translation rules satisfying the word alignment can be extracted. In phrase-based SMT, we may extract a huge number of phrasal translation rules for one source phrase. It becomes a key issue to choose the most appropriate translation rules during decoding. Conventionally, translation rule selection is usually performed according to the rule's translation probability which is calculated using the co-occurrence statistics in the bilingual training data (Koehn et al. 2003). For example, the conditional probability $p(y_i^{i+l}|x_j^{j+k})$ for the phrasal translation rule $\langle x_j^{j+k}, y_i^{i+l}\rangle$ is computed with maximum likelihood estimation (MLE):

$$p(y_i^{i+l}|x_j^{j+k}) = \frac{count(x_j^{j+k}, y_i^{i+l})}{count(x_j^{j+k})}. \tag{6.22}$$

The MLE method is prone to encounter the data sparsity problem and the estimated probability will be incorrect for infrequent phrasal translation rules. Furthermore, the MLE method cannot capture the deep semantics of the phrasal rules and explore the larger contexts beyond the phrase of interest. In recent years, deep learning-based methods are proposed to better estimate the quality of a translation rule using distributed semantic representations and more contextual information.

For a phrasal translation rule $\langle x_j^{j+k}, y_i^{i+l}\rangle$, Gao et al. (2014) attempt to calculate the translation score $score(x_j^{j+k}, y_i^{i+l})$ in a low-dimensional vector space. The main idea of the method is shown in Fig. 6.4.

A feed-forward neural network with two hidden layers is employed to map the word string (phrase) into an abstract vector representation. Take the source phrase x_j^{j+k}, for example, it starts with bag-of-words one-hot representation $\mathbf{h}_x^{(0)}$, followed by two hidden layers:

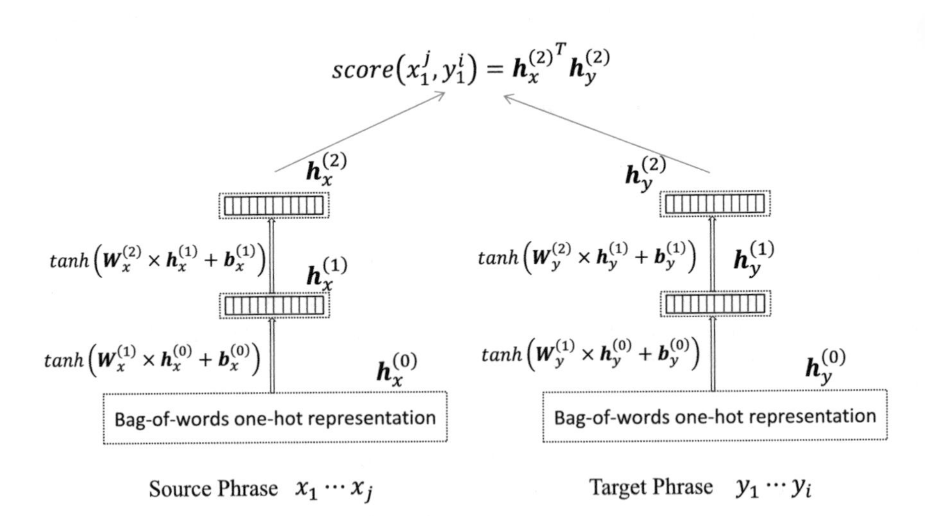

Fig. 6.4 Bag-of-words distributed phrase representations for phrasal translation rules and the goal is to learn evaluation metric (BLEU) sensitive phrase embeddings. The dot product similarity between source and target phrase is employed as the translation score in SMT

$$\mathbf{h}_x^{(1)} = f(\mathbf{W}_x^{(1)}\mathbf{h}_x^{(0)} + \mathbf{b}_x^{(1)}), \tag{6.23}$$

$$\mathbf{h}_x^{(2)} = f(\mathbf{W}_x^{(2)}\mathbf{h}_x^{(1)} + \mathbf{b}_x^{(2)}), \tag{6.24}$$

where the activation function is set $f(\cdot) = \tanh(\cdot)$. $\mathbf{h}_y^{(2)}$ for the target phrase y_i^{i+l} can be learned in the same manner. Then, the dot product between the source and target phrase representations is used as the translation score, namely, $score(x_j^{j+k}, y_i^{i+l}) = \mathbf{h}_x^{(2)T}\mathbf{h}_y^{(2)}$. The network parameters, such as word embeddings and weight matrix, are optimized to maximize the score of the phrase pairs which can lead to better translation quality (e.g., BLEU) in the validation set.

The distributed representation for phrases alleviates the data sparsity problem to large extent, and the learned phrase presentations are sensitive to evaluation metrics. However, it is worth noting that, due to bag-of-words modeling, this method cannot capture the word order information of a phrase, which is very important to determining the meaning of a phrase. For example, *cat eats fish* is totally different from *fish eats cat* even though they share the same bag-of-words.

Accordingly, Zhang et al. (2014a, b) propose to model the word order in a phrase and capture the semantics of the phrase by using bilingually constrained recursive autoencoders (BRAE). The basic idea behind is that a source phrase and its correct target translation share the same meaning, and should share the same semantic vector representation. The framework of this method is illustrated in Fig. 6.5. Two recursive autoencoders are employed to learn the initial embeddings $(\mathbf{x}_1^3, \mathbf{y}_1^4)$ of the source and target phrases for the rule $\langle x_1^3, y_1^4\rangle$. A recursive autoencoder applies the same autoencoder for each node in the binary tree. The autoencoder takes two vector representations (e.g., $\mathbf{x}_1$ and $\mathbf{x}_2$) as inputs, and generates the phrase representation $(\mathbf{x}_1^2)$ as follows:

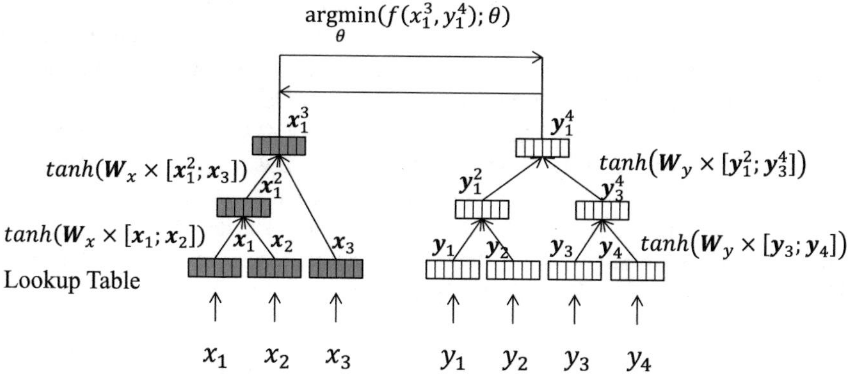

Fig. 6.5 Bilingually constrained phrase embeddings using recursive autoencoders that take word order into consideration. The goal is to learn semantic representation of a phrase

$$\mathbf{x}_1^2 = f(\mathbf{W}_x[\mathbf{x}_1; \mathbf{x}_2] + \mathbf{b}_x). \tag{6.25}$$

From $\mathbf{x}_1^2$, the autoencoder tries to reconstruct the inputs:

$$[\mathbf{x}'_1, \mathbf{x}'_2] = f(\mathbf{W}'_x \mathbf{x}_1^2 + \mathbf{b}'_x). \tag{6.26}$$

The network parameters are optimized to minimize the following reconstruction error:

$$E_{rec}[\mathbf{x}_1, \mathbf{x}_2] = \frac{1}{2}\|[\mathbf{x}_1, \mathbf{x}_2] - [\mathbf{x}'_1, \mathbf{x}'_2]\|^2. \tag{6.27}$$

In the recursive autoencoder, the network parameters are trained to minimize the sum of the reconstruction errors at each node. To capture the semantics of a phrase, besides the reconstruction error, the objective is also designed to minimize the semantic distance between translation equivalents and maximize the semantic distance between non-translation pairs simultaneously. After the network parameters and the word embeddings are optimized, the method can learn semantic vector representations of any source and target phrase. The similarity between two phrases in the semantic vector space (e.g., cosine similarity) is used as the translation confidence of the corresponding phrasal translation rule. With the help of semantic similarities, translation rule selection is much more accurate. Su et al. (2015) and Zhang et al. (2017a) propose to enhance the BRAE model and further improve the translation quality.

The above two methods focus on the phrasal translation rule itself and do not consider much more contexts. Devlin et al. (2014) propose a joint neural network model aiming at modeling both of the source and target-side contexts to predict translation probability. The idea is very simple: for a target word y_i to predict, we can track its corresponding source-side word (central source word x_j) according to the translation rule.[2] Then, the source context in a window centering x_j can be obtained, $x_{j-w} \cdots x_j \cdots x_{j+w}$ (e.g., $w = 5$). The vector representations of the source context and the target history translation $y_{i-3}y_{i-2}y_{i-1}$ are concatenated as the input of a feed-forward neural network as shown in Fig. 6.6. Following two hidden layers, a softmax function outputs the probability of the word y_i. Since much more contextual information is captured, the predicted translation probability becomes much more reliable.

However, the source-side context depends on the fix-sized window and cannot capture the global information. To solve this problem, Zhang et al. (2015) and Meng et al. (2015) try to learn the semantic representation for the source-side sentence and use the global sentence embedding as the additional input to augment the above joint network model. This kind of methods can perform better when a target word translation needs the sentence-level knowledge to disambiguate.

[2]For example, if a phrasal rule ⟨you bangjiao, have diplomatic relations⟩ matches the source sentence "*aozhou shi yu beihan you bangjiao de shaoshu guojia zhiyi*", the central source word will be *bangjiao* for predicting the target word *relations*.

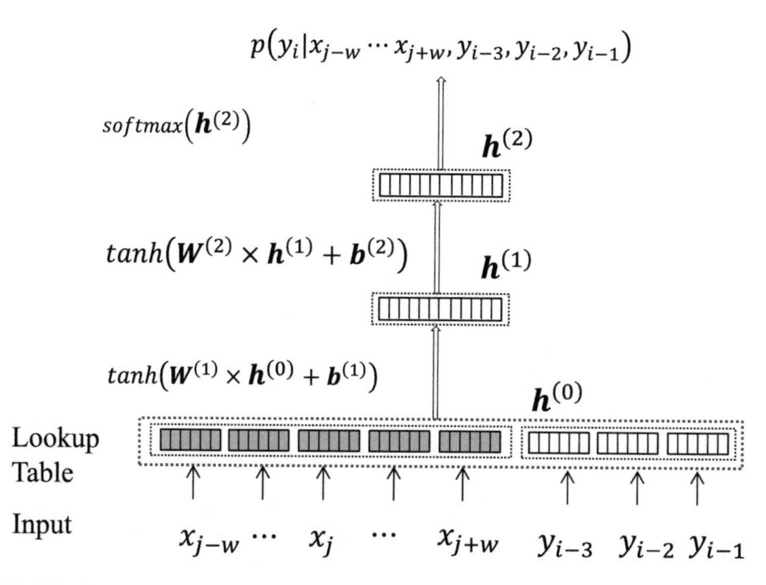

Fig. 6.6 Joint learning for target translation word prediction with a feed-forward neural network. The input includes the source-side context surrounding the central word and the target-side history. The output is the predicted conditional probability of next target word

6.3.3 Deep Learning for Reordering Phrases

For a source language sentence $\mathbf{x} = x_1^J$, phrasal translation rules match the sentence, segment the word sequence x_1^J into phrase sequence, and map each source phrase into the target language phrase using the neural rule selection model discussed in the previous section. The next task needs to rearrange the target phrases to produce a well-formed translation. This phrase reordering task is usually casted as a binary classification problem for any two neighboring target phrases: keep the two phrases in order (monotone) or swap the two phrases. For the two neighboring source phrases x^0 = yu beihan, x^1 = you bangjiao, and their translation candidates $y^0 = \mathit{with\ North\ Korea}$ and $y^1 = \mathit{have\ the\ diplomatic\ relations}$, the reordering model utilizes only the boundary discrete words of the four phrases as features and adopts a maximum entropy model to predict the reordering probability (Xiong et al. 2006):

$$p(o|x^0, x^1, y^0, y^1) = \frac{\sum_i \{\lambda_i f_i(x^0, x^1, y^0, y^1, o)\}}{\sum_o' \sum_i \{\lambda_i f_i(x^0, x^1, y^0, y^1, o')\}}, \tag{6.28}$$

where $f_i(x^0, x^1, y^0, y^1, o)$ and λ_i denote the discrete word features and their corresponding feature weights. o indicates the reordering type, $o = mono$ or $o = swap$. The reordering model using discrete symbols as features faces a serious issue of data sparseness. Furthermore, it cannot make full use of the whole phrase information and fails to capture the similar reordering patterns.

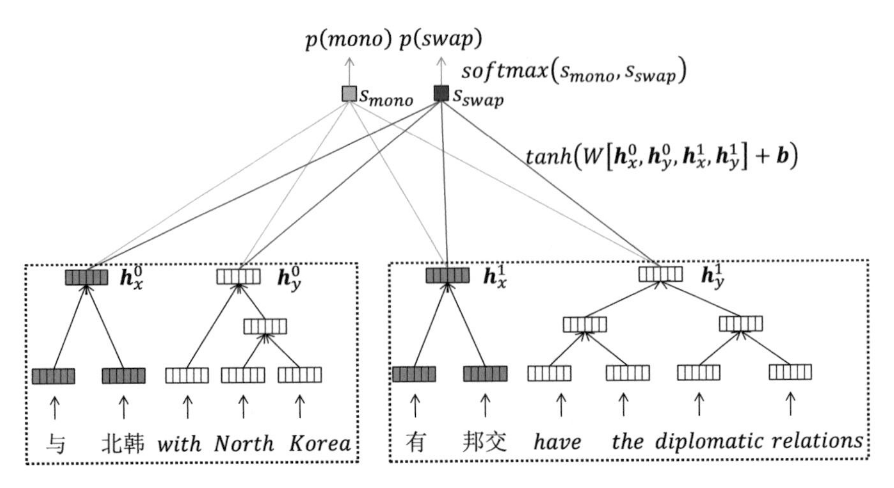

Fig. 6.7 A neural phrase reordering model in which four phrases in two phrasal translation rules are mapped into distributed representations using recursive autoencoders, and a feed-forward network is employed to predict the probability of reordering

Learning feature representations of the phrases in the real-valued vector space can alleviate the data sparsity problem and fully exploit the whole phrase information for reordering. Li et al. (2013, 2014) propose a neural phrase reordering model as shown in Fig. 6.7. The neural phrase reordering model first applies the recursive autoencoders to learn the distributed representations of the four phrases, $\mathbf{x}_0$, $\mathbf{y}_0$, $\mathbf{x}_1$, $\mathbf{y}_1$. Then, a feed-forward neural network is employed to convert the four vectors into a score vector consisting of two elements s_{mono} and s_{swap} using the following equation:

$$[s_{mono}, s_{swap}] = \tanh(\mathbf{W}[\mathbf{x}_0, \mathbf{y}_0, \mathbf{x}_1, \mathbf{y}_1] + \mathbf{b}). \tag{6.29}$$

Finally, a softmax function is leveraged to normalize the two scores s_{mono} and s_{swap} into two probabilities $p(mono)$ and $p(swap)$. The network parameters and word embeddings in the neural reordering model are optimized to minimize the following semi-supervised objective function:

$$Err = \alpha E_{rec}(x^0, x^1, y^0, y^1) + (1 - \alpha) E_{reorder}((x^0, y^0), (x^1, y^1)). \tag{6.30}$$

In which, $E_{rec}(x^0, x^1, y^0, y^1)$ is the sum of the reconstruction errors of recursive autoencoders for the four phrases and $E_{reorder}((x^0, y^0), (x^1, y^1))$ is the phrase reordering loss which is calculated with cross-entropy error function. α is employed to balance these two kinds of errors. This semi-supervised recursive autoencoder demonstrates that it can automatically group the phrases sharing the similar reordering patterns and leads to much better translation quality.

6.3.4 Deep Learning for Language Modeling

During phrase reordering, any two neighboring partial translations (target phrases) are composed into a bigger partial translation. The language model performs the task to measure whether the (partial) translation hypothesis is more fluent than others. The conventional SMT employs the most popular count-based n-gram language model whose conditional probability is calculated as follows:

$$p(y_i|y_{i-n+1}^{i-1}) = \frac{y_{i-n+1}^{i}}{y_{i-n+1}^{i-1}}. \tag{6.31}$$

Similar to the rule probability estimation and the reordering model, the string-match-based n-gram language model faces the severe data sparsity problem and cannot take full advantage of the semantically similar but surface different contexts. To alleviate this problem, deep learning-based language models are introduced to estimate the probability of a word conditioned on the history context in the continuous vector space.

Bengio et al. (2003) designed a feed-forward neural network as shown in Fig. 6.8a to learn the n-gram model in the continuous vector space. Vaswani et al. (2013) integrate this neural n-gram language model into SMT. During SMT decoding (phrase reordering and composition in phrase-based SMT), it is easy to find the partial history context (e.g., four words y_{i-4}, y_{i-3}, y_{i-2}, y_{i-1}) before the current word y_i in each decoding step. Thus, the neural n-gram model can be incorporated into the SMT decoding stage. Just as Fig. 6.8a illustrates, the fix-sized history words are first mapped into real-valued vectors which are then combined to feed the following two hidden layers. Finally, the softmax layer outputs the probability of the current word given the history

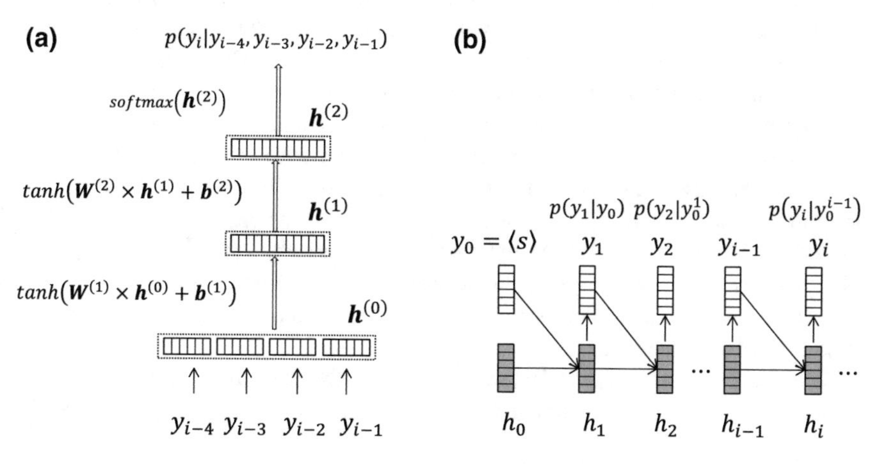

Fig. 6.8 Two popular neural language models: **a** the feed-forward neural network for language model which exploits a fix-sized window context; **b** the recurrent neural network for language model that takes full advantage of all the history context before the current word

context $p(y_i|y_{i-4}^{i-1})$. Large-scale experiments indicate that the neural n-gram language model could significantly improve the translation quality.

The n-gram language model assumes that generation of the current word depends only on the previous $n-1$ words, which is not the case in practice. In order to relax this assumption, recurrent neural network (including LSTM and GRU) tries to model all the history information when predicting the current word. As shown in Fig. 6.8b, a sentence start symbol $y_0 = <s>$ and the initial history context $\mathbf{h}_0$[3] are input into a recurrent neural network unit. It gets a new history context $\mathbf{h}_1$ which is used to predict the probability of y_1 using the following equation:

$$\mathbf{h}_1 = RNN(\mathbf{h}_0, \mathbf{y}_0). \tag{6.32}$$

In addition to the simple function (e.g., $\tanh(\mathbf{W}_h\mathbf{h}_0 + \mathbf{W}_y\mathbf{y}_0 + \mathbf{b})$), $RNN(\cdot)$ can use LSTM or GRU. $\mathbf{h}_1$ and y_1 are then employed to obtain the new history $\mathbf{h}_2$ that is believed to remember y_0 to y_1. $\mathbf{h}_2$ is utilized to predict $p(y_2|y_0^1)$. This process iterates. When predicting the probability of y_i, all the history context y_0^{i-1} can be used. Since the recurrent neural language model needs the entire history to predict a word while it is very difficult to record all the history during SMT decoding, this language model is usually employed to rescore the final n-best translation hypotheses. Auli and Gao (2014) try to integrate the recurrent neural language model into the SMT decoding stage with additional efforts and some improvements can be achieved compared to only rescoring.

6.3.5 Deep Learning for Feature Combination

Suppose that we have two phrasal translation rules[4] (x^1, y^1) and (x^2, y^2), and they happen to exactly match two neighboring source phrases x_i^k and x_{k+1}^j in a test sentence. Then, these two rules can be composed using the phrase reordering model to obtain the translation candidate for the longer source phrase x_i^j. In this case, we need to determine whether the monotone composition y^1y^2 is better than the swapped composition y^2y^1. Based on the above introductions in the previous sections, the two translation candidates can be evaluated with at least three sub-models: the rule probability estimation model, the phrase reordering model, and the language model. We will have three scores[5] for each of the translation candidate: $s_t(y^1y^2)$, $s_r(y^1y^2)$, $s_l(y^1y^2)$ and $s_t(y^2y^1)$, $s_r(y^2y^1)$, $s_l(y^2y^1)$. The final task needs to design a feature combination mechanism that maps the three model scores into one overall score so that the translation candidates can be compared with each other.

In the last decade, the log-linear model dominates the SMT community. It combines all the sub-model scores in a linear way as shown in Fig. 6.9a. The log-linear model

[3] $\mathbf{h}_0$ is usually set to all zeros.

[4] For example, the two phrasal translation rules are, respectively, (yu beihan, *with North Korea*) and (you bangjiao, *have the diplomatic relations*).

[5] The scores are usually log-probabilities.

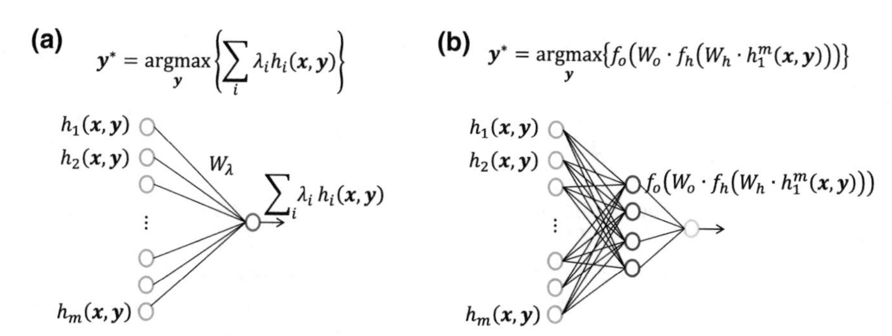

Fig. 6.9 Different framework for model feature combination: **a** log-linear model that combines the model features in a linear way; **b** nonlinear neural model which makes full use of the model features in a nonlinear space

assumes that all the sub-model features interact linearly with each other and thus limits the expressive power of the SMT model. In order to capture the complex interactions between different sub-model features, Huang et al. (2015) propose a neural network model to combine the feature scores in a nonlinear manner as illustrated in Fig. 6.9b.

Compared to the log-linear model, the neural combination model maps all the sub-model scores into one overall score using the following equation:

$$s_{neural}(e) = f_o(\mathbf{W}_o \cdot f_h(\mathbf{W}_h \cdot h_1^m(\mathbf{x}, \mathbf{y}))). \quad (6.33)$$

In which, we omit the bias term in the hidden layer and the output layer for simplification. $h_1^m(\mathbf{x}, \mathbf{y})$ denotes m sub-model feature scores, such as translation probability, reordering model probability and language model probability. $\mathbf{W}_h$ and $\mathbf{W}_o$ are weight matrices for the hidden layer and the output layer, respectively. $f_h(\cdot)$ and $f_o(\cdot)$ are activation functions for the hidden layer and the output layer. It is found best to set $f_h(\cdot) = sigmoid(\cdot)$ and set $f_o(\cdot)$ to be a linear function.

Parameter optimization of the neural combination model is much more difficult than that of the log-linear model. In the log-linear model, the weights of the sub-models can be efficiently tuned with the MERT (minimum error rate training) method (Och 2003), which searches the best weights by enumerating the model scores of all the translation candidates and utilizing the interactions between the linear functions to generate a well-formed search space. However, it is infeasible to obtain the interactions of the nonlinear functions employed by the neural combination model. To solve this problem, Huang et al. (2015) resort to the ranking-based training criteria, and the objective function is designed as follows:

$$\text{argmin}_\theta \tfrac{1}{N} \textstyle\sum_{x \in D} \sum_{(y_1, y_2) \in T(x)} \delta(x, y_1, y_2; \theta) + \lambda \cdot \|\theta\|_1 \quad (6.34)$$

$$\delta(x, y_1, y_2; \theta) = max\{s_{neural}(x, y_2; \theta) - s_{neural}(x, y_1; \theta) + 1, 0\}. \quad (6.35)$$

In the above equation, D is the sentence-aligned training data. (y_1, y_2) is the core of this training algorithm and denotes the training hypothesis pair, in which y_1 is a better translation hypothesis than y_2 according to sentence-level BLEU+1 evaluation. The model aims at optimizing the network parameters so as to guarantee that better translation hypotheses get higher network scores. $T(x)$ is the hypothesis pair set for each training sentence x and N is the total number of hypothesis pairs in the training data D.

For a training sentence x, it remains unclear how to efficiently sample the hypothesis pairs (y_1, y_2). Ideally, y_1 should be the correct translation (or reference translation), and y_2 is any other translation candidate. However, the correct translation does not exist in the search space of SMT in most cases due to many reasons, such as beam size limit, reordering distance constraints, and unknown words. Accordingly, Huang et al. (2015) attempt to sample (y_1, y_2) in the n-best translation list T_{nbest} using three methods: (1) **Best versus Rest**: y_1 is chosen to be the best candidate in T_{nbest} and y_2 can be any of the rest; (2) **Best versus Worst**: y_1 and y_2 are chosen to be the best and worst candidates in T_{nbest}, respectively; (3) **Pairwise**: sample two hypotheses from T_{nbest}, y_1 is set to be the better candidate and y_2 is the worse one.

Extensive experiments on Chinese-to-English translation demonstrate that the neural nonlinear model feature combination significantly outperforms the log-linear framework in translation quality.

6.4 End-to-End Deep Learning for Machine Translation

6.4.1 The Encoder–Decoder Framework

Research on component-wise deep learning for SMT is very active from 2013 to 2015. The log-linear model facilitates any integration of the deep learning-based translation features. Various kinds of neural network structures have been designed to improve different sub-modules, and the overall SMT performance has been upgraded significantly, for example, the joint neural model proposed by Devlin et al. (2014) achieved a surprising improvement of more than six BLEU points on Arabic-to-English translation. However, although deep learning is used to improve key components, SMT still uses linear modeling that is unable to deal with nonlinearities in textual data. In addition, the global dependency required by newly introduced neural features makes it impossible to design efficient dynamic programming training and decoding algorithms for SMT. Therefore, it is necessary to find new ways to improve machine translation using deep learning.

End-to-end neural machine translation (NMT) (Sutskever et al. 2014; Bahdanau et al. 2015) aims to directly map natural languages using neural networks. The major difference from conventional statistical machine translation (SMT) (Brown et al. 1993; Och and Ney 2002; Koehn et al. 2003; Chiang 2007) is that NMT is capable of learning representations from data, without the need to design features to capture translation regularities manually.

Given a source language sentence $\mathbf{x} = x_1, \ldots, x_i, \ldots, x_I$ and a target language sentence $\mathbf{y} = y_1, \ldots, y_j, \ldots, y_J$, standard NMT decomposes the sentence-level translation probability as a product of context-dependent word-level translation probabilities:

$$P(\mathbf{y}|\mathbf{x}; \boldsymbol{\theta}) = \prod_{j=1}^{J} P(y_j|\mathbf{x}, \mathbf{y}_{<j}; \boldsymbol{\theta}), \tag{6.36}$$

where $\mathbf{y}_{<j} = y_1, \ldots, y_{j-1}$ is a partial translation.

The word-level translation probability can be defined as

$$P(y_j|\mathbf{x}, \mathbf{y}_{<j}; \boldsymbol{\theta}) = \frac{\exp\left(g(\mathbf{x}, y_j, \mathbf{y}_{<j}, \boldsymbol{\theta})\right)}{\sum_y \exp\left(g(\mathbf{x}, y, \mathbf{y}_{<j}, \boldsymbol{\theta})\right)}, \tag{6.37}$$

where $g(\mathbf{x}, y_j, \mathbf{y}_{<j}, \boldsymbol{\theta})$ is a real-valued score that indicates how well the j-th target word y_j is given the source context $\mathbf{x}$ and target context $\mathbf{y}_{<j}$.

A major challenge is that the source and target contexts are highly sparse, especially for long sentences. To address this problem, Sutskever et al. (2014) propose to use a recurrent neural network (RNN), which is referred to as an *encoder*, to encode the source context $\mathbf{x}$ into a vector representation.

Figure 6.10 illustrates the basic idea of an encoder. Given a two-word source sentence $\mathbf{x} = x_1, x_2$, an end-of-sentence token $\langle\text{EOS}\rangle$ is appended to control the length of its translation. After obtaining vector representations of source words, the recurrent neural network runs to generate hidden states:

$$\mathbf{h}_i = f(\boldsymbol{x}_i, \mathbf{h}_{i-1}, \boldsymbol{\theta}), \tag{6.38}$$

where $\mathbf{h}_i$ is the i-th hidden state, $f(\cdot)$ is a nonlinear activation function, $\boldsymbol{x}_i$ is the vector representation of the i-th source word x_i.

For the nonlinear activation function $f(\cdot)$, long short-term memory (LSTM) (Hochreiter and Schmidhuber 1997) and gated recurrent units (GRUs) (Cho et al. 2014) are widely used to address the gradient vanishing or explosion problem. This leads to a significant advantage of NMT over conventional SMT in predicting global word reordering thanks to the capability of LSTM or GRUs to handle long-distance dependencies.

As there is an end-of-sentence symbol "EOS" appended to the source, the length of the source sentence is $I + 1$, and the last hidden state $\mathbf{h}_{I+1}$ is considered to encode the entire source sentence $\mathbf{x}$.

On the target side, Sutskever et al. (2014) use another RNN, which is called a *decoder*, for generating translations in a word-by-word manner. As shown in Fig. 6.10, each target-side hidden state that represents the target context $\mathbf{y}_{<j}$ is calculated as

$$\mathbf{s}_j = \begin{cases} \mathbf{h}_{I+1} & \text{if } j = 1 \\ f(\boldsymbol{y}_{j-1}, \mathbf{s}_{j-1}, \boldsymbol{\theta}) & \text{otherwise.} \end{cases} \tag{6.39}$$

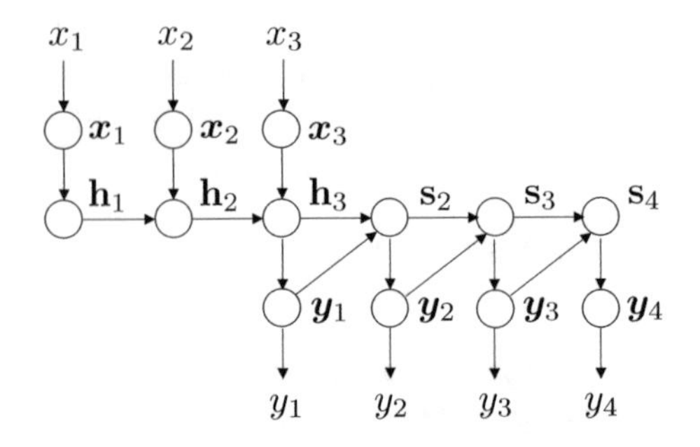

Fig. 6.10 The encoder–decoder framework for end-to-end neural machine translation. Given a source sentence $\mathbf{x} = x_1, x_2$, an end-of-sentence token (i.e., x_3) is appended to help predict when to terminate during generating target words. After mapping source words to their vector representations (i.e., $\boldsymbol{x}_1$, $\boldsymbol{x}_2$, and $\boldsymbol{x}_3$), a recurrent neural network (i.e., encoder) is used to compute the source-side hidden states $\mathbf{h}_1$, $\mathbf{h}_2$, and $\mathbf{h}_3$. Then, another recurrent neural network (i.e., decoder) runs to generate the target sentence word by word. The last source hidden state $\mathbf{h}_3$ is used to initialize the first target hidden state $\mathbf{s}_1$, from which the first target word y_1 and its vector representation $\boldsymbol{y}_1$ are determined. The first target hidden state $\mathbf{s}_1$ and word vector $\boldsymbol{y}_1$ are used to generate the second hidden state $\mathbf{s}_2$. This process iterates until an end-of-sentence token (i.e., y_4) is generated

Note that the source sentence representation $\mathbf{h}_{I+1}$ is only used to initialize the first target-side hidden state $\mathbf{s}_1$.

With the target-side hidden state $\mathbf{s}_j$, the scoring function $g(\mathbf{x}, y_j, \mathbf{y}_{<j}, \boldsymbol{\theta})$ can be simplified to $g(y_j, \mathbf{s}_j, \boldsymbol{\theta})$ that is calculated by another neural network. Please refer to (Sutskever et al. 2014) for more details.

Given a set of parallel sentences $\{\langle \mathbf{x}^{(s)}, \mathbf{y}^{(s)} \rangle\}_{s=1}^{S}$, the standard training objective is to maximize the log-likelihood of the training data:

$$\hat{\boldsymbol{\theta}} = \underset{\boldsymbol{\theta}}{\operatorname{argmax}} \Big\{ L(\boldsymbol{\theta}) \Big\}, \tag{6.40}$$

where the log-likelihood is defined as

$$L(\boldsymbol{\theta}) = \sum_{s=1}^{S} \log P(\mathbf{y}^{(s)} | \mathbf{x}^{(s)}; \boldsymbol{\theta}). \tag{6.41}$$

Standard mini-batch stochastic gradient descent algorithms can be used to optimize model parameters.

Given learned model parameters $\hat{\boldsymbol{\theta}}$, the decision rule for translating an unseen source sentence $\mathbf{x}$ is given by

$$\hat{\mathbf{y}} = \underset{\mathbf{y}}{\operatorname{argmax}} \Big\{ P(\mathbf{y} | \mathbf{x}; \hat{\boldsymbol{\theta}}) \Big\}. \tag{6.42}$$

6.4.2 Neural Attention in Machine Translation

In the original encoder–decoder framework (Sutskever et al. 2014), the encoder needs to represent the entire source sentence as a fixed-length vector regardless of the sentence length, which is used to initialize the first target-side hidden state. Bahdanau et al. (2015) indicate that this may make it difficult for neural networks to deal with long-distance dependencies. Empirical results reveal that the translation quality of the original encoder–decoder framework decreases significantly with the increase of sentence length (Bahdanau et al. 2015).

To address this problem, Bahdanau et al. (2015) introduce an *attention* mechanism to dynamically select relevant source context for generating a target word. As shown in Fig. 6.11, the attention-based encoder leverages bidirectional RNNs to capture global contexts:

$$\overrightarrow{\mathbf{h}}_i = f(\boldsymbol{x}_i, \overrightarrow{\mathbf{h}}_{i-1}, \boldsymbol{\theta}) \tag{6.43}$$

$$\overleftarrow{\mathbf{h}}_i = f(\boldsymbol{x}_i, \overleftarrow{\mathbf{h}}_{i+1}, \boldsymbol{\theta}), \tag{6.44}$$

where $\overrightarrow{\mathbf{h}}_i$ denotes the forward hidden state of the i-th source word x_i that captures the context on the left, $\overleftarrow{\mathbf{h}}_i$ denotes the backward hidden state of x_i that captures the context on the right. Therefore, the concatenation of forward and backward hidden states $\mathbf{h}_i = [\overrightarrow{\mathbf{h}}_i; \overleftarrow{\mathbf{h}}_i]$ is capable of capturing sentence-level context.

The basic idea of attention is to find *relevant* source context for target word generation. This is done by first calculating attention weight:

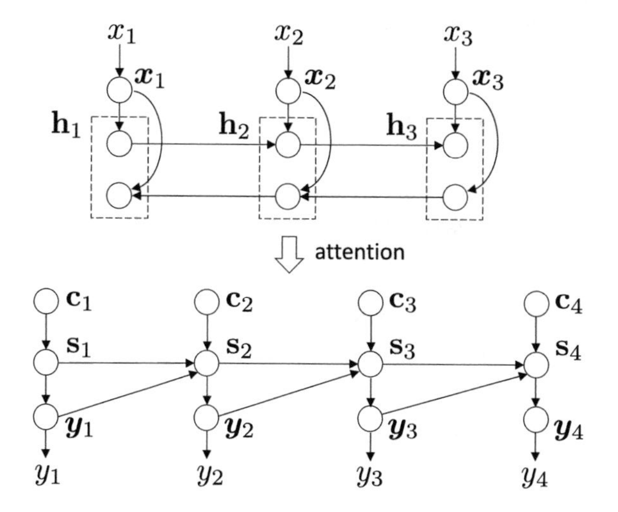

Fig. 6.11 Attention-based neural machine translation. Unlike the original encoder–decoder framework, the new encoder exploits bidirectional RNNs to compute forward and backward hidden states, which are concatenated to serve as context-dependent representations of each source word. Then, the attention mechanism is used to calculate a dynamic source context $\mathbf{c}_j$ ($j = 1, \ldots, 4$) for each target word, which involves the generation of corresponding target hidden state $\mathbf{s}_j$

$$\alpha_{j,i} = \frac{\exp\left(a(\mathbf{s}_{j-1}, \mathbf{h}_i, \boldsymbol{\theta})\right)}{\sum_{i'=1}^{I+1} \exp\left(a(\mathbf{s}_{j-1}, \mathbf{h}_{i'}, \boldsymbol{\theta})\right)}, \tag{6.45}$$

where the alignment function $a(\mathbf{s}_{j-1}, \mathbf{h}_i, \boldsymbol{\theta})$ evaluates how well the inputs around position i and the output at position i are related.

Then, the source context vector $\mathbf{c}_j$ is computed as a weighted sum of source hidden states:

$$\mathbf{c}_j = \sum_{i=1}^{I+1} \alpha_{j,i}\mathbf{h}_i. \tag{6.46}$$

As a result, the target hidden state can be calculated as

$$\mathbf{s}_j = f(\mathbf{y}_{j-1}, \mathbf{s}_{j-1}, \mathbf{c}_j, \boldsymbol{\theta}). \tag{6.47}$$

The major difference of attention-based NMT (Bahdanau et al. 2015) from the original encoder–decoder framework (Sutskever et al. 2014) is the way to calculating source context. The original framework only uses the last hidden state to initialize the first target hidden state. It is unclear how the source context controls the generation of target words, especially for words near the end of long sentences. In contrast, the attention mechanism enables each source word to contribute to the generation of a target word according to attention weight regardless of the position of the target word. This strategy proves to be very effective in improving translation quality, especially for long sentences. Therefore, the attention-based approach has become the *de facto* approach in neural machine translation.

6.4.3 Addressing Technical Challenges of Large Vocabulary

Although end-to-end NMT (Sutskever et al. 2014; Bahdanau et al. 2015) has delivered state-of-the-art translation performance across a variety of language pairs, one of the major challenges NMT faces is how to address the efficiency problem caused by target language vocabulary.

As the word-level translation probability requires normalization over all target words [(see Eq. (6.37)], the log-likelihood of the training data $\langle \mathbf{x}^{(s)}, \mathbf{y}^{(s)} \rangle$ is given by

$$L(\boldsymbol{\theta}) = \sum_{s=1}^{S} \log P(\mathbf{y}^{(s)}|\mathbf{s}^{(s)}; \boldsymbol{\theta}) \tag{6.48}$$

$$= \sum_{s=1}^{S} \sum_{j=1}^{J^{(s)}} \log P(y_j^{(s)}|\mathbf{x}^{(s)}, \mathbf{y}_{<j}^{(s)}; \boldsymbol{\theta}) \tag{6.49}$$

$$= \sum_{s=1}^{S} \sum_{j=1}^{J^{(s)}} \left(g(\mathbf{x}^{(s)}, y_j^{(s)}, \mathbf{y}_{<j}^{(s)}, \boldsymbol{\theta}) - \log \sum_{y \in V_y} \exp\left(g(\mathbf{x}^{(s)}, y, \mathbf{y}_{<j}^{(s)}, \boldsymbol{\theta})\right)\right), \tag{6.50}$$

where $J^{(s)}$ denotes the length of the s-th target sentence, and V_y denotes the target vocabulary.

Training NMT models requires to compute the gradients of log-likelihood:

$$\nabla L(\boldsymbol{\theta}) = \sum_{s=1}^{S} \sum_{j=1}^{J^{(s)}} \Big(\nabla g(\mathbf{x}^{(s)}, y_j^{(s)}, \mathbf{y}_{<j}^{(s)}, \boldsymbol{\theta}) - \sum_{y \in V_y} P(y|\mathbf{x}^{(s)}, \mathbf{y}_{<j}^{(s)}; \boldsymbol{\theta}) \nabla g(\mathbf{x}^{(s)}, y, \mathbf{y}_{<j}^{(s)}, \boldsymbol{\theta}) \Big). \tag{6.51}$$

It is clear that calculating the gradients involves the enumeration of all target words in V_y, which makes training NMT model prohibitively slow. In addition, predicting a target word at position j during decoding also requires enumerating all target words:

$$\hat{y}_j = \operatorname*{argmax}_{y \in V_y} \Big\{ P(y|\mathbf{x}, \mathbf{y}_{<j}; \boldsymbol{\theta}) \Big\}. \tag{6.52}$$

Therefore, Sutskever et al. (2014) and Bahdanau et al. (2015) have to use a subset of the full vocabulary, which is restricted to contain 30,000 to 80,000 frequent target words due to the limit of GPU memory. This significantly deteriorates the translation quality of source sentences that contain rare words falling out of the subset or out-of-vocabulary (OOV) words. Hence, it is important to address the large vocabulary problem to improve the efficiency of NMT.

To address this problem, Luong et al. (2015) propose to identify the correspondence between OOV words in source and target sentences and translate OOV words in a post-processing step. Table 6.1 shows an example. Given a source sentence "meiguo daibiaotuan baokuo laizi shidanfu de zhuanjia", two words "daibiaotuan" and "shidanfu" are identified as OOV (row 1). Therefore, the two words are replaced with "OOV"s (row 2). The source sentence with OOV words is translated to a target sentence with OOV words "the us OOV_1 consists of experts from OOV_3" (row 3), where the subscripts indicate the relative positions of corresponding source-side OOVs. In this example, the third target word "OOV_1" corresponds to the second source word "OOV" (i.e., $3 - 1 = 2$) and the eighth target word "OOV_3" is aligned to the fifth source word "OOV" (i.e., $8 - 3 = 5$). Finally, "OOV_1" is replaced with "delegation", which is a translation of "daibiaotuan". This can be done by using external translation knowledge sources such as bilingual dictionaries.

An alternative approach is to exploit sampling to address the large target vocabulary problem (Jean et al. 2015). As the major challenge for calculating the gradients is how to efficiently compute the expected gradient of the energy function [(i.e., the second term in Eq. (6.51)], Jean et al. (2015) propose to approximate the expectation by importance sampling with a small number of samples. Given a predefined proposal distribution $Q(y)$ and a set of V' samples from $Q(y)$, their approximation is given by

Table 6.1 Addressing out-of-vocabulary (OOV) words in neural machine translation. Identified as OOV, infrequent source words such as "*daibiaotuan*" and "*shidanfu*" are translated in a post-processing step using external knowledge sources such as bilingual dictionaries. The subscripts of target OOVs indicate the correspondence between source and target OOVs

source w/o OOV	meiguo *daibiaotuan* baokuo laizi *shidanfu* de zhuanjia
source w/ OOV	meiguo OOV baokuo laizi OOV de zhuanjia
target w/ OOV	the us OOV_1 consists of experts from OOV_3
target w/o OOV	the us *delegation* consists of experts from *stanford*

$$\sum_{y \in V_y} P(y|\mathbf{x}^{(s)}, \mathbf{y}^{(s)}_{<j}; \boldsymbol{\theta}) \nabla g(\mathbf{x}^{(s)}, y, \mathbf{y}^{(s)}_{<j}, \boldsymbol{\theta})$$

$$\approx \sum_{y \in V'} \frac{\exp\left(g(\mathbf{x}^{(s)}, y, \mathbf{y}^{(s)}_{<j}, \boldsymbol{\theta}) - \log Q(y)\right)}{\sum_{y' \in V'} \exp\left(g(\mathbf{x}^{(s)}, y', \mathbf{y}^{(s)}_{<j}, \boldsymbol{\theta}) - \log Q(y')\right)} \nabla g(\mathbf{x}^{(s)}, y, \mathbf{y}^{(s)}_{<j}, \boldsymbol{\theta}). \quad (6.53)$$

As a result, computing the normalization constant during training only requires to sum over a small subset of the target vocabulary, which significantly reduces computational complexity for each parameter update.

Another important direction is to model neural machine translation at character (Chung et al. 2016; Luong and Manning 2016; Costa-jussà and Fonollosa 2016) or sub-word (Sennrich et al. 2016b) levels. The intuition is that using characters or sub-words as the basic unit of translation significantly reduces the vocabulary size since there are always much fewer characters and sub-words as compared with words.

6.4.4 End-to-End Training to Optimize Evaluation Metric Directly

The standard training objective for neural machine translation is maximum likelihood estimation (MLE), which aims to find a set of model parameters maximizing the log-likelihood of the training data [(see Eqs. (6.40) and (6.41)]. Ranzato et al. (2016) identify two drawbacks of MLE. First, translation models are only exposed to gold-standard data during training. In other words, when generating a word in training, all context words are from ground-truth target sentences. However, during decoding, the context words are predicted by models, which are inevitably erroneous. This discrepancy between training and decoding has a negative effect on translation quality. Second, MLE only uses a loss function defined at the word level while machine translation evaluation metrics such as BLEU (Papineni et al. 2002) and TER (Snover et al. 2006) are often defined at corpus and sentence levels. This discrepancy between training and evaluation also hinders neural machine translation.

To address this problem, Shen et al. (2016) introduce minimum risk training (MRT) (Och 2003; Smith and Eisner 2006; He and Deng 2012) into neural machine translation. The basic idea is to use evaluation metrics as loss functions to measure the difference between model predictions and ground-truth translation and find a set of model parameters to minimize the expected loss (i.e., risk) on the training data.

Formally, the new training objective is given by

$$\hat{\boldsymbol{\theta}} = \underset{\boldsymbol{\theta}}{\text{argmin}} \Big\{ R(\boldsymbol{\theta}) \Big\}. \tag{6.54}$$

The risk on the training data is defined as

$$R(\boldsymbol{\theta}) = \sum_{s=1}^{S} \sum_{\mathbf{y} \in \mathcal{Y}(\mathbf{x}^{(s)})} P(\mathbf{y}|\mathbf{x}^{(s)}; \boldsymbol{\theta}) \Delta(\mathbf{y}, \mathbf{y}^{(s)}) \tag{6.55}$$

$$= \sum_{s=1}^{S} \mathbb{E}_{\mathbf{y}|\mathbf{x}^{(s)};\boldsymbol{\theta}} \Big[\Delta(\mathbf{y}, \mathbf{y}^{(s)}) \Big], \tag{6.56}$$

where $\mathcal{Y}(\mathbf{x}^{(s)})$ is a set of all possible translations of $\mathbf{x}^{(s)}$, $\mathbf{y}$ is a model prediction, $\mathbf{y}^{(s)}$ is a ground-truth translation, and $\Delta(\mathbf{y}, \mathbf{y}^{(s)})$ is a loss function calculated using sentence-level evaluation metrics such as BLEU.

Shen et al. (2016) argue that MRT has the following advantages over MLE. First, MRT is capable of directly optimizing model parameters with respect to evaluation metrics. This has proven to effectively improve translation quality by minimizing the discrepancy between training and evaluation (Och 2003). Second, MRT accepts arbitrary sentence-level loss functions, which are not necessarily differentiable. Third, MRT is transparent to model architectures and can be applied to arbitrary neural networks and artificial intelligence tasks.

However, a major challenge for MRT is that calculating the gradients requires enumerating all possible target sentences:

$$\nabla R(\boldsymbol{\theta}) = \sum_{s=1}^{S} \sum_{\mathbf{y} \in \mathcal{Y}(\mathbf{x}^{(s)})} \nabla P(\mathbf{y}|\mathbf{x}^{(s)}; \boldsymbol{\theta}) \Delta(\mathbf{y}, \mathbf{y}^{(s)}). \tag{6.57}$$

To alleviate this problem, Shen et al. (2016) propose to only use a subset of the full search space to approximate the posterior distribution $P(\mathbf{y}|\mathbf{x}^{(s)}; \boldsymbol{\theta})$ as

$$Q(\mathbf{y}|\mathbf{x}^{(s)}; \boldsymbol{\theta}, \beta) = \frac{P(\mathbf{y}|\mathbf{x}^{(s)}; \boldsymbol{\theta})^{\beta}}{\sum_{\mathbf{y}' \in \mathcal{S}(\mathbf{x}^{(s)})} P(\mathbf{y}'|\mathbf{x}^{(s)}; \boldsymbol{\theta})^{\beta}}, \tag{6.58}$$

where $\mathcal{S}(\mathbf{x}^{(s)}) \subset \mathcal{Y}(\mathbf{x}^{(s)})$ is a subset of the full search space that can be constructed by sampling, and β is a hyper-parameter for controlling the sharpness of the distribution.

Then, the new training objective is defined as

$$\tilde{R}(\boldsymbol{\theta}) = \sum_{s=1}^{S} \sum_{\mathbf{y} \in \mathcal{S}(\mathbf{x}^{(s)})} Q(\mathbf{y}|\mathbf{x}^{(s)}; \boldsymbol{\theta}, \beta) \Delta(\mathbf{y}, \mathbf{y}^{(s)}). \tag{6.59}$$

Ranzato et al. (2016) also propose an approach very similar with MRT. They cast the sequence generation problem in the reinforcement learning framework (Sutton and Barto 1988). The generative model can be viewed as an agent, which takes actions to predict the next word in the sequence at each time step. The agent receives a reward when it has reached the end of the sequence. Wiseman and Rush (2016) introduce a beam search training scheme to avoid biases associated with local training and unifies the training loss with the test-time usage.

In summary, as these evaluation metrics-oriented training criteria are capable of minimizing the discrepancy between training and evaluation, they have proven to be very effective in practical NMT systems (Wu et al. 2016).

6.4.5 Incorporating Prior Knowledge

Another important topic in neural machine translation is how to integrate prior knowledge into neural networks. As a data-driven approach, NMT acquires all translation knowledge from parallel corpora. It is difficult to integrate prior knowledge into neural networks due to the difference in representations. Neural networks use continuous real-valued vectors to represent all language structures involved in the translation process. While these vector representations prove to be capable of capturing translation regularities implicitly (Sutskever et al. 2014), it is hard to interpret each hidden state in neural networks from a linguistic perspective. In contrast, prior knowledge in machine translation is usually represented in discrete symbolic forms such as dictionaries and rules (Nirenburg 1989) that explicitly encode translation regularities. It is challenging to transform prior knowledge represented in discrete forms to continuous representations required by neural networks.

Therefore, a number of authors have endeavored to integrate prior knowledge into NMT in recent years. The following prior knowledge sources have exploited to improve NMT:

1. *Bilingual dictionary*: a set of source and target word pairs that are translationally equivalent (Arthur et al. 2016);
2. *Phrase table*: a set of source and target phrase pairs that are translationally equivalent (Tang et al. 2016);
3. *The coverage constraint*: each source phrase should be translated into exactly one target phrase (Tu et al. 2016; Mi et al. 2016);
4. *The agreement constraint*: the attention weight on which source-to-target and target-to-source translation models agree is reliable (Cheng et al. 2016a; Cohn et al. 2016);
5. *The structural biases*: position bias, Markov condition, and fertility that capture the structural divergence between source and target languages (Cohn et al. 2016);

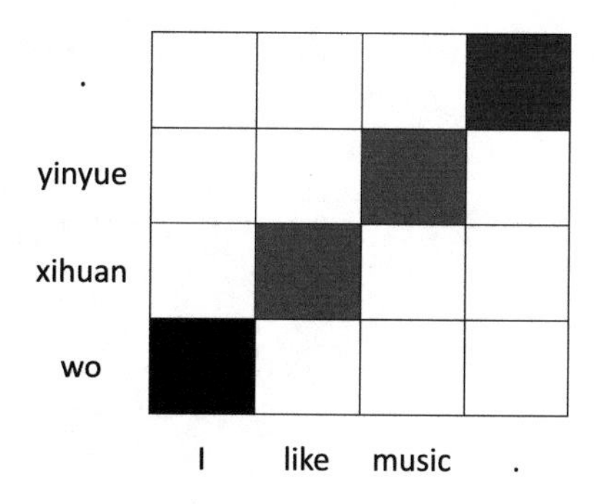

Fig. 6.12 Position bias for neural machine translation. Translation equivalents tend to have similar relative positions in source and target sentences. This prior knowledge source can be used to guide the learning of attentional NMT models

6. *Linguistic syntax*: exploiting syntactic trees to guide the learning process of neural machine translation (Eriguchi et al. 2016; Li et al. 2017; Wu et al. 2017; Chen et al. 2017a).

There are two broad categories of approaches to prior knowledge integration in neural networks. The first category is to modify model architectures. For example, as shown in Fig. 6.12, the position bias is based on the observation that a word at a given relative position in the source tends to be align to a word at a similar relative position in the target (i.e., $i/I \approx j/J$), especially for closely related language pairs such as English and French. In other words, aligned source and target words tend to occur near the diagonal of the alignment matrix.

To include this bias into NMT, Cohn et al. (2016) append a bias term to the alignment function:

$$a(\mathbf{h}_i, \mathbf{s}_{j-1}, \boldsymbol{\theta}) = \mathbf{v}^\top f\big(\mathbf{W}_1\mathbf{h}_i + \mathbf{W}_2\mathbf{s}_{j-1} + \mathbf{W}_3 \underbrace{\psi(j, i, I)}_{\text{position bias}}\big), \tag{6.60}$$

where $\mathbf{v}$, $\mathbf{W}_1$, $\mathbf{W}_2$, and $\mathbf{W}_3$ are model parameters.

The position bias term is defined as a function of the positions in the source and target sentences and the source length:

$$\psi(j, i, I) = \Big[\log(1 + j), \log(1 + i), \log(1 + I)\Big]^\top. \tag{6.61}$$

Note that the target length J is excluded because it is unknown during decoding.

Although modifying model architectures to inject prior knowledge into neural networks has shown its effectiveness in improving NMT, it is still hard to combine multiple overlapping, arbitrary prior knowledge sources. This is because neural networks usually impose strong independence assumptions between hidden states. As a result, extending a neural model requires that the interdependence of information sources be modeled explicitly.

This problem can be partly alleviated by appending additional additive terms to training objectives (Cheng et al. 2016a; Cohn et al. 2016), which keeps the NMT models unchanged. For example, Cheng et al. (2016a) introduce a new training objective to encourage source-to-target and target-to-source translation models to agree on attention weight matrices:

$$J(\overrightarrow{\boldsymbol{\theta}}, \overleftarrow{\boldsymbol{\theta}}) = \sum_{s=1}^{S} \log P(\mathbf{y}^{(s)}|\mathbf{x}^{(s)}; \overrightarrow{\boldsymbol{\theta}}) + \sum_{s=1}^{S} \log P(\mathbf{x}^{(s)}|\mathbf{y}^{(s)}; \overleftarrow{\boldsymbol{\theta}}) - \lambda \sum_{s=1}^{S} \underbrace{\Delta\Big(\mathbf{x}^{(s)}, \mathbf{y}^{(s)}, \overrightarrow{\alpha}^{(s)}(\overrightarrow{\boldsymbol{\theta}}), \overleftarrow{\alpha}^{(s)}(\overleftarrow{\boldsymbol{\theta}})\Big)}_{\text{agreement}}, \tag{6.62}$$

where $\overrightarrow{\boldsymbol{\theta}}$ is a set of source-to-target translation model parameters, $\overleftarrow{\boldsymbol{\theta}}$ is a set of target-to-source translation model parameters, $\overrightarrow{\alpha}^{(s)}(\overrightarrow{\boldsymbol{\theta}})$ is the source-to-target attention weight matrix for the s-th sentence pair, $\overleftarrow{\alpha}^{(s)}(\overleftarrow{\boldsymbol{\theta}})$ is the target-to-source attention weight matrix for the s-th sentence pair, and $\Delta(\cdot)$ measures the disagreement between two attention weight matrices.

However, the terms appended to training objectives have been restricted to a limited number of simple constraints because it is hard to manually tune the weight of each term.

More recently, Zhang et al. (2017b) have proposed a general framework for incorporating arbitrary knowledge sources based on posterior regularization (Ganchev et al. 2010). The central idea is to encode prior knowledge sources into a probability distribution, which guides the learning process of translation models by minimizing the KL divergence between two distributions:

$$J(\boldsymbol{\theta}, \boldsymbol{\gamma}) = \lambda_1 \sum_{s=1}^{S} \log P(\mathbf{y}^{(s)}|\mathbf{x}^{(s)}; \boldsymbol{\theta}) - \lambda_2 \sum_{s=1}^{S} \mathrm{KL}\Big(Q(\mathbf{y}|\mathbf{x}^{(s)}; \boldsymbol{\gamma})|P(\mathbf{y}^{(s)}|\mathbf{x}^{(s)}; \boldsymbol{\theta})\Big), \tag{6.63}$$

where prior knowledge sources are encoded in a log-linear model:

$$Q(\mathbf{y}|\mathbf{x}^{(s)}; \boldsymbol{\gamma}) = \frac{\exp\big(\boldsymbol{\gamma} \cdot \boldsymbol{\phi}(\mathbf{x}^{(s)}, \mathbf{y})\big)}{\sum_{\mathbf{y}'} \exp\big(\boldsymbol{\gamma} \cdot \boldsymbol{\phi}(\mathbf{x}^{(s)}, \mathbf{y}')\big)}. \tag{6.64}$$

Note that prior knowledge sources are represented as features $\boldsymbol{\phi}(\cdot)$ in conventional discrete symbolic forms.

6.4.6 Low-Resource Language Translation

Parallel corpora, which are collections of parallel texts, play a critical role in training NMT models because they are the main source for translation knowledge acquisition.

It is widely accepted that the quantity, quality, and coverage of parallel corpora directly influence the translation quality of NMT systems.

Although NMT has delivered state-of-the-art performance for resource-rich language pairs, the unavailability of large-scale, high-quality, and wide-coverage parallel corpora still remains a major challenge for NMT, especially for low-resource language translation. For most language pairs, parallel corpora are nonexistent. Even for the top handful of resource-rich languages, the available parallel corpora are usually unbalanced because the major sources are restricted to government documents or news articles. Due to the large parameter space, neural models usually learn poorly from low-count events, making NMT a poor choice for low-resource language pairs. Zoph et al. (2016) indicate that NMT obtains much worse translation quality than conventional statistical machine translation on low-resource languages.

To address this problem, a straightforward solution is to exploit abundant monolingual data. Gulcehre et al. (2015) propose two methods, which are referred to as shallow fusion and deep fusion, to integrate a language model into NMT. The basic idea is to use the language model trained on large-scale monolingual data to score the candidate words proposed by the neural translation model at each time step or concatenating the hidden states of the language model and the decoder. Although their approach leads to significant improvements, one possible downside is that the network architecture has to be modified to integrate the language model.

Alternatively, Sennrich et al. (2016a) propose two approaches to exploiting monolingual corpora that are transparent to network architectures. The first approach pairs monolingual sentences with dummy input. Then, the parameters of encoder and attention model are fixed during training on these pseudo-parallel sentence pairs. The second approach first trains a neural translation model on the parallel corpus and then uses the learned model to translate a monolingual corpus. The monolingual corpus and its translations constitute an additional pseudo-parallel corpus. Similar methods are investigated by (Zhang and Zong 2016) to exploit the source-side monolingual data.

Cheng et al. (2016b) introduce a semi-supervised learning approach to using monolingual data for NMT. As shown in Fig. 6.13, given a source sentence in a monolingual corpus, Cheng et al. (2016b) use source-to-target and target-to-source translation mod-

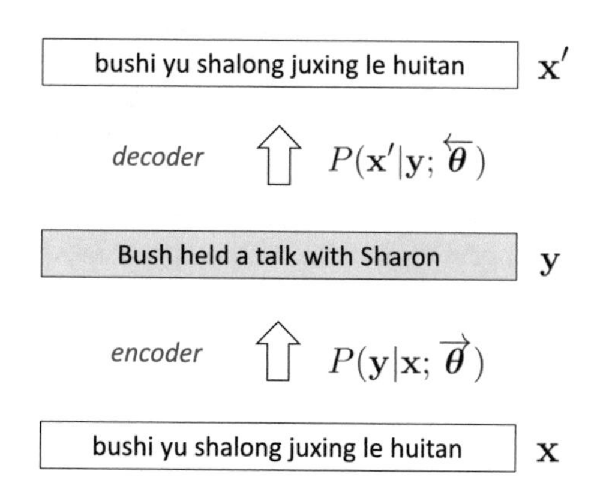

Fig. 6.13 Autoencoders for exploiting monolingual corpora for NMT. Given a source sentence, a source-to-target NMT model transforms it into a latent target sentence, from which a target-to-source model is used to recover the input source sentence

els to build an autoencoder that recovers the input source sentence via a latent target sentence. More formally, the reconstruction probability is given by

$$P(\mathbf{x}'|\mathbf{x}; \overrightarrow{\boldsymbol{\theta}}, \overleftarrow{\boldsymbol{\theta}}) = \sum_{\mathbf{y}} P(\mathbf{y}|\mathbf{x}; \overrightarrow{\boldsymbol{\theta}}) P(\mathbf{x}'|\mathbf{y}; \overleftarrow{\boldsymbol{\theta}}). \tag{6.65}$$

Therefore, both parallel and monolingual corpora can be used for semi-supervised learning. Let $\{\langle \mathbf{x}^{(s)}, \mathbf{y}^{(s)} \rangle\}_{s=1}^{S}$ be a parallel corpus, $\{\mathbf{x}^{(m)}\}_{m=1}^{M}$ be a monolingual corpus of source language, and $\{\mathbf{y}^{(n)}\}_{n=1}^{N}$ be a monolingual corpus of target language. The new training objective is given by

$$\begin{aligned} J(\overrightarrow{\boldsymbol{\theta}}, \overleftarrow{\boldsymbol{\theta}}) = & \underbrace{\sum_{s=1}^{S} \log P(\mathbf{y}^{(s)}|\mathbf{x}^{(s)}; \overrightarrow{\boldsymbol{\theta}})}_{\text{source-to-target likelihood}} + \underbrace{\sum_{s=1}^{S} \log P(\mathbf{x}^{(s)}|\mathbf{y}^{(s)}; \overleftarrow{\boldsymbol{\theta}})}_{\text{target-to-source likelihood}} + \\ & \underbrace{\sum_{m=1}^{M} \log P(\mathbf{x}'|\mathbf{x}^{(m)}; \overrightarrow{\boldsymbol{\theta}}, \overleftarrow{\boldsymbol{\theta}})}_{\text{source autoencoder}} + \underbrace{\sum_{n=1}^{N} \log P(\mathbf{y}'|\mathbf{y}^{(n)}; \overrightarrow{\boldsymbol{\theta}}, \overleftarrow{\boldsymbol{\theta}})}_{\text{target autoencoder}}. \end{aligned} \tag{6.66}$$

Another interesting direction is to exploit multilingual data for NMT (Firat et al. 2016; Johnson et al. 2016). Firat et al. (2016) present a multi-way, multilingual model with shared attention to achieve zero-resource translation. They fine-tune the attention part using pseudo-bilingual sentences for the zero-resource language pair. Johnson et al. (2016) develop a universal NMT model in multilingual scenarios. They use parallel corpora of multiple languages to train one single model, which is then able to translate a language pair without parallel corpora available.

Using a pivot language to bridge source and target languages has also been investigated in neural machine translation (Nakayama and Nishida 2016; Cheng et al. 2017). The basic idea is to use source-pivot and pivot-target parallel corpora to train source-to-pivot and pivot-to-target translation models. During decoding, a source sentence is first translated into a pivot sentence using the source-to-pivot model, which is then translated to a target sentence using the pivot-to-target model. Nakayama and Nishida (2016) achieve zero-resource machine translation by utilizing image as a pivot and training multimodal encoders to share common semantic representation. Cheng et al. (2017) propose pivot-based NMT by simultaneously improving source-to-pivot and pivot-to-target translation quality in order to improve source-to-target translation quality. As pivot-based approaches face the error propagation problem resulted from indirect modeling, direct modeling approaches such as teacher–student framework (Chen et al. 2017b) and maximum expected likelihood estimation (Zheng et al. 2017) have been proposed recently.

6.4.7 Network Structures in Neural Machine Translation

Recurrent models, such as LSMT and GRU, dominate the network structure design for encoder and decoder in neural machine translation. Recently, convolution networks (Gehring et al. 2017) and self-attention networks (Vaswani et al. 2017) are fully investigated and promising improvements are achieved.

Gehring et al. (2017) argue that parallel computation is inefficient in sequence modeling using the recurrent network since it needs to maintain a hidden state of the entire history. In contrast, convolution networks learn representations for fixed-length context and do not depend on the computations of all the history information. Thus, each element in the sequence can be calculated in parallel for both encoding and decoding (during training). Furthermore, the convolution layers can be deeply stacked to capture the long-distance dependency relationship. Figure 6.14a illustrates the translation process of the convolutional sequence to sequence model. Kernel size of the convolution is set $k = 3$. For encoder, multiple convolution and nonlinearity layers (only one is displayed in Fig. 6.14a for simplicity) are adopted to create the hidden state of each input position. When decoder tries to generate the fourth target word y_4, multiple convolution and nonlinearity layers are employed to obtain the hidden representation of previous k words. Then, standard attention is applied to predict y_4.

Recurrent networks require $O(n)$ operations to model the dependency between the first and n-th word, while convolution models need $O(log_k(n))$ stacked convolution operations. Without using any recurrence and convolution, Vaswani et al. (2017) propose to directly model the relationship between any word pair with a self-attention mechanism, as shown in Fig. 6.14b. To learn the hidden state of each input position (e.g., second word) in encoder, self-attention model and feed-forward network calculate the relevance between the second word and other words and obtain a hidden state.

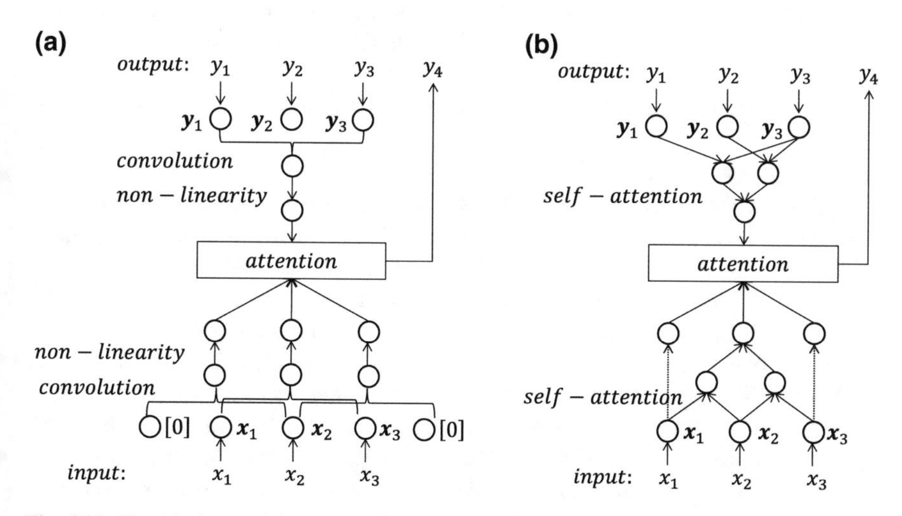

Fig. 6.14 Convolution model (**a**) and self-attention model (**b**) for neural machine translation

We can stack multiple self-attention and feed-forward layers to yield the highly abstract representation of the second position. To decode y_4, another self-attention model is employed to capture the dependency between the current target position and any previous ones. Then, we use conventional attention mechanism to model the relationship between source and target words, and predict the next target word y_4. Due to the highly parallelizable network structure and direct connection between any two positions, this translation model significantly speeds up training process and remarkably improves the translation performance. However, decoding is not efficient when translating unseen sentences since parallelism cannot be used in the target side.

Currently, there is no consensus on which network structure is best for neural machine translation. Network structure design will still be a hot research topic in the future.

6.4.8 *Combination of SMT and NMT*

Although NMT is superior to SMT in translation quality (especially in translation fluency), NMT sometimes lacks reliability in translation adequacy and generates translations having different meanings with the source sentences, particularly when rare words occur in the input. In comparison, SMT can usually produce adequate but influent translations. It is, therefore, a promising direction to combine the merits of both NMT and SMT.

Recent 2 years witnessed the great efforts to take both advantages of NMT and SMT (He et al. 2016; Niehues et al. 2016; Wang et al. 2017; Zhou et al. 2017). He et al. (2016) and Wang et al. (2017) attempt to enhance the NMT system with SMT features or SMT translation recommendations. For example, Wang et al. (2017) utilize SMT to generate a recommendation vocabulary V_{smt} by using the partial translation of NMT as prefix. Then, the following formula is employed to predict the next target word:

$$P(y_t|\mathbf{y}_{<t}, \mathbf{x}) = (1 - \alpha_t)P_{nmt}(y_t|\mathbf{y}_{<t}, \mathbf{x}) + \alpha_t P_{smt}(y_t|\mathbf{y}_{<t}, \mathbf{x}). \tag{6.67}$$

In which $P_{smt}(y_t|\mathbf{y}_{<t}, \mathbf{x}) = 0$ if $y_t \notin V_{smt}$.

Niehues et al. (2016) adopt an SMT system to pre-translate the input into the target language sentence. Then, a neural machine translation system is developed to take as input the pre-translation or the combination of pre-translation and source sentence.

Zhou et al. (2017) argue that this kind of methods can make use of only one SMT system. Accordingly, they propose a neural system combination method that can take advantages of multiple SMT and NMT systems. As illustrated in Fig. 6.15, the outputs of SMT and NMT systems serve as the input to the neural system combination framework. Then, the hierarchical attention mechanism is designed to determine which part of which system should be paid more attention to when predicting the next target word. Due to efficient combination, this method leads to promising gains in translation quality. However, translation n-best list cannot be used in this framework and we believe that there is much room to explore in the direction of system combination.

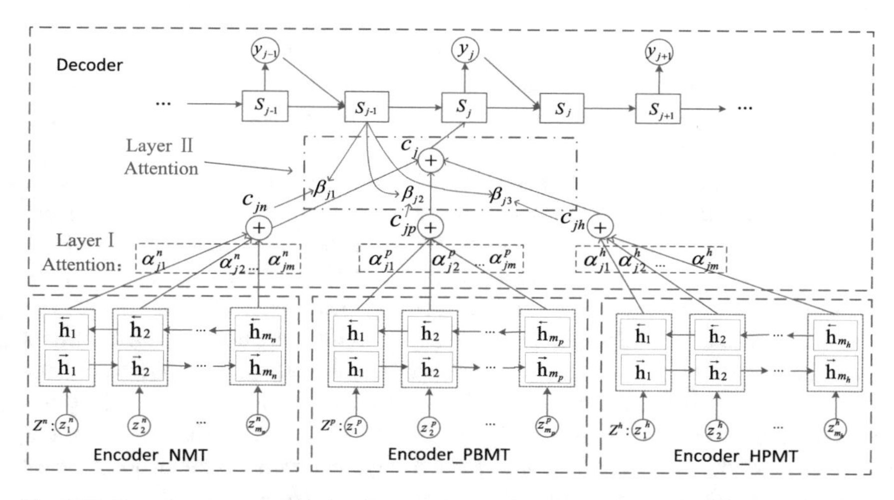

Fig. 6.15 Neural system combination framework for machine translation, in which multiple SMT and NMT systems can be combined using a hierarchical attention model to generate better translations

6.5 Summary

In this chapter, we have introduced how deep learning is used to improve machine translation. As traditional statistical machine translation faces the data sparsity and feature engineering problems, early efforts have focused on using deep learning to improve key components of linear translation models such as rule translation probabilities (Gao et al. 2014), reordering models (Li et al. 2013), and language models (Vaswani et al. 2013). Since 2014, end-to-end neural machine translation (Sutskever et al. 2014; Bahdanau et al. 2015) that aims to directly map between natural languages using neural networks has become increasingly popular in the MT community. NMT has made remarkable progress in the last 2 years and quickly replaced SMT to be the new *de facto* technology of commercial translation systems.

Although deep learning has proven to revolutionize machine translation, there are still a number of key limitations of current NMT methods. First, it is hard to interpret the internal workings of neural networks and design linguistically motivated neural translation models. While recent work on using layer-wise relevance propagation to quantify the connection between two arbitrary neurons in a network (Ding et al. 2017), it is still hard to associate hidden states in neural networks with interpretable language structures. As a result, it is also challenging to incorporate prior knowledge using symbolic representations into neural networks using continuous representations.

Another major challenge is data scarcity. NMT is a data-hungry approach while there is only limited or even no parallel data available for most language pairs in the world. How to make a better use of limited labeled data and abundant unlabeled data continues to be a hot topic in the future. The universal NMT model proposed by Johnson et al. (2016) is an interesting direction for addressing the data scarcity problem. Although their experiments show promising results for the many-to-one direction (i.e., multiple source languages and a single target language), there are no consistent and significant improvements for one-to-many and many-to-many directions. It is still unclear how to represent and exploit the common knowledge of all natural languages in NMT from a linguistic point of view.

Finally, most existing NMT systems are still restricted to dealing with textual data. Fortunately, the use of continuous representations makes it possible to combine text, speech, and vision information to develop multimodal NMT models. Duong et al. (2016) propose to develop speech translation systems without transcription. This can be done by enabling the NMT model to take the continuous representations of source language speech as input. However, they fail to report significant improvement in terms of translation quality. Calixto et al. (2017) introduce a doubly attentive decoder to incorporate both text and image to improve NMT. However, the training data for their system only contains 30 K images and five descriptions for each image. Therefore, building large-scale multimodal parallel corpora and designing new multimodal neural translation models is also an interesting future direction to explore.

References

Arthur, P., Neubug, G., & Nakamura, S. (2016). Incorporating discrete translation lexicons into neural machine translation. *In Proceedings of EMNLP.* arXiv:1606.02006v2.

Auli, M., & Gao, J. (2014). Decoder integration and expected bleu training for recurrent neural network language models. In *Proceedings of ACL* (pp. 136–142).

Bahdanau, D., Cho, K., & Bengio, Y. (2015). Neural machine translation by jointly learning to align and translate. In *Proceedings of ICLR*.

Bengio, Y., Ducharme, R., Vincent, P., & Jauvin, C. (2003). A neural probabilistic language model. *Journal of Machine Learning Research*, *3*, 1137–1155.

Brown, P. F., Della Pietra, S. A., Della Pietra, V. J., & Mercer, R. L. (1993). The mathematics of statistical machine translation: Parameter estimation. *Computational Linguistics*.

Calixto, I., Liu, Q., & Campbell, N. (2017). Doubly-attentive decoder for multi-modal neural machine translation. In *Proceedings of ACL*.

Chen, H., Huang, S., Chiang, D., & Chen, J. (2017a). Improved neural machine translation with a syntax-aware encoder and decoder. In *Proceedings of ACL 2017*.

Chen, S., & Goodman, J. (1999). An empirical study of smoothing techniques for language modeling. *Computer Speech and Language*.

Chen, Y., Liu, Y., Cheng, Y., & Li, V. O. (2017b). A teacher-student framework for zero-resource neural machine translation. In *Proceedings of the 55th Annual Meeting of the Association for Computational Linguistics* (Vol. 1: Long Papers, pp. 1925–1935). Vancouver, Canada: Association for Computational Linguistics.

Cheng, Y., Shen, S., He, Z., He, W., Wu, H., Sun, M., & Liu, Y. (2016a). Agreement-based learning of parallel lexicons and phrases from non-parallel corpora. In *Proceedings of IJCAI*.

Cheng, Y., Xu, W., He, Z., He, W., Wu, H., Sun, M., & Liu, Y. (2016b). Semi-supervised learning for neural machine translation. In *Proceedings of the 54th Annual Meeting of the Association for Computational Linguistics* (Vol. 1: Long Papers, pp. 1965–1974). Berlin, Germany: Association for Computational Linguistics.

Cheng, Y., Yang, Q., Liu, Y., Sun, M., & Xu, W. (2017). Joint training for pivot-based neural machine translation. In *Proceedings of IJCAI*.

Chiang, D. (2007). Hierarchical phrase-based translation. *Computational Linguistics*.

Chiang, D., Knight, K., & Wang, W. (2009). 11,001 new features for statistical machine translation. In *Proceedings of NAACL*.

Cho, K., van Merrienboer, B., Gulcehre, C., Bahdanau, D., Bougares, F., Schwenk, H., & Bengio, Y. (2014). Learning phrase representations using RNN encoder–decoder for statistical machine translation. In *Proceedings of EMNLP*.

Chung, J., Cho, K., & Bengio, Y. (2016). A character-level decoder without explicit segmentation for neural machine translation. In *Proceedings of the 54th Annual Meeting of the Association for Computational Linguistics* (Vol. 1: Long Papers, pp. 1693–1703). Berlin, Germany: Association for Computational Linguistics.

Cohn, T., Hoang, C. D. V., Vymolova, E., Yao, K., Dyer, C., & Haffari, G. (2016). Incorporating structural alignment biases into an attentional neural translation model. In *Proceedings of NAACL*.

Costa-jussà, M. R., & Fonollosa, J. A. R. (2016). Character-based neural machine translation. In *Proceedings of the 54th Annual Meeting of the Association for Computational Linguistics* (Vol. 2: Short Papers, pp. 357–361). Berlin, Germany: Association for Computational Linguistics.

Devlin, J., Zbib, R., Huang, Z., Lamar, T., Schwartz, R. M., & Makhoul, J. (2014). Fast and robust neural network joint models for statistical machine translation. In *Proceedings of ACL* (pp. 1370–1380).

Ding, Y., Liu, Y., Luan, H., & Sun, M. (2017). Visualizing and understanding neural machine translation. In *Proceedings of ACL*.

Duong, L., Anastasopoulos, A., Chiang, D., Bird, S., & Cohn, T. (2016). An attentional model for speech translation without transcription. In *Proceedings of NAACL*.

Eriguchi, A., Hashimoto, K., & Tsuruoka, Y. (2016). Tree-to-sequence attentional neural machine translation. In *Proceedings of the 54th Annual Meeting of the Association for Computational Linguistics* (Vol. 1: Long Papers, pp. 823–833). Berlin, Germany: Association for Computational Linguistics.

Firat, O., Cho, K., & Bengio, Y. (2016). Multi-way, multilingual neural machine translation with a shared attention mechanism. In *HLT-NAACL*.

Galley, M., & Manning, C. (2008). A simple and effective hierarchical phrase reordering model. In *Proceedings of EMNLP*.

Ganchev, K., Graça, J., Gillenwater, J., & Taskar, B. (2010). Posterior regularization for structured latent variable models. *Journal of Machine Learning Research*.

Gao, J., He, X., Yih, W.-t., & Deng, L. (2014). Learning continuous phrase representations for translation modeling. In *Proceedings of ACL* (pp. 699–709).

Gehring, J., Auli, M., Grangier, D., Yarats, D., & Dauphin, Y. N. (2017). Convolutional sequence to sequence learning. In *Proceedings of ICML 2017*.

Gulcehre, C., Firat, O., Xu, K., Cho, K., Barrault, L., Lin, H. C. et al. (2015). On using monolingual corpora in neural machine translation. *Computer Science*.

He, W., He, Z., Wu, H., & Wang, H. (2016). Improved neural machine translation with SMT features. In *Proceedings of AAAI 2016* (pp. 151–157).

He, X., & Deng, L. (2012). Maximum expected bleu training of phrase and lexicon translation models. In *Proceedings of ACL*.

Hochreiter, S., & Schmidhuber, J. (1997). Long short-term memory. *Neural Computation*.

Huang, S., Chen, H., Dai, X., & Chen, J. (2015). Non-linear learning for statistical machine translation. In *Proceedings of ACL*.

Jean, S., Cho, K., Memisevic, R., & Bengio, Y. (2015). On using very large target vocabulary for neural machine translation. In *ACL*.

Johnson, M., Schuster, M., Le, Q. V., Krikun, M., Wu, Y., Chen, Z. et al. (2016). Google's multilingual neural machine translation system: Enabling zero-shot translation. *CoRR*, abs/1611.04558.

Koehn, P., Och, F. J., & Marcu, D. (2003). Statistical phrase-based translation. In *Proceedings of NAACL*.

Li, J., Xiong, D., Tu, Z., Zhu, M., Zhang, M., & Zhou, G. (2017). Modeling source syntax for neural machine translation. In *Proceedings of ACL 2017*.

Li, P., Liu, Y., & Sun, M. (2013). Recursive autoencoders for ITG-based translation. In *Proceedings of EMNLP*.
Li, P., Liu, Y., Sun, M., Izuha, T., & Zhang, D. (2014). A neural reordering model for phrase-based translation. In *Proceedings of COLING* (pp. 1897–1907).
Luong, M.-T., & Manning, C. D. (2016). Achieving open vocabulary neural machine translation with hybrid word-character models. In *Proceedings of ACL*.
Luong, T., Sutskever, I., Le, Q. V., Vinyals, O., & Zaremba, W. (2015). Addressing the rare word problem in neural machine translation. In *ACL*.
Meng, F., Lu, Z., Wang, M., Li, H., Jiang, W., & Liu, Q. (2015). Encoding source language with convolutional neural network for machine translation. In *Proceedings of ACL*.
Mi, H., Sankaran, B., Wang, Z., & Ittycheriah, A. (2016). Coverage embedding models for neural machine translation. In *Proceedings of EMNLP*.
Nakayama, H., & Nishida, N. (2016). Zero-resource machine translation by multimodal encoder-decoder network with multimedia pivot. Machine Translation 2017. *CoRR*, abs/1611.04503.
Niehues, J., Cho, E., Ha, T.-L., & Waibel, A. (2016). Pre-translation for neural machine translation. In *Proceedings of COLING 2016*.
Nirenburg, S. (1989). Knowledge-based machine translation. *Machine Translation*.
Och, F. J. (2003). Minimum error rate training in statistical machine translation. In *Proceedings of the 41st Annual Meeting on Association for Computational Linguistics* (Vol. 1, pp. 160–167).
Och, F. J., & Ney, H. (2002). Discriminative training and maximum entropy models for statistical machine translation. In *Proceedings of ACL*.
Papineni, K., Roukos, S., Ward, T., & Zhu, W.-J. (2002). BLEU: A method for automatic evaluation of machine translation. In *ACL*.
Ranzato, M., Chopra, S., Auli, M., & Zaremba, W. (2016). Sequence level training with recurrent neural networks. In *CoRR*.
Sennrich, R., Haddow, B., & Birch, A. (2016a). Improving neural machine translation models with monolingual data. In *Proceedings of the 54th Annual Meeting of the Association for Computational Linguistics* (Vol. 1: Long Papers, pp. 86–96). Berlin, Germany: Association for Computational Linguistics.
Sennrich, R., Haddow, B., & Birch, A. (2016b). Neural machine translation of rare words with subword units. In *Proceedings of the 54th Annual Meeting of the Association for Computational Linguistics* (Vol. 1: Long Papers, pp. 1715–1725). Berlin, Germany: Association for Computational Linguistics.
Shen, S., Cheng, Y., He, Z., He, W., Wu, H., Sun, M., & Liu, Y. (2016). Minimum risk training for neural machine translation. In *Proceedings of ACL*.
Smith, D. A., & Eisner, J. (2006). Minimum risk annealing for training log-linear models. In *Proceedings of ACL*.
Snover, M., Dorr, B., Schwartz, R., Micciulla, L., & Makhoul, J. (2006). A study of translation edit rate with targeted human annotation. In *Proceedings of AMTA*.
Su, J., Xiong, D., Zhang, B., Liu, Y., Yao, J., & Zhang, M. (2015). Bilingual correspondence recursive autoencoder for statistical machine translation. In *Proceedings of EMNLP* (pp. 1248–1258).
Sutskever, I., Vinyals, O., & Le, Q. V. (2014). Sequence to sequence learning with neural networks. In *Proceedings of NIPS*.
Sutton, R. S., & Barto, A. G. (1988). *Reinforcement Learning: An Introduction*. Cambridge, MA: MIT Press.
Tamura, A., Watanabe, T., & Sumita, E. (2014). Recurrent neural networks for word alignment model. In *Proceedings of ACL*.
Tang, Y., Meng, F., Lu, Z., Li, H., & Yu, P. L. H. (2016). Neural machine translation with external phrase memory. arXiv:1606.01792v1.
Tu, Z., Lu, Z., Liu, Y., Liu, X., & Li, H. (2016). Modeling coverage for neural machine translation. In *Proceedings of ACL*.
Vaswani, A., Shazeer, N., Parmar, N., Uszkoreit, J., Jones, L., Gomez, A. N. et al. (2017). Attention is all you need. arXiv preprint arXiv:1706.03762.

Vaswani, A., Zhao, Y., Fossum, V., & Chiang, D. (2013). Decoding with large-scale neural language models improves translation. In *Proceedings of EMNLP* (pp. 1387–1392).
Vogel, S., Ney, H., & Tillmann, C. (1996). HMM-based word alignment in statistical translation. In *Proceedings of COLING*.
Wang, X., Lu, Z., Tu, Z., Li, H., Xiong, D., & Zhang, M. (2017). Neural machine translation advised by statistical machine translation. In *Proceedings of AAAI 2017* (pp. 3330–3336).
Wiseman, S., & Rush, A. M. (2016). Sequence-to-sequence learning as beam-search optimization. In *Proceedings of the 2016 Conference on Empirical Methods in Natural Language Processing* (pp. 1296–1306). Austin, TX: Association for Computational Linguistics.
Wu, S., Zhang, D., Yang, N., Li, M., & Zhou, M. (2017). Sequence-to-dependency neural machine translation. In *Proceedings of ACL 2017* (Vol.e 1, pp. 698–707).
Wu, Y., Schuster, M., Chen, Z., Le, Q. V., Norouzi, M., Macherey, W. et al. (2016). Google's neural machine translation system: Bridging the gap between human and machine translation. arXiv:1609.08144v2.
Xiong, D., Liu, Q., & Lin, S. (2006). Maximum entropy based phrase reordering model for statistical machine translation. In *Proceedings of ACL-COLING* (pp. 505–512).
Yang, N., Liu, S., Li, M., Zhou, M., & Yu, N. (2013). Word alignment modeling with context dependent deep neural network. In *Proceedings of ACL*.
Zhang, B., Xiong, D., & Su, J. (2017a). BattRAE: Bidimensional attention-based recursive autoencoders for learning bilingual phrase embeddings. In *Proceedings of AAAI*.
Zhang, J., Liu, S., Li, M., Zhou, M., & Zong, C. (2014a). Bilingually-constrained phrase embeddings for machine translation. In *Proceedings of ACL* (pp. 111–121).
Zhang, J., Liu, S., Li, M., Zhou, M., & Zong, C. (2014b). Mind the gap: Machine translation by minimizing the semantic gap in embedding space. In *AAAI* (pp. 1657–1664).
Zhang, J., Liu, Y., Luan, H., Xu, J., & Sun, M. (2017b). Prior knowledge integration for neural machine translation using posterior regularization. In *Proceedings of the 55th Annual Meeting of the Association for Computational Linguistics* (Vol. 1: Long Papers, pp. 1514–1523). Vancouver, Canada: Association for Computational Linguistics.
Zhang, J., Zhang, D., & Hao, J. (2015). Local translation prediction with global sentence representation. In *Proceedings of IJCAI*.
Zhang, J., & Zong, C. (2016). Exploiting source-side monolingual data in neural machine translation. In *Proceedings of EMNLP 2016* (pp. 1535–1545).
Zheng, H., Cheng, Y., & Liu, Y. (2017). Maximum expected likelihood estimation for zero-resource neural machine translation. In *Proceedings of IJCAI*.
Zhou, L., Hu, W., Zhang, J., & Zong, C. (2017). Neural system combination for machine translation. In *Proceedings of ACL 2017*.
Zoph, B., Yuret, D., May, J., & Knight, K. (2016). Transfer learning for low-resource neural machine translation. In *EMNLP*.

Chapter 7
Deep Learning in Question Answering

Kang Liu and Yansong Feng

Abstract Question answering (QA) is a challenging task in natural language processing. Recently, with the remarkable success of deep learning on many natural language processing tasks, including semantic and syntactic analysis, machine translation, relation extraction, etc., more and more efforts have also been devoted to the task of question answering. This chapter briefly introduces the recent advances in deep learning methods on two typical and popular question answering tasks. (1) Deep learning in question answering over knowledge base (KBQA) which mainly employs deep neural networks to understand the meaning of the questions and try to translate them into structured queries, or directly translate them into distributional semantic representations compared with candidate answers in the knowledge base. (2) Deep learning in machine comprehension (MC) which manages to construct an end-to-end paradigm based on novel neural networks for directly computing the deep semantic matching among question, answers and the given passage.

7.1 Introduction

Web search is on the cusp of a profound change, from simple document retrieval to natural language question answering (QA) (Etzioni 2011). It needs to precisely understand the meaning of the users' natural language questions, extract useful facts from various information on the web, and select appropriate answers. Similar to other natural language processing (NLP) tasks, such as part-of-speech tagging, parsing, and machine translation, most traditional QA methods were based on symbolic representation. In such paradigm, all elements in questions and answers, including words, phrases, clauses, sentences, documents, etc., are usually processed through NLP basic modules and then converted into certain structured or unstructured formats,

K. Liu (✉)
Institute of Automation, Chinese Academy of Sciences, Beijing, China
e-mail: kliu@nlpr.ia.ac.cn

Y. Feng
Peking University, Beijing, China
e-mail: fengyansong@pku.edu.cn

L. Deng and Y. Liu (eds.), *Deep Learning in Natural Language Processing*, https://doi.org/10.1007/978-981-10-5209-5_7

like bag-of-words, parsing trees, logical forms, etc. Then, in the given documents or webpages, the semantic similarity or relatedness between the question and candidate answer is computed and the candidate with the highest score will be the final answer. Unfortunately, the key weakness of such paradigm is the so-called "*semantic gap*", that text spans with similar meanings may have different symbolic representations.

In neural networks, texts are usually represented into distributed vectors. Then, the exact matching between text spans could be replaced by the operations among distributed vectors. In this way, the *semantic gap* problem in traditional approaches could be alleviated to a certain extent.

This chapter briefly introduces those deep learning-based QA efforts. Moreover, there are several branches in question answering, including retrieval-based QA (IRQA), community QA (cQA), question answering over knowledge base (KBQA), and machine comprehension (MC). Here, we mainly focus on KBQA and MC, since these two QA tasks demand more semantic analysis and understanding of texts, from questions to documents. In the rest of this chapter, we will first discuss recent advances in KBQA from two perspectives, and further review the deep learning efforts targeting MC, as well as the resources involved.

7.2 Deep Learning in Question Answering over Knowledge Base

There have been many successful attempts to extend novel neural network models to improve the performance of question answering systems over knowledge bases (KBQA). Various novel neural network components or architectures and their variants, e.g., CNN, RNN (LSTM, BLSTM), attention mechanism, and memory networks, have been examined within the task. These efforts can be categorized into two main paradigms, either **the information extraction style** (IE), or **the semantic parsing style** (SP). The former usually retrieves a set of candidate answers from KB using various relation extraction techniques, which are then compared with the questions in a condensed feature space. While the semantic parsing style methods manage to distill the formal/symbolic representations or structured queries from the sentence with the help of novel components or network structures (Fig. 7.1).

From another point of view, in the context of deep learning-inspired paradigms, recent works in applying deep learning methods to facilitate knowledge-based question answering (KBQA) can also be categorized into two types, using novel neural network models to improve specific components within the traditional KBQA framework, and formalizing the task in a unified neural network architecture. The former view mainly focuses on utilizing advanced neural network models to improve existing components, e.g., feature extraction, relation identification, semantic matching or similarity computation, etc., while the latter puts their emphasis on using novel deep learning frameworks to project natural language questions and candidate answers into the same low-dimensional semantic space. Consequently, this KBQA task can

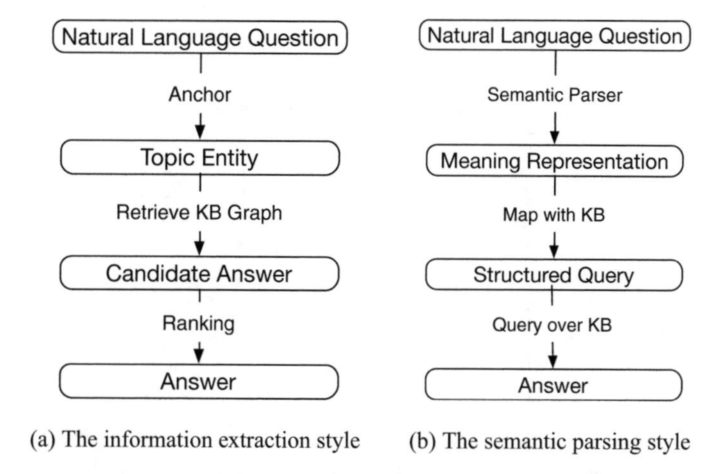

(a) The information extraction style (b) The semantic parsing style

Fig. 7.1 The illustrations for the information extraction- styled KBQA framework (**a**) and the semantic parsing-styled framework (**b**)

be converted into the problem of similarity computation between the embeddings of the questions and candidate answers in this space, often in an information extraction style.

In the following, we will review the recent deep learning-based KBQA efforts in two mainstreams, the information extraction style and the semantic parsing based. Note that there are not strict differences between them, and most advances actually benefit from both paradigms. We will try to highlight the advantages of different components or specific treatments.

7.2.1 The Information Extraction Style

The mainstream of works using deep learning methods put their emphasis on finding better ways to embed natural language questions and candidate answers from a KB in the same, condensed, semantic space. These works usually formalize the solution in a *retrieval–embedding–comparing* pipeline, within a unified neural network architecture.

7.2.1.1 Simple Vector Representation

The pioneer work out of the information extraction style approaches is contributed by Bordes et al. (2014a, b). Instead of mapping categories, entity mentions, and relation patterns to corresponding types, entities, and predicates in the KB individually, Bordes et al. (2014b) propose a more straightforward approach: they design a joint

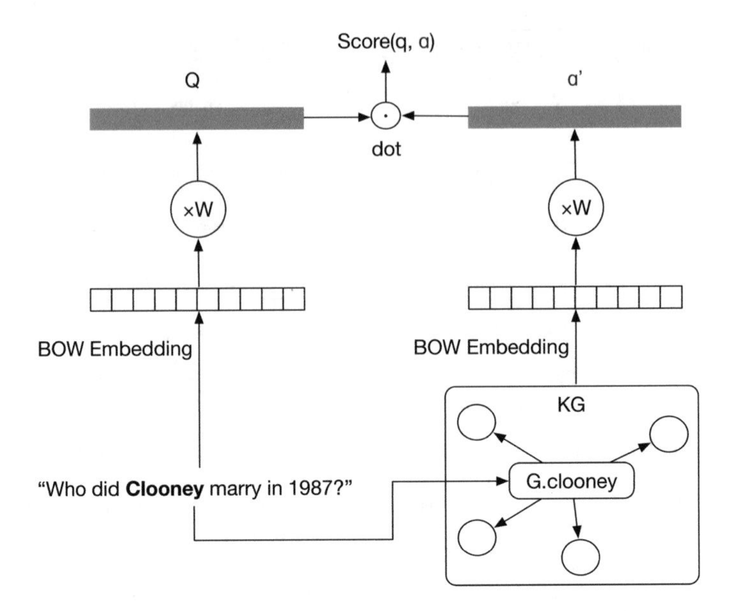

Fig. 7.2 Illustration for simple vector representation in (Bordes et al. 2014a)

embedding framework to learn vector representations of words, entities, relations, and other semantic items in a structured knowledge base, and manage to map a natural language question with a subgraph in the KB (Fig. 7.2).

When the natural language questions and the candidate subgraphs are represented by low-dimensional embeddings, one can easily compute the similarity between a question and a subgraph. The model requires annotated question–answer pairs as training data, but they are also designed to automatically collect more training instances through simple patterns and a multitask paradigm. By simultaneously optimizing over other resources or related side tasks, e.g., a paraphrase task, the model intends to ensure similar utterances with higher similarities, thus relieves the requirement of human effort.

This framework follows a simple and clear pipeline structure, i.e., *retrieval–embedding–comparing*, without relying on human-crafted features, extra syntactic analysis, or empirical rules as traditional extraction models do, and achieve competitive performances on benchmark datasets.

However, for ease of implementation, both natural language questions are first represented by bag-of-words and then go through a condensing process, which ignores the syntactic structures within questions. Similar approaches also apply to the candidate answers, where a subgraph is simply represented by a multi-hot representation of its involved entities and relations. This simplification prevents the model from utilizing more sources of clues, in either natural language utterances or the KB itself, e.g., relational phrase or answer type indicators in the questions, or entity-predicate consistencies in the KB.

Moreover, current treatments from neural network models are unable to properly deal with semantic compositionality and various constraints beyond the bag-of-words or bag-of-entities-relations representations, e.g., *Tom's father's mother's son* versus *Tom's mother's father's son*. This, to some extent, could be handled by more deep syntactic analysis for the questions, or frequent structure mining for the KB.

7.2.1.2 Embedding Features with CNN

Different from treating everything within a bag (Bordes et al. 2014a), Yih et al. (2014) propose to target single-relation questions with the help of convolutional neural networks (CNN). They use a CNN-based semantic model (CNNSM) to construct two different mapping models, one for identifying the entities in the questions and the other one for mapping relations to the KB relations. Note that the target questions are assumed to contain one entity and one relation only, which indeed take a large proportion of various KBQA benchmark datasets. And the structured queries for such questions are relatively simple, only one <*subject, predicate, object*> triple involved, thus one does not need a structure prediction process to recover the inherent query structure among multiple entities and relations.

The key idea is, similar to (Bordes et al. 2014a), that relational patterns expressed in the natural language questions and the relations/predicates in the structured KB can be projected into the same low-dimensional semantic space through CNNs. Similarly, the surface form of an entity in the KB is treated as the same as the entity mention in a question, and can be captured by a CNN. Thus, the CNNSM can provide the similarity between a natural language question and candidate triples in the KB and select the top scoring one as the final answer.

This solution benefits from convolutional neural network models, which is superior to the simple bag-of-words format, and works with letter-trigrams vectors as input to deal with the out-of-vocabulary (OOV) issue, to some extent. But, this reminds us two important issues in the KBQA task, entity linking and relation identification. Both of them are challenging enough by themselves, and require sufficient training data, i.e., mention–entity pairs and natural language pattern–KB relation pairs, to train the model. Especially, there have been a large volume of entities and relations in current large-scale knowledge bases, e.g., Freebase, making it more challenging to handle questions with multiple entities and relations.

On the other hand, Dong et al. (2015) propose to use CNNs to encode different kinds of features between a question and a candidate answer. They propose a multicolumn convolutional neural networks (MCCNNs) model to capture different aspects for a question, and further score a pair of question and answer through three channels, **answer path**, **answer context**, and **answer type**.

Comparing with simple vector representation (Bordes et al. 2014a), MCCNNs use CNNs to extract different features, which can explicitly capture the path between the topic entity in the question and a candidate answer on the KB, and also the expected answer types. These two are shown to be more important in evaluating a candidate answer. The framework is also easy to extend more kinds of features by adding required columns to the networks.

Again, entity linking is still an open question for these feature-based models. The encoding of answer path helps MCCNNs to be able to perform shallow inference along the path, to some extent. However, due to the nature of typical *retrieval–embedding –comparing* framework, MCCNNs are still unable to find better solutions to deal with comparisons among candidate answers, e.g., *the highest mountain*, or *his first son*.

7.2.1.3 Embedding Features with Attention

Moreover, Hao et al. (2017) employ a bidirectional RNN model to capture the semantics of a given question. They believe a question should be represented differently according to the different focuses of various answer aspects (An answer aspect could be the answer entity itself, the answer type, the answer context, etc.). Take the question "Who is the president of France?" and one of its candidate answers "Francois Hollande" as an example. When dealing with the answer entity `Francois Holland`, "president" and "France" in the question is more focused, and the question representation should bias toward the two words. While facing the answer type `/business/board_member`, "Who" should be the most prominent word. Meanwhile, some questions may value answer type more than other answer aspects. While in some other questions, answer relation may be the most important information we should consider, which is dynamic and flexible corresponding to different questions and answers. Obviously, this requires an attention mechanism, which reveals the mutual influences between the representation of questions and the corresponding answer aspects.

Instead of representing questions using three CNNs with different parameters (Dong et al. 2015) when dealing with different answer aspects including answer path, answer context, and answer type, Hao et al. (2017) proposed a cross-attention-based neural network to perform KBQA.

The cross-attention model, which stands for the mutual attention between the question and the answer aspects, contains two parts: the answer-towards-question attention part and the question-towards-answer attention part. The former could help learn flexible and adequate question representation, and the latter helps adjust the question–answer weight. Finally, the similarity scores between the question and each corresponding candidate answer on different aspects are calculated, and the final score for each candidate is combined by all similarity scores according to the corresponding question–answer weights. Then the candidates with the highest score will be selected as the final answers.

7.2.1.4 Question Answering with Memory

Memory network is a novel learning framework that is designed around a memory mechanism that can be read and modified/appended during a specific task (Weston et al. 2015b). There have been a few attempts to investigate the task of knowledge-

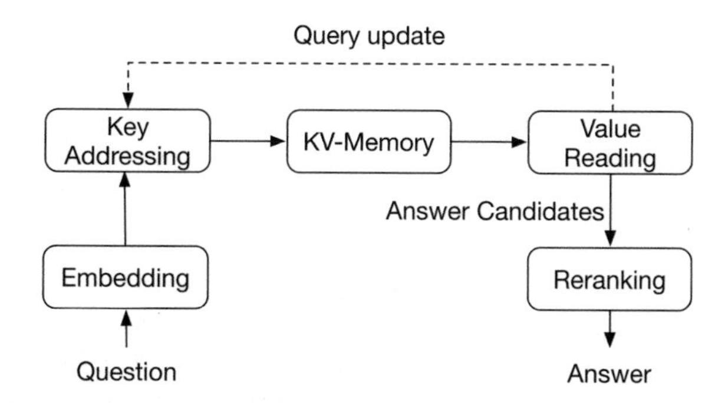

Fig. 7.3 Illustration for the key-value memory networks in (Miller et al. 2016)

based question answering under the memory network paradigm, mostly following the information extraction style's *retrieval–comparing* routine.

The first attempt (Bordes et al. 2015) focuses on simple questions, which can be answered with one <subject, relation, object> triple. In the input component, structured knowledge bases, in a *bag-of-symbols* form, are read and stored in the memory, and questions are processed into a *bag-of-ngrams* form. The output component will then compare the *bag-of-ngrams* question with the entries in the memory to find candidate triples, which are in turn evaluated with the input question. The object of the top scoring triple will be provided as the answer by the response component. This should be considered as a straightforward application of memory networks in the KBQA task, but actually shows the potential of memory networks in managing large scale of KB entries, even from multiple resources (Fig. 7.3).

Miller et al. (2016) further extend the idea by investigating various forms of **Key-Value** knowledge in the memory. The improved model also allows multiple *address-and-read* from the memory to collect evidence/context to dynamically update the *question* for the final answers. An advantage of the Key-Value design is to make the memory mechanism more flexible to store various knowledge, from KB triples ($subject + relation$ as the key, and obj as the value), to documents (sentences or a window of words as the key or value), which potentially supports to answer more complicated questions with heterogeneous resources.

7.2.2 *The Semantic Parsing Style*

The *retrieval–embedding–comparing* framework benefits from various neural network components to capture the question–answer similarity and works better in simple questions, where the entities and relations are within a simple subgraph in the KB. But, they are not good at solving complex composition of semantics, since there is no explicit mechanism for information extraction- styled approaches to capture such composition when understanding a question. By contrast, other mainstream of works in KBQA, the **semantic parsing styled** models, try to formally represent the

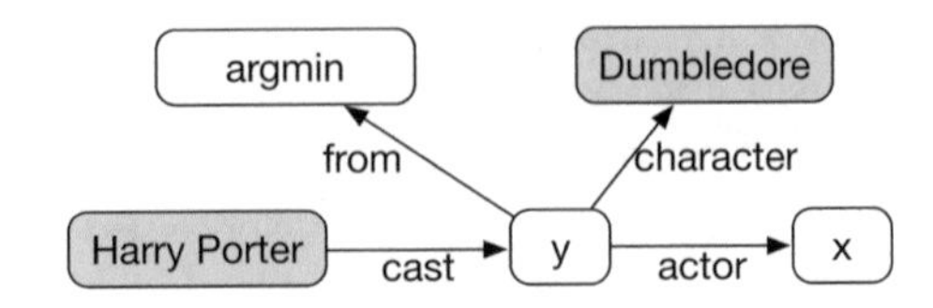

Fig. 7.4 Illustration for the staged query graph generation model (STAGG) in (Yih et al. 2015)

meaning of a question, and then instantiate with a KB to construct a structured query over the KB, which makes it possible to explicitly capture complex queries.

The core components of such models are, therefore, recovering the formal meaning representations, e.g., logical forms or structured query representations, from the natural language questions, and further find answers from a KB by mapping the representations with KB components and querying over the KB. The efforts of deep learning methods are mostly designed to improve certain components of the framework.

The CNNSM model (Yih et al. 2014) discussed in Sect. 7.2.1.2 can also be considered as a semantic parsing- styled method, which can only produce one *<subject, predicate, object>* triple as the query, where the CNNs are used to perform entity linking and relation identification. But it does not work for slightly complex questions, e.g., involving multiple entities and relations, let alone constraints. The main reason is that the neural network components are only responsible for mapping with KB components, either entities or relations, but there is no explicit mechanism responsible for identifying the inherent structure among multiple entities or relations. In fact, such structures have been intensively investigated in traditional semantic parsing-based models via either PCCG, PCFG, dependency structures, or other syntactic/semantic parsing paradigms (Cai and Yates 2013; Kwiatkowski et al. 2013; Berant and Liang 2014; Reddy et al. 2014; Kun et al. 2014).

7.2.2.1 STAGG: Semantic Parsing While Searching and Pruning

Beyond single-relation questions, Yih et al. (2015) propose to use a **query graph** to represent the meaning of a question, which contains four kinds of nodes: *grounded entities, existential variables, lambda variables* and *constraints/functions*. Here, lambda variables are ungrounded entities and expected to be the final answers. The existential variables could refer to middle nodes, e.g., the **father** in the utterance *Tom's* ***father****'s mother*, or abstract nodes, e.g., a compound value-typed (CVT) node[1] in Freebase. And the constraints or functions are designed to filter a set of entities according to certain numerical properties, e.g., arg min. In the query graph, nodes are connected by directed edges, indicating the relationship between two nodes, which is expected to be mapped with KB predicates (Fig. 7.4).

[1] A CVT node is usually not a real-world entity, but often refers to an event, e.g., a marriage event, or a presidency event, which can represent an entry of data with multiple fields.

Then, the task becomes how to convert a natural language question into such a query graph. Yih et al. (2015) propose a staged query graph generation model (STAGG) to leverage KB from beginning to incrementally prune the search space, and construct the structured queries.

The key components of STAGG include linking topical entity, identifying the core inferential chain, and finally augmenting with constraints and functions, which is basically a parsing-and-ranking process with step-by-step searching. Here, the core inferential chain captures the relationship between the topical entity and the lambda variable, and provides the backbone for a query. Yih et al. (2015) use a deep convolutional neural network model to semantically match a question and a sequence of predicates (up to length of 2 with a CVT node in the middle).

Despite its success on the benchmark dataset, we can learn several lessons from the design of STAGG.

Topic Entities: Finding topic entities and linking to the KB is the very first but a crucial step. STAGG uses S-MART (Yang and Chang 2015), a statistical model for entity linking in short text, which plays an important role to the upcoming steps and then the overall performance. When changing to Freebase API for topical entity linking, STAGG will have an absolute 4.1% drop in the overall F1 score.

Identifying the Core Inferential Chain: Basically, this is a relation extraction step to capture how one can get to the lambda variable starting from the topical entity on the graph of the KB, which is captured through CNNs, similar to the CNNSM (Yih et al. 2014). Given the huge space of all candidate relations, STAGG only considers those related to the topical entities and captures how a question is semantically matched with a sequence of KB relation around the topical entity. This process of identifying core inferential chain thus becomes a match-and-rank step, while avoiding a large-size multi-class classification style.

Augmenting Constraints and Aggregations: STAGG considers other entities or time expressions in the questions as constraint nodes to the core inferential chain, and also introduces certain functions to further filter the answers, e.g., converting *first*, *smallest* into arg min. And it is promising to see a KBQA system formally introducing aggregation functions as part of the formal representations, though through a set of rules.

Understanding Superlative Expressions: As also discussed in (Berant and Liang 2014; Zhang et al. 2015), superlative utterances are common to see in questions. Most KBQA works adopt templates or rules to analyze superlative expressions, by simply choosing from arg min or arg max (Berant and Liang 2014; Yih et al. 2015). However, formally analyzing a superlative expression into a structured comparative construction against a KB will help a KBQA system to better handle not only superlative utterances, but also those with ordinal constraints. Zhang et al. (2015) design a neural network model to learn the underlying correspondence between a superlative utterance and KB relations, which serve as the comparison dimension within the comparative construction. For example, from *the longest river* into a tuple <`river.length`, *descending*, **1**>, we expect that all rivers are compared upon a KB predicate `river.length`, sorted in descending and the top ranked is the target.

7.2.2.2 Improving Relation Identification

As being discussed in many previous semantic parsing-styled works (Kwiatkowski et al. 2013; Berant et al. 2013; Berant and Liang 2014), identifying the KB relations/predicates from a question is the key to success, where traditional feature-based models are hard to capture the mismatch between sentences and KB relations and also the variance among natural language utterances. There have been many attempts in applying deep learning methods for relation extraction using either CNN or RNN models to explore lexical or syntactic features (Zeng et al. 2014; Liu et al. 2015; Xu et al. 2015).

The relation extraction component in KBQA is designed to deal with KB-based relations within a short context, where there could be up to thousands of candidates. One possible solution is to perform a semantic match between a natural language utterance and a KB relation through CNN (Yih et al. 2014, 2015), avoiding direct classification over hundreds of relations (Fig. 7.5).

Xu et al. (2016) propose a multichannel convolutional neural networks (MCCNNs) model to learn compact and robust relation representations from both lexical

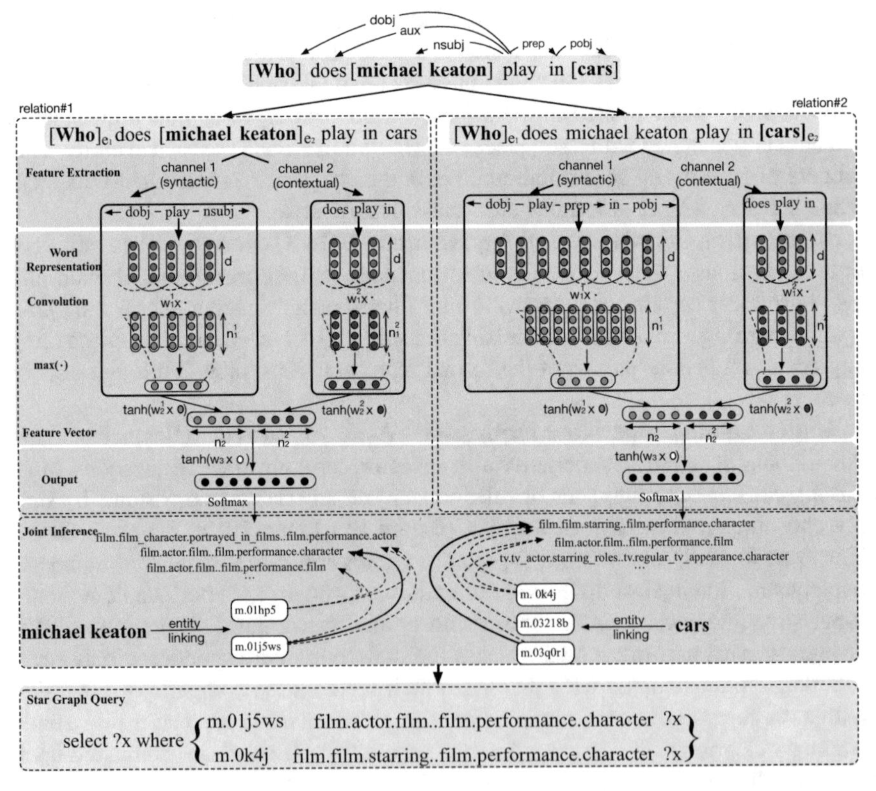

Fig. 7.5 Illustration for the multichannel convolutional neural networks model in (Xu et al. 2016)

and syntactic views. These approaches better suit the open-domain KBQA scenario for good a reason. There are often thousands of relations in an open-domain KB, and traditional feature-based models inevitably suffer from the data sparsity issue and their poor generalization ability on unseen words.

7.2.2.3 Neural Symbolic Machines

There is another interesting line of semantic parsing-styled work trying to combine advantages of neural networks and symbolic reasoning to improve question answering. Liang et al. (2017) introduce a neural symbolic machine (NSM) that is equipped with a neural network component responsible for mapping from natural language representations to executable code, and a symbolic component to execute the code to prune the search space or find the answers.

Specifically, the neural network component, called *programmer*, is basically a sequence-to-sequence model which maintains a key-variable memory to deal with intermediate results when generating a sequence of program. However, the mixture design of neural network component and symbolic interpreter will make the whole framework hard to train, which is then cast and solved as a reinforcement learning problem.

7.2.3 The Information Extraction Style Versus the Semantic Parsing Style

Given the discussion above, it is easy to find that we do not need to draw a clear distinction between *the information extraction style* and *the semantic parsing style*. The two streams indeed have their own advantages. The IE style efforts take more advantages of novel neural network models and architectures to better represent questions and candidate answers in a condensed semantic space, and are easier to incorporate various feature representations in the model structures. On the other hand, deep learning models provide SP models with more accurate relation or constraint identification/mapping, and enable or support more accurate/complex meaning representations and derivations.

In fact, many previous works can be considered to be with both sides, especially those targeting simple questions or benefiting from both the styles. For example, the STAGG follows the traditional semantic parsing style to construct the structured query from a question, but its staged rank-and-prune helps prune the search space, leading to better query construction and overall performance. We believe that several novel paradigms with merits from both IE and SP styles, such as memory network models and neural symbolic frameworks, are flexible enough to be adapted to more complicated questions.

7.2.4 Datasets

There have been several datasets available for evaluating knowledge-based question answering systems.

WebQuestions: The widely used `WebQuestions` dataset is originally constructed by Berant et al. (2013), containing 5,810 question–answer pairs, which are crawled via Google Suggest service, and annotated with Freebase answers through Amazon Mechanical Turk. There is a publicly available training/testing split as well as evaluation script for comparisons.[2]

Yih et al. (2016) further augment the `WebQuestions` dataset with semantic parse tags, `WebQuestionsSP`, where, despite of original answers, each *answerable* question is annotated using SPARQL queries with Freebase entity identifiers, 4,737 questions in total.[3]

Free917: The `Free917` dataset is constructed by Cai and Yates (2013), containing 917 questions, each annotated with a logical form, where the entities and relations are grounded to Freebase. Kun et al. (2014) further annotate each question with its ungrounded semantic parse, where the entity phrases, relational phrases, categories, and variables as well as the dependency structure among them are explicitly labeled.

SimpleQuestions: `SimpleQuestions` is constructed by Bordes et al. (2015), containing 108,442 questions, each of which is manually annotated with a <subject, relation, object> triple from Freebase. The questions in `SimpleQuestions` are relatively simple, and one can answer such a question by retrieving and utilizing only one triple from the KB, e.g., *What do Jamaican people speak ?* paired with a KB query (*jamaica*, *language_spoken*, ?).

WikiMovies: `WikiMovies` is contributed by Miller et al. (2016), containing around 100k questions in the movie domain. This dataset is designed to be answered by either Wikipedia documents (containing the movies' Wikipedia pages), human cured structured KB (carefully created from the Open Movie Database[4] and MovieLens[5]), or KB triples automatically obtained with OpenIE tools. Each question is guaranteed to be equally answered with Wikipedia documents or the cured KB.

QALD: Question answering over linked data (QALD) challenge is a series of open evaluations on question answering over linked data since 2011.[6] The theme of QALD evaluations is to properly represent users' natural language questions into standard, executable queries, e.g., SPARQL queries, which can be executed over large-scale knowledge bases, e.g., DBpedia.

There have been several classical KBQA tasks with several hundreds of question-answer pairs, including the *multilingual question answering* task which contains pairs of natural language questions (in multiple languages, e.g., English, French, German, etc.) and DBpedia answers or corresponding SPARQL queries which can

---

[2]More details can be found in https://nlp.stanford.edu/software/sempre/.

[3]Obtained through https://www.microsoft.com/en-us/download/details.aspx?id=52763.

[4]Obtained through http://beforethecode.com/projects/omdb/download.aspx.

[5]Obtained through http://grouplens.org/datasets/movielens/.

[6]http://qald.sebastianwalter.org/.

Table 7.1 Example questions from the currently popular KBQA datasets

WebQuestions
Which country in Europe has the largest land area
Who did Shaq first play for
What is the largest city in the county in which Faulkner spent most of his life
Free917
Who was nominated for the academy award for best director in 2011
How many countries use euros
What was the strongest storm in the 1992 Atlantic Hurricane season
SimpleQuestions
What American cartoonist is the creator of Andy Lippincott?
Which forest is fires creek in?
what is an active ingredient in children's earache relief?
WikiMovies
What movies did Harrison Ford star in?
Can you describe movie blade runner in a few words?
Which films can be described by dystopian?
QALD
What countries do Queen Elizabeth II reign
What is the best sandals resort in St. Lucia
What currency do they use in Switzerland

be executed over DBpedia. Another related task is the *hybrid question answering* task, where each natural language question should be answered with both a structured knowledge base, DBpedia, and free text, e.g., DBpedia abstracts (Table 7.1).

7.2.5 Challenges

Along with the development of knowledge-based question answering systems, there have been several issues that are mostly concerned or discussed, especially in the context of employing deep learning models.

7.2.5.1 Compositionality

Traditional semantic parsing-based KBQA works usually rely on combinatory categorial grammar (CCG) (Steedman 2000) or probabilistic CCG to derive its meaning representation from a question (Cai and Yates 2013; Kwiatkowski et al. 2013), which is relatively hard to explicitly capture in a unified model without considerations for such syntactic structures, e.g., information extraction-styled methods. Therefore, many existing works rely on manually defined rules or templates to handle compositionality (Yih et al. 2015). However, the neural symbolic framework may provide a new direction to augment neural network models with the ability for shallow symbolic reasoning (Liang et al. 2017).

7.2.5.2 Gap Between Natural Language and Knowledge Base

We have discussed that entity linking and relation extraction are two main hurdles when we try to retrieve candidate answers or match natural language utterances with knowledge base items. The main reason is the mismatch between natural language and knowledge bases, including limited or omission of context in the language side, sub-lexical compositionality, or even the defect in the KB design. Various neural network models have been proposed to improve the relation matching or extraction, but far less attention is given to the entity linking task, which is the fundamental step in a KBQA system.

7.2.5.3 Training Data

Training data has been a long-standing problem in various machine learning-based methods, especially for neural network models which are assumed to require more training data than traditional methods. And in the question answering scenario, it is very expensive to collect question–answer pairs, let alone any fine annotations, e.g., logical forms, structured queries, or even entity and relation annotations. Possible solutions include using question–answer pairs as indirect supervision to collect pseudo labels (Yih et al. 2015; Xu et al. 2016), or automatically collecting training data with noisy labels or templates (Miller et al. 2016; Bordes et al. 2014a).

Question answering over knowledge bases is a challenging task, which requires many NLP or IR techniques, such as lexical analysis, syntactic analysis, information extraction, entity linking, reasoning, and so on. Recent advances in deep learning provide helpful tools or novel frameworks to improve knowledge-based question answering, which are admitted on the earlier stage. We believe that deep fusion of neural network modeling and question answering will bring more opportunities in this field.

7.3 Deep Learning in Machine Comprehension

7.3.1 Task Description

Machine comprehension (MC) is a recently proposed application that has gained significant popularity over the past few years within the natural language processing and artificial intelligence communities. MC tests the ability of the machine to read the text, process it, and understand its meaning. MC follows the traditional QA setup, but still remains some differences as follows:

- In traditional QA, for a given question, the answer may come from various resources, such as knowledge base (KBQA), web searching results, and even some question answering platforms (also known as community QA). However, in MC, the context knowledge is restricted to a single given document.
- Compared to traditional QA, especially for IRQA and KBQA, MC mainly focused on those questions which could not be answered directly and need to be reasoned according to multiple entities or events in the given document. In the way, the reasoning ability is more required in MC.
- Compared to traditional QA, the answer types in MC are more diverse and vary from a single word to multiple sentences. In addition, the question forms of the MC are also diverse, such as multiple-choice questions (the answer candidates are provided previously) and cloze-style questions (candidates are not provided and the answer should be generated from the system).

7.3.1.1 Datasets

MCTest: The task of machine comprehension started from NLP community. In 2013, Microsoft researchers proposed *MCTest* (Richardson et al. 2013) dataset to evaluate the comprehension ability of a machine. In MCtest, each document (stories) is associated with four questions. For each question, four candidate answers are provided and the system is required to select the correct one. An example of the MCTest is shown in Fig. 7.6.

Obviously, MCTest is a standard reading comprehension dataset in which the stories are fictitious and some questions could be answered from several sentences (labeled as *multiple*). The author divides the dataset into two subsets, including MC160 and MC500, which contains 160 and 500 stories, respectively. However, the size of this dataset is too small which sometimes only serves as a *test* setup. Many recent researchers usually resort to external linguistic tools to extract features and then make the inference based on them. Started from MCtest, several MC datasets have been released. Here, we mainly introduce four standard resources as follows.

bAbi: bAbi (Weston et al. 2015a) is a MC dataset that is AI-complete according to the author's description. In total, bAbi contains 20 subtasks, where each subtask requires different answer skills. Some subtask examples are shown in Fig. 7.7.

As this dataset has been divided into different categories, the performance on different subtasks could expose the advantages or disadvantages of one model on different question types. Moreover, the whole dataset is auto-synthesized and auto-generated with several human-designed rules. Although the rules are supposed to be unlimited, in fact, the generation rules are merely based on no more than 100 words. As a result, some questions or documents in this dataset are duplicated. Furthermore, as bAbi is auto-synthesized by rules, the exploited algorithms or systems are more likely to approximate the used generation rules.

Alyssa got to the beach after a long trip. She's from Charlotte. She traveled from Atlanta. She' s now in Miami. She went to Miami to visit some friends

...

The girls went to a restaurant for dinner. The restaurant had a special on catfish. Alyssa enjoyed the restaurant's special. Ellen ordered a salad. Kristin had soup. Rachel had a steak.

1: one: Why did Alyssa go to Miami?
A) swim
B) travel
*C) visit friends
D) laying out

2: multiple: What did Alyssa eat at the restaurant?
A) steak
B) soup
C) salad
*D) catfish

Fig. 7.6 An example of MCTest stories and questions

John is in the playground.
Bob is in the office.
Where is John? *playground*

single supporting fact

John is in the playground.
Daniel picks up the milk.
Is John in the classroom? *no*
Does Daniel have the milk? *yes*

Yes or No questions

John is in the playground.
Bob is in the office.
John picked up the football.
Bob went to the kitchen.
Where is the football? *playground*
Where was Bob before the kitchen? *office*

two supporting facts

The office is north of the bedroom.
The bedroom is north of the bathroom.
What is north of the bedroom? *office*
What is the bedroom north of? *bathroom*

Subject vs. Object

Fig. 7.7 Examples of bAbi questions

SQuAD: SQuAD (Rajpurkar et al. 2016) denotes Stanford question answering dataset, which is a recently released human-created large machine comprehension dataset. This dataset contains nearly 100,000 document–question pairs. The documents are derived from Wikipedia pages, then crowdsourcing annotators are asked to propose some questions based on these documents and label the corresponding answer in the document. Note that, in SQuAD, no candidate answer is provided. And the system could '*generate*' the answer by predicting the start and end position of the answer in the document. An example of the question is shown in Fig. 7.8.

In meteorology, precipitation is any product of the condensation of atmospheric water vapor that falls under gravity. The main forms of precipitation include drizzle, rain, sleet, snow, graupel and hail... Precipitation forms as smaller droplets coalesce via collision with other rain drops or ice crystals within a cloud. Short, intense periods of rain in scattered locations are called "showers".

Where do water droplets collide with ice crystals to form precipitation?

Fig. 7.8 An example of SQuAD questions

CONTEXT:
(@entity4) if you feel a ripple in the force today , it may be the news that the official @entity6 is getting its first gay character . according to the sci-fi website @entity9 , the upcoming novel " @entity11 " will feature a capable but flawed @entity13 official named @entity14 who " also happens to be a lesbian . " comics and books approved by @entity6 franchise owner @entity22 -- according to @entity24 , editor of " @entity6 " books at @entity28 imprint @entity26 .

QUESTION:
characters in " @placeholder " movies have gradually become more diverse
ANSWER:
@entity6

Fig. 7.9 An example of CNN questions

Moreover, recently, several MC datasets with similar scales and similar form to SQuAD have also been released, such as NewsQA[7] and Marco.[8]

Cloze-Style Machine Comprehension Dataset: Besides the aforementioned QA forms in MC, the Cloze-style queries (Taylor 1953) is one of the fundamental forms. Such type shares most of the characteristics of reading comprehension, but the answer is a single word in the document. Recently many datasets have been proposed for this type, such as CNN/Daily Mail (Hermann et al. 2015) and CBT (Hill et al. 2015). In CNN/Daily Mail, the authors proposed a semiautomatic method to generate the cloze from two news corpora. Each news story is accompanied with a headline or summary. The authors remove one specific noun in the headline, and the system is required to fill this placeholder based on the given document. In order to avoid the impact of language modeling or real-world knowledge beyond text comprehension, the authors anonymized all entities in the documents and queries. In CBT, each document contains 20 consecutive sentences in book story. One word in the twenty-first sentence is removed. To avoid the usage of the methods based on language modeling in reading comprehension, the answers are restricted to proper nouns. An example of CNN/Daily Mail is shown in Fig. 7.9.

---

[7]https://datasets.maluuba.com/NewsQA.

[8]http://www.msmarco.org.

Skills	Descriptions or examples
List/Enumeration	Tracking, retaining, and list/enumeration of entities or states
Mathematical operations	Four arithmetic operations and geometric comprehension
Coreference resolution	Detection and resolution of coreference
Logical reasoning	Induction, deduction, conditional statement, and quantifier
Analogy	Trope in figures of speech, e.g., metaphor
Spatiotemporal relations	Spatial and/or temporal relations of events
Causal relations	Relations of events expressed by why, because, the reason for, and so on
Commonsense reasoning	Taxonomic knowledge, qualitative knowledge, action and event changes
Schematic/Rhetorical clause relations	Coordination or subordination of clauses in a sentence
Special sentence structure	Scheme in figures of speech, constructions, and punctuation marks in a sentence

Fig. 7.10 Machine comprehension skills

7.3.1.2 Knowledge Requirement to Achieve Machine Comprehension

Machine comprehension is a comprehensive inference task that requires the deep understanding of natural languages. In psychology, comprehension comes from the interaction between the words how they trigger knowledge inside/outside the given passages/documents. And it is a creative, multifaceted process dependent upon four language skills: phonology, syntax, semantics, and pragmatics. For machine comprehension problem, achieving genuine comprehension even requires the abilities to understand the relations among multiple clauses. For example, the skill of understanding temporal relations between events implicitly requires the recognition of expressions such as conjunctions (when, as, since, …), time indexicals (morning, evening, …), tense, and aspects (went, is going, will go, …). Moreover, other inference skills are needed. For example, mathematical operations are required to answer the question related to arithmetic problem such as '*Tom has four pencils and he gave his desk-mate 2 of them, how much pencils did he have at hand ?*.' Systems required to answer this type of questions should inference the equation '*4 – 2 = 2*'. Sugawara et al. (2017) proposed 10 roughly skills that are required for MC which are listed in Fig. 7.10.

General speaking, machine comprehension involves dealing with many linguistic patterns, such as lexical, syntactical, or high-level discourse, paraphrase. To model these features, according to the methodology perspective, the current methods could be divided into two parts: feature engineering-based methods and deep learning- based methods. We briefly introduce them as follows.

7.3.2 Feature Engineering-Based Methods in Machine Comprehension

The existing feature engineering-based methods often model the text comprehension task as a task of computing semantical similarity between the given question and the document or passage. These methods try to model the semantics of sentence and document through several shallow linguistic features, including POS tag-based features, dependency parsing features, coreference, reference etc. Based on different features, different types of semantics are captured, such as lexical level semantic, discourse-level semantics, and so on.

7.3.2.1 Lexical Matching

The lexical matching method is a simple yet effective approach for machine comprehension task. This kind of approach usually adopts a sliding window-based algorithm that could rank the answer candidates by forming a bag-of-words vector for each answer paired with the question text. Then each candidate is scored according to their overlap with the story text, and the candidate with the highest score will be figured out. More concretely, this algorithm passes a sliding window over the whole story texts, and the size of such window is equal to the number of words in the question–answer pair. The highest overlap score between a story text window and the question–answer pair is taken as the corresponding score for the answer. The algorithm detail is illustrated in Algorithm 2.

Algorithm 2 Sliding window

Require: Passage P, set of passage words PW, i^{th} word in passage P_i, set of words in question Q, set of words in hypothesized answer $A_{1..4}$, and a set of stop words U.
Define: $C(W) = \sum_i \mathbb{I}(P_i = w)$
Define: $IC(W) = log(1 + \frac{1}{c(w)})$

1: **for** $i = 1$ to 4 **do**
2: $\quad S = A_i \cup Q$
3: $\quad sw_i = \max_{j=1..|s|} \sum_{w=1..|S|} \begin{cases} IC(P_{j+w}) & if\, P_{j+w} \in S \\ 0 & otherwise \end{cases}$
4: **end forReturn:** $sw_{1..4}$

However, the used text window in above algorithm is fixed. Smith et al. (2015) score each answer by making multiple passes and summing the obtained scores. In specific, they start from window size 2 and increase it to 30 tokens. Then they combine these scores with the overall number of matches for question–answer pair across the story. As they declared, this solution could enable the system to capture the long-distance relations in the story. The compared results of the original and enhanced sliding window lexical matching methods on MCTest is shown in Table 7.2.

Table 7.2 The performance of lexical matching method on MCtest

	Sliding window (%)	Enhanced sliding window (%)
MC160	69.43	72.65
MC500	63.01	63.57

7.3.2.2 Discourse Relations

Moreover, the required and relevant information for answering a question may distribute across multiple sentences. Understanding the semantic relation(s) among these sentences is important to find the correct answer. Take the example in Fig. 7.11 for instance. To answer the question about 'why Sally put on her shoes', we need to infer that 'She put on her shoes' and 'She went outside to walk' are connected by a *causality* relation.

Some prior works have demonstrated the values of discourse relations in related applications such as question answering (Jansen et al. 2014). Narasimhan and Barzilay (2015) proposed three models to incorporate the discourse relation into the MC system.

Denote the sentence in a document as z, and questions as q, answer as a.

Model 1:

$$P(a, z|q_j) = P(z|q_j)P(a|z, q_j) \tag{7.1}$$

Equation 7.1 defines a joint probability as a product of two distributions. The first is the conditional distribution of sentences in the paragraph given in the question. This is to help identify those sentences which are required to answer the question. The second component models the conditional probability of selecting an answer by given a question q and a sentence z. We can use exponential family for these component probabilities, that is : $P(z|q) \propto \exp^{\theta_1\phi_1(q,z)}$ and $P(z|a, q) \propto \exp^{\theta_2\phi_2(q,a,z)}$, where ϕ is the feature vector and θ represent the associate weight. Sums over all sentences z_n in document and we can get the probability for a specific answer a_j:

Sally liked going outside. She put on her shoes. She went outside to walk. [...] Missy the cat meowed to Sally. Sally waved to Missy the cat [...] Sally hears her name. "Sally, Sally, come home", Sally's mom calls out. Sally runs home to her Mom. Sally liked going outside.

Why did Sally put on her shoes?

A) To wave to Missy the cat
B) To hear her name
C) Because she wanted to go outside
D) To come home

Fig. 7.11 An example of question that requires multiple sentence inference

$$P(a_j|q_j) = \sum_n P(a_j, z_n|q_j). \tag{7.2}$$

In this way, a likelihood objective function could be written as

$$L_1(\theta) = log \sum_j \sum_n P(a_j, z_n|q_j). \tag{7.3}$$

Model 2: The above model could only consider one support sentence (i.e., z) once. Naturally, we could extend it for the multi-sentence case where we make use of more than a single relevant sentence for the given question. In this scenario, a joint model is defined as follows.

$$P(a, z_1, z_2|q) = P(z_1, |q)P(z_2|z_1, q)P(a|z_1, z_2, q). \tag{7.4}$$

Given a question q, we first predict the first sentence support sentence z_1 that relate to q with probability $P(z_1, |q)$, then given q and z_1, the second support sentence z_2 is inferenced. Finally, the answer a is predicted.

Model 3: This model tries to directly specify the discourse relation among questions and then employs this relation to inference other related sentences in a document. In specific, Model 3 adds a hidden variable $r \in \mathscr{R}$ to represent the relation type. It incorporates features that tie in the question type with the relation type. It also employes the type of relation to compute the lexical and syntactic similarities between sentences. Relation set $\mathscr{R}$ consists of the following relations: (1) *Causal*: Causes of events or reasons for facts. (2) *Temporal*: Time-ordering of events (3) *Explanation*: Predominantly dealing with how-type questions. (4) *Other*: A relation other than the above three relations (including non-relation).

Now the joint probability from Eq. 7.4 is modified through adding relation type r:

$$P(a, r, z_1, z_2|q) = P(z_1|q)P(r|q)P(z_2|z_1, r, q)P(a|z_1, z_2, r, q). \tag{7.5}$$

The extra component $P(r|q)$ is the conditional probability of the relation type r depending on the question. Thus, this model could learn, for instance, that *why*-questions correspond to the causal relation.

The results of three models are shown in Table 7.3.

Table 7.3 Accuracy on MCtest of three models. *Single* refers to the questions that only need one support sentence to answer, and *Multi* refers to questions requiring multiple sentences to answer

	MC160			MC500		
	Single (%)	Multi (%)	All (%)	Single (%)	Multi (%)	All (%)
Model 1	78.45	60.57	68.47	70.58	57.77	63.58
Model 2	74.68	60.07	66.52	66.17	59.9	62.75
Model 3	72.79	60.07	65.69	68.38	59.9	63.75

7.3.2.3 Answer-Entailing Structures

Some previous works of NLP have benefited from learning the latent structure between two text snippets. For example, in recognizing textual entailment (RTE), the hypothesis could be inferred from its premise by some latent alignment between them. In MC, we could also incorporate this entailing structures information into account. For instance, in the example of Fig. 7.6, to answer the second question *What did c ?*, we can use some syntactic rules to transform the question and a candidate answer into statements. For example, one of the candidates' answer is *catfish*, we combine it with the query to form a statement: *Alyssa eat catfish at the restaurant.* Deem this statement as a hypothesis, and the document as the premise, we can infer the probability of this entailment. The structure is illustrated in Fig. 7.12.

The answer-entailing structures considered here could align multiple sentences in the text to the hypothesis. The sentences in the text considered for alignment are not restricted to occur contiguously in the text. To allow such a discontinuous alignment, Sachan et al. (2015) make use of the document structure. In particular, they take help from rhetorical structure theory which could capture the event or entity coreference links across sentences. They specifically trained a max-margin fashion using a latent structural SVM (LSSVM) where the answer-entailing structures are latent. The experiment results of this answer-entailing model on MC500 is shown in Table 7.4.

7.3.2.4 Challenges in Feature Engineering-Based Methods

Feature engineering-based methods are efficient and explicit ways to dealing with machine comprehension problem, they usually utilize several linguistic features to model the semantical relations between the given document and question. Then, it makes an inference based on these features. The process is clear and easy to follow

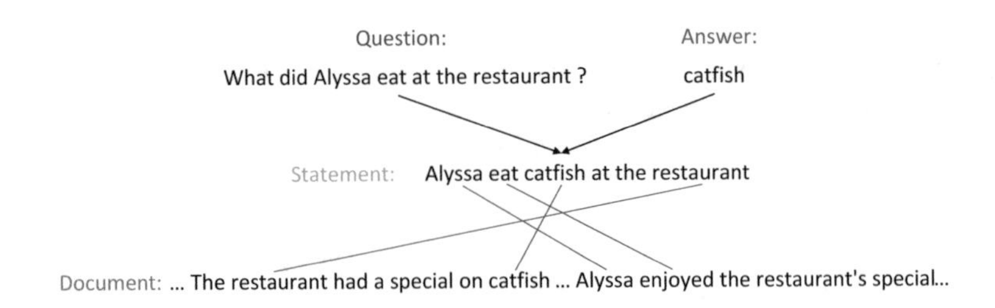

Fig. 7.12 An answer-entailing structure example from MCTest

Table 7.4 Answer-entailing model accucary on MC500

	Single	Multiple	All
Accuracy (%)	67.65	67.99	67.83

what is going wrong in the method. However, these linguistic features sometimes need to be extracted by empirical or heuristic experiences. And they may not cover much deeper semantic information. Moreover, they heavily rely on standalone linguistic tools, such as part-of-speech tagging, parser etc., which may introduce noises to the system. Therefore, the feature engineering methods usually focus on MC data with such as MCTest. And for some large-scale MC datasets such as SQuAD and bAbi, it is difficult for existing feature engineering-based methods to design and extract effective features from texts. Recently, as deep learning gained great successes on computer vision and speech recognition, more and more researchers began to focus on deep learning-based techniques for MC task.

7.3.3 Deep Learning Methods in Machine Comprehension

In this section, for MC task, we would like to introduce several prevalent deep learning-based methods on different datasets. Formally, given a document d and a question q, the probability of selecting the answer a could be modeled as follows:

$$P(a|d,q) \propto \exp(W(a)g(d,q)), \tag{7.6}$$

where $W(a)$ means the embedding of the answer candidate a and $g(d,q)$ denotes the embedding of document d under the given question q. The critical part is to compute the function of $g(d,q)$, where several deep neural network could be applied, such as RNN, LSTM, and Memory Network (Weston et al. 2015b).

7.3.3.1 LSTM-Based Encoder

Long short-term memory networks (LSTM) have been proved to be effective for modeling sequence data into vectors. Thus, to model function $g(d,q)$, Hermann et al. (2015) feed documents on word at a time into a LSTM-based encoder. Then the question q is also fed into this encoder after a delimiter. In this way, the pair of the given document d and the question q could be as a long single sequence as shown in Fig. 7.13. The details are omitted here and could refer to (Hermann et al. 2015).

7.3.3.2 Bidirectional Attention Encoder

Unidirectional LSTM is difficult to propagate dependencies over long distances. As a result, information would decay in the transport from a component to another and the semantics of the document could not be encoded precisely. Thus, more and more researchers adopt bidirectional LSTM model to encode sequential data. Moreover, not all sentences or contexts in a document d have related to the given question q. For example, d is "Michael Jordan abruptly retired from Chicago Bulls before the

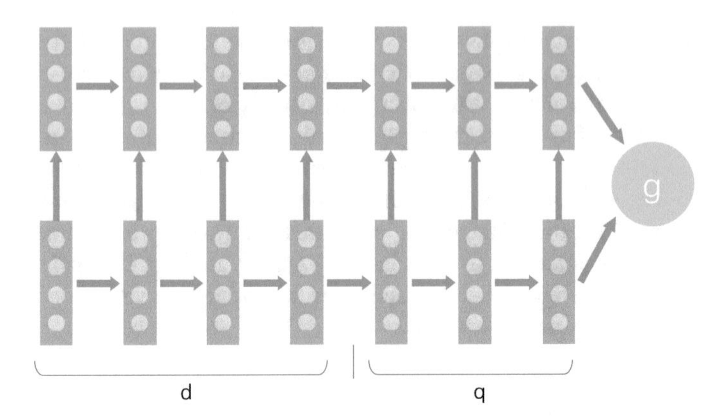

Fig. 7.13 An answer-entailing structure example from MCTest

beginning of the 1993–94 NBA season to pursue a career in baseball." When q is "When did Michael Jordan retired from NBA?", the focused contexts in d should be "before the beginning of the 1993–94 NBA." When q is "Which sports does Michael Jordan participate after his retirement from NBA?," the focused contexts in d should be "pursue a career in baseball." That is to say that we should pay different attention of different parts in d when dealing with different questions. Therefore, it is natural to introduce attention mechanism into the deep neural network. Chen et al. (2016) proposed a bidirectional encoding model with attention mechanism (BiDEA), which did get a promising performance on CNN/Daily Mail dataset.

The structure of this model is quite intuitive. Its procedure of predicting a answer mainly contains the following three steps:

1. **Encoding**: After all the words being mapped to d-dimensional vectors, the passage p (d) and query q can be represented as $p_1, p_2, \ldots, p_m$ and $q_1, q_2, \ldots, q_l$ respectively. Therefore, the contextual information of p can be calculated as

$$\overrightarrow{h_i} = LSTM(\overrightarrow{h_{i-1}, p_i}), i = 1, \ldots, m$$
$$\overleftarrow{h_i} = LSTM(\overleftarrow{h_{i+1}, p_i}), i = m, \ldots, 1$$
$$\widetilde{p_i} = concat(\overrightarrow{h_i}, \overleftarrow{h_i}).$$

   Meanwhile, the question can be embedding into q (a single vector) by another LSTM layer in the same way.

2. **Attention**: All textual information in $\widetilde{p_i}$ can be combined into output vector o in the following way:

$$\alpha_i = softmax_i q^{\top} W_s \widetilde{p_i}$$
$$o = \sum_i \alpha_i \widetilde{p_i}$$

Table 7.5 Results of BiDEA and other models on CNN/Daily mail

	CNN		Daily mail	
	Val	Test	Val	Test
Attentive reader (Hermann et al. 2015)	61.6	63.0	70.5	69.0
MemNN (Sukhbaatar et al. 2015)	63.4	6.8	–	–
AS reader (Hermann et al. 2015)	68.6	69.5	75.0	73.9
Stanford AR (Chen et al. 2016)	68.6	69.5	75.0	73.9
DER network (Kobayashi et al. 2016)	71.3	72.9	–	–
Iterative attention (Sordoni et al. 2016)	72.6	73.3	–	–
EpiReader (Trischler et al. 2016)	73.4	74.0	–	–
GAReader (Dhingra et al. 2016)	73.0	73.8	76.7	75.7
AoA reader (Cui et al. 2017)	73.1	74.4	–	–
ReasoNet (Shen et al. 2017)	72.9	74.7	77.6	76.6
BiDAF (Seo et al. 2016)	76.3	76.9	80.3	79.6
BiDEA (Chen et al. 2016)	72.4	72.4	76.9	75.8

In the above equations, $W_s \in \mathbb{R}^{h \times h}$ is used for measuring similarity between the question q and a word in passage p_i.

3. **Prediction**: The predicted answer a is computed as follows:

$$a = \operatorname{argmax}_{a \in p \cap E} W_a^{\top} o,$$

where E is the embedding matrix and W_a is the measurement matrix between output o and candidate word a.

Although the computation of aforementioned model is quite straightforward, it gains pretty promising performance on CNN/Daily Mail (experimental results are listed in Table 7.5). According to the analysis of Chen et al. (2016), the effectiveness of the proposed model are caused by (i) reasoning and inference level in CNN/Daily Mail are still simple enough to be handled by a simple model; (ii) all kinds of models have reached a performance ceiling on CNN/Daily Mail, and this corpus may even be handled well by an information retrieve system.

Moreover, in order to represent the context at different granularities and achieve a query-aware context representation without early summarization, Seo et al. (2016) adopt a multistage hierarchical process and also propose a bidirectional attention flow networks (BiDAF) for MC task.

As showed in Fig. 7.14, the proposed model is mainly composed by the following 6 layers:

1. **Character Embedding Layer**: a character-level CNNs that can map characters in a word to a continuous vector.
2. **Word Embedding Layer**: a pre-trained word embedding matrix.

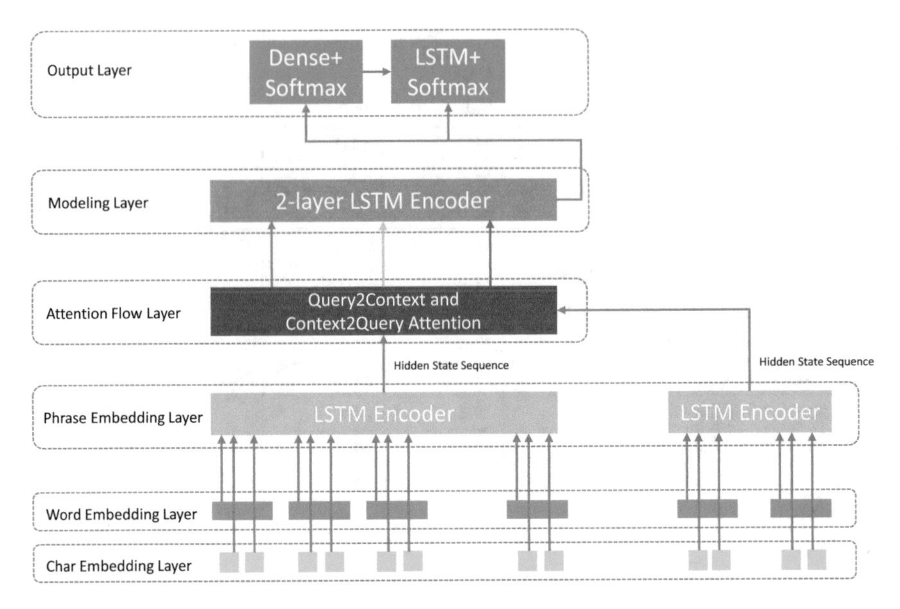

Fig. 7.14 Diagram of bi-directional attention flow model

3. **Phrase Embedding Layer**: a bidirectional LSTM layer that can catch contextual information of a word.
4. **Attention Flow Layer**: a similarity matrix, S, measures similarity between words of context and query from two directions, i.e., Context-to-Query and Query-to-Context.
5. **Modeling Layer**: a two-layer bidirectional LSTM containing contextual information about all the words.
6. **Output Layer**: two logistic regression models that capture the start-index and end-index respectively.

The experimental results, as showed in Table 7.6, on SQuAD indicates that BiDAF's idea brings an improvement in performance, which may be caused by BiDAF's capability on finding start and end of support evidence in a hierarchical level.

7.3.3.3 Memory Networks

Memory Networks (MemNNs) (Weston et al. 2015b) are proposed to address the decaying of information in sequential neural networks. And it can reason with inference components combined with a long-term memory component (actually a matrix or tensor, where its name came from). In general, it contains four major important components:

I (input feature map) converts input vectors to internal feature representation.
G (generalization) updates the existing memory according to the new input.

Table 7.6 Results of BiDAF and other models on SQuAD test set

	Single model		Ensemble	
	EM	F1	EM	F1
Logistic regression baseline (Rajpurkar et al. 2016)	40.4	51.0	–	–
Dynamic chunk reader (Yu et al. 2016)	62.5	71.0	–	–
Fine-grained gating (Yang et al. 2016)	62.5	73.3	–	–
Match LSTM (Wang and Jiang 2016)	64.7	73.7	67.9	77.0
Multi-perspective matching (Wang et al. 2016)	65.5	75.1	68.2	77.2
Dynamic coattention networks (Xiong et al. 2016)	66.2	75.9	71.6	80.4
R-Net (Wang et al. 2017)	68.4	77.5	72.1	79.7
BiDAF (Seo et al. 2016)	68.0	77.3	73.3	81.1

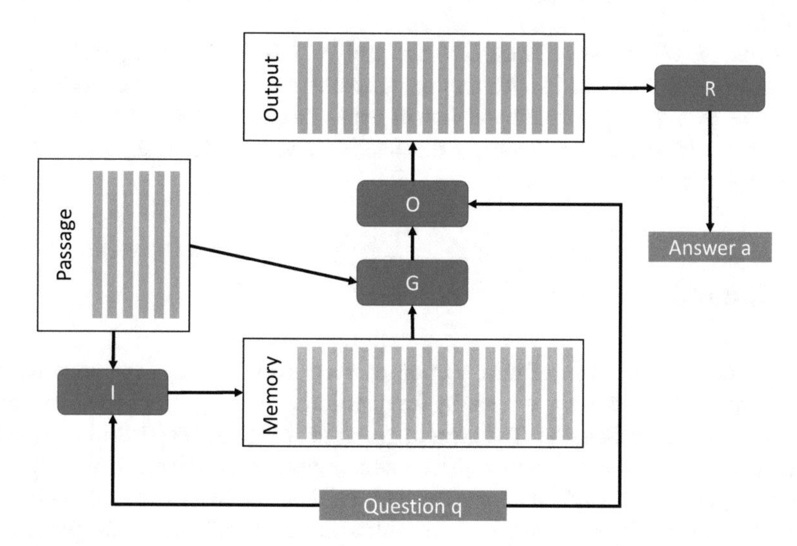

Fig. 7.15 Diagram of memory networks for bAbI task

O (output feature map) calculates a new output based on the new input and current memory states.
R (response) converts the output to desired response format.

The diagram of MemNNs is shown in Fig. 7.15. An important form of MemNNs is End2End Memory Networks and we will abbreviate it as MemN2Ns. One advantage of MemN2Ns is that it can be trained in an end-to-end way, which means it requires less-supervised information and is more generally applicable in realistic settings. The following equations are computation in I, G, O, and R respectively:

I $p_i = Softmax(u^T m_i)$,
where $m_i = Ax_i$ (x_i is the embedding vectors of the input sentences) and $u = Bq$(q is the input query).

G in MemN2Ns, memory has not been updated.
O $o = \sum_i p_i c_i$ where $c_i = Cx_i$.
R $\hat{a} = Softmax(W(o + u))$.

Also, it is easy to insert layers into MemN2Ns in the following way:

- u of $(k+1)th$ layer can be calculated as $u^{k+1} = u^k + o^k$.
- Each layer has its own A^k and C^k.
- Prediction can be computed by $\hat{a} = Softmax(Wu^{K+1})$

MemN2N is originally applied to 20 tasks in bAbI task. And, as demonstrated by experiments in Table 7.7, the best MemN2N performance are close to the supervised models, the position encoding(PE) representation improves over bag-of-words (BoW), the linear start (LS) to training seems help avoid local minima and joint training helps on all tasks.

Beyond the MemNNs or MemN2Ns themselves, it is worthy noting that the computation in G is actually a kind of attention mechanism. And memory networks are the first model that keep external knowledge in a specific matrix, which has a significant effect on the development of memory mechanism in various deep models for natural language processing.

7.4 Summary

This chapter presents a brief introduction of deep learning-based methods for the tasks of question answering, especially for question answering over knowledge base and machine comprehension. The advantages of the usage of deep learning is that it could convert all text spans, including documents, questions, and the potential answers into vector embeddings. As a result, all texts could be handled in a unified semantic space. Thus, the existing semantic gap problem in traditional QA approaches based on symbolic representation could be alleviated to a certain degree. Moreover, such paradigm makes that the QA system could be constructed in a end-to-end way. As a result, the existing complicated pipeline-based QA process could be replaced by a more straightforward or easy way. The results are expected to be improved.

Nevertheless, there are many challenges in deep learning-based QA models. For example, the existing neural networks, such as RNN and CNN, could still not precisely capture the semantic meaning of the given questions. Especially for the document, the topical or logical structure in a document could not be easily modeled by a neural network. Moreover, there are still no effective methods for embedding items in knowledge base. And, the reasoning process in QA is difficult to be modeled by simply numerical operations between vectors. These problems are the key challenges for QA task and should be paid more attention in the future.

Table 7.7 Test error rates (%) on the 20 QA tasks for models using 1k training examples (mean test errors for 10k training examples are shown at the bottom)

Task	Baseline			MemN2N								
	Strongly super-vised MemNN	LSTM	MemNN WSH	BoW	PE	PE LS	PE LS RN	1 hop PE LS joint	2 hop PE LS joint	3 hop PE LS joint	PE LS RN joint	PE LS LW joint
1 supporting fact	0.0	50.0	0.1	0.6	0.1	0.2	0.0	0.8	0.0	0.1	0.0	0.1
2 supporting facts	0.0	80.0	42.8	17.6	21.6	12.8	8.3	62.0	15.6	14.0	11.4	18.8
3 supporting facts	0.0	80.0	76.4	71.0	64.2	58.8	40.3	76.9	31.6	33.1	21.9	31.7
2 argument relations	0.0	39.0	40.3	32.0	3.8	11.6	2.8	22.8	2.2	5.7	13.4	17.5
3 argument relations	2.0	30.0	16.3	18.3	14.1	15.7	13.1	11.0	13.4	14.8	14.4	12.9
Yes/no questions	0.0	52.0	51.0	8.7	7.9	8.7	7.6	7.2	2.3	3.3	2.8	2.0
Counting	15.0	51.0	36.1	23.5	21.6	20.3	17.3	15.9	25.3	17.9	18.3	10.1
Lists/sets	9.0	55.0	37.8	11.4	12.6	12.7	10.0	13.2	11.7	10.1	9.3	6.1
Simple negation	0.0	36.0	35.9	21.1	23.3	17.0	13.2	5.1	2.0	3.1	1.9	1.5
Indefinite knowledge	2.0	56.0	68.7	22.9	17.4	18.6	15.1	10.6	5.0	6.6	6.5	2.6
Basic coreference	0.0	38.0	30.0	4.1	4.3	0.0	0.9	8.4	1.2	0.9	0.3	3.3
Conjunction	0.0	26.0	10.1	0.3	0.3	0.1	0.2	0.4	0.0	0.3	0.1	0.0
Compound coreference	0.0	6.0	19.7	10.5	9.9	0.3	0.4	6.3	0.2	1.4	0.2	0.5
Time reasoning	1.0	73.0	18.3	1.3	1.8	2.0	1.7	36.9	8.1	8.2	6.9	2.0
Basic deduction	0.0	79.0	64.8	24.3	0.0	0.0	0.0	46.4	0.5	0.0	0.0	1.8
Basic induction	0.0	77.0	50.5	52.0	52.1	1.6	1.3	47.4	51.3	3.5	2.7	51.0
Positional reasoning	35.0	49.0	50.9	45.4	50.1	49.0	51.0	44.4	41.2	44.5	40.4	42.6

(continued)

Table 7.7 (continued)

Task	Baseline			MemN2N								
	Strongly supervised MemNN	LSTM	MemNN WSH	BoW	PE	PE LS	PE LS RN	1 hop PE LS joint	2 hop PE LS joint	3 hop PE LS joint	PE LS RN joint	PE LS LW joint
Size reasoning	5.0	48.0	51.3	48.1	13.6	10.1	11.1	9.6	10.3	9.2	9.4	9.2
Path finding	64.0	92.0	100.0	89.7	87.4	85.6	82.8	90.7	89.9	90.2	88.0	90.6
Agent's motivation	0.0	9.0	3.6	0.1	0.0	0.0	0.0	0.0	0.1	0.0	0.0	0.2
Mean error (%)	6.7	51.3	40.2	25.1	20.3	16.3	13.9	25.8	15.6	13.3	12.4	15.2
Filed tasks (err.>5%)	4	20	18	15	13	12	11	17	11	11	11	10
On 10k training data												
Mean error (%)	3.2	36.4	39.2	15.4	9.4	7.2	6.6	24.5	10.9	7.9	7.5	11.0
Filed tasks (err.>5%)	2	16	17	9	6	4	4	16	7	6	6	6

Key *BoW* bag-of-words representation; *PE* position encoding representation; *LS* linear start training; *RN* random injection of time index noise; *LW* RNN-style layer-wise weight tying (if not stated, adjacent weight tying is used); *joint* joint training on all tasks (as opposed to per-task training)

References

Berant, J., Chou, A., Frostig, R., & Liang, P. (2013). Semantic parsing on freebase from question-answer pairs. In *EMNLP*.
Berant, J., & Liang, P. (2014). Semantic parsing via paraphrasing. In *ACL*.
Bordes, A., Chopra, S., & Weston, J. (2014a). Question answering with subgraph embeddings. In *EMNLP*.
Bordes, A., Usunier, N., Chopra, S., & Weston, J. (2015). Large-scale simple question answering with memory networks. In *arXiv*.
Bordes, A., Weston, J., & Usunier, N. (2014b). Open question answering with weakly supervised embedding models. In *ECML*.
Cai, Q., & Yates, A. (2013). Large-scale semantic parsing via schema matching and lexicon extension. In *ACL*.
Chen, D., Bolton, J., & Manning, C. D. (2016). A thorough examination of the CNN/Daily Mail reading comprehension task. In *Association for Computational Linguistics (ACL)*.
Cui, Y., Chen, Z., Wei, S., Wang, S., Liu, T., & Hu, G. (2017). Attention-over-attention neural networks for reading comprehension. In *ACL*.
Dhingra, B., Liu, H., Yang, Z., Cohen, W. W., & Salakhutdinov, R. (2016). Gated-attention readers for text comprehension. arXiv preprint arXiv:1606.01549.
Dong, L., Wei, F., Zhou, M., & Xu, K. (2015). Question answering over freebase with multi-column convolutional neural networks. In *ACL-IJCNLP*.
Etzioni, O. (2011). Search needs a shake-up. *Nature*, *476*(7358), 25–26.
Hao, Y., Zhang, Y., Liu, K., He, S., Liu, Z., Wu, H., & Zhao, J. (2017). An end-to-end model for question answering over knowledge base with cross-attention combining global knowledge. In *Association for Computational Linguistics (ACL)*.
Hermann, K. M., Kocisky, T., Grefenstette, E., Espeholt, L., Kay, W., Suleyman, M., & Blunsom, P. (2015). Teaching machines to read and comprehend. In *Advances in Neural Information Processing Systems* (pp. 1693–1701).
Hill, F., Bordes, A., Chopra, S., & Weston, J. (2015). The goldilocks principle: Reading children's books with explicit memory representations. arXiv preprint arXiv:1511.02301.
Jansen, P., Surdeanu, M., & Clark, P. (2014). Discourse complements lexical semantics for non-factoid answer reranking. In *Proceedings of the 52nd Annual Meeting of the Association for Computational Linguistics* (Vol. 1: Long Papers, pp. 977–986). Association for Computational Linguistics.
Kobayashi, S., Tian, R., Okazaki, N., & Inui, K. (2016). Dynamic entity representation with max-pooling improves machine reading. In *Proceedings of the 2016 Conference of the NAACL*.
Kun, X., Sheng, Z., Yansong, F., & Dongyan, Z. (2014). Answering natural language questions via phrasal semantic parsing. In *Proceedings of the 2014 Conference on Natural Language Processing and Chinese Computing (NLPCC)*.
Kwiatkowski, T., Choi, E., Artzi, Y., & Zettlemoyer, L. S. (2013). Scaling semantic parsers with on-the-fly ontology matching. In *EMNLP*.
Liang, C., Berant, J., Le, Q., Forbus, K. D., & Lao, N. (2017). Neural Symbolic Machines: Learning Semantic Parsers on Freebase with Weak Supervision. In *Proceedings of the Association for Computational Linguistics (ACL 2017)*. Canada: Association for Computational Linguistics.
Liu, Y., Wei, F., Li, S., Ji, H., Zhou, M., & Wang, H. (2015). A dependency-based neural network for relation classification. In *ACL*.
Miller, A., Fisch, A., Dodge, J., Karimi, A.-H., Bordes, A., & Weston, J. (2016). Key-value memory networks for directly reading documents. In *Proceedings of the 2016 Conference on Empirical Methods in Natural Language Processing* (pp. 1400–1409). Austin, TX: Association for Computational Linguistics.
Narasimhan, K., & Barzilay, R. (2015). Machine comprehension with discourse relations. In *Proceedings of the 53rd Annual Meeting of the Association for Computational Linguistics and the*

7th International Joint Conference on Natural Language Processing (Vol. 1: Long Papers, pp. 1253–1262). Association for Computational Linguistics.
Rajpurkar, P., Zhang, J., Lopyrev, K., & Liang, P. (2016). Squad: 100, 000+ questions for machine comprehension of text. *CoRR*, abs/1606.05250.
Reddy, S., Lapata, M., & Steedman, M. (2014). Large-scale semantic parsing without question-answer pairs. *Transactions of the Association of Computational Linguistics* (pp. 377–392).
Richardson, M., Burges, J. C., & Renshaw, E. (2013). Mctest: A challenge dataset for the open-domain machine comprehension of text. In *Proceedings of the 2013 Conference on Empirical Methods in Natural Language Processing* (pp. 193–203). Association for Computational Linguistics.
Sachan, M., Dubey, K., Xing, E., & Richardson, M. (2015). Learning answer-entailing structures for machine comprehension. In *Proceedings of the 53rd Annual Meeting of the Association for Computational Linguistics and the 7th International Joint Conference on Natural Language Processing* (Vol. 1: Long Papers, pp. 239–249). Association for Computational Linguistics.
Seo, M. J., Kembhavi, A., Farhadi, A., & Hajishirzi, H. (2016). Bidirectional attention flow for machine comprehension. *CoRR*, abs/1611.01603.
Shen, Y., Huang, P.-S., Gao, J., & Chen, W. (2017). Reasonet: Learning to stop reading in machine comprehension. In *Proceedings of the 23rd ACM SIGKDD International Conference on Knowledge Discovery and Data Mining*, KDD '17 (pp. 1047–1055). New York, USA: ACM.
Smith, E., Greco, N., Bosnjak, M., & Vlachos, A. (2015). A strong lexical matching method for the machine comprehension test. In *Proceedings of the 2015 Conference on Empirical Methods in Natural Language Processing* (pp. 1693–1698). Association for Computational Linguistics.
Sordoni, A., Bachman, P., Trischler, A., & Bengio, Y. (2016). Iterative alternating neural attention for machine reading. arXiv preprint arXiv:1606.02245.
Steedman, M. (2000). *The Syntactic Process*. Cambridge, MA: The MIT Press.
Sugawara, S., Yokono, H., & Aizawa, A. (2017). Prerequisite skills for reading comprehension: Multi-perspective analysis of mctest datasets and systems.
Sukhbaatar, S., Weston, J., Fergus, R., et al. (2015). End-to-end memory networks. In *Advances in Neural Information Processing Systems* (pp. 2440–2448).
Taylor, W. L. (1953). cloze procedure: a new tool for measuring readability. *Journalism Bulletin*, *30*(4), 415–433.
Trischler, A., Ye, Z., Yuan, X., & Suleman, K. (2016). Natural language comprehension with the epireader. arXiv preprint arXiv:1606.02270.
Wang, S., & Jiang, J. (2016). Machine comprehension using match-lstm and answer pointer. *CoRR*, abs/1608.07905.
Wang, W., Yang, N., Wei, F., Chang, B., & Zhou, M. (2017). Gated self-matching networks for reading comprehension and question answering. In *Proceedings of the 55th Annual Meeting of the Association for Computational Linguistics* (Vol. 1: Long Papers, pp. 189–198). Association for Computational Linguistics.
Wang, Z., Mi, H., Hamza, W., & Florian, R. (2016). Multi-perspective context matching for machine comprehension. *CoRR*, abs/1612.04211.
Weston, J., Bordes, A., Chopra, S., Rush, A. M., van Merriënboer, B., Joulin, A., & Mikolov, T. (2015a). Towards ai-complete question answering: A set of prerequisite toy tasks. arXiv preprint arXiv:1502.05698.
Weston, J., Chopra, S., & Bordes, A. (2015b). Memory networks. In *ICLR*.
Xiong, C., Zhong, V., & Socher, R. (2016). Dynamic coattention networks for question answering. *CoRR*, abs/1611.01604.
Xu, K., Feng, Y., Huang, S., & Zhao, D. (2015). Semantic relation classification via convolutional neural networks with simple negative sampling. In *EMNLP*.
Xu, K., Reddy, S., Feng, Y., Huang, S., & Zhao, D. (2016). Question Answering on Freebase via Relation Extraction and Textual Evidence. In *Proceedings of the Association for Computational Linguistics (ACL 2016)*. Berlin, Germany: Association for Computational Linguistics.
Yang, Y., & Chang, M.-W. (2015). S-mart: Novel tree-based structured learning algorithms applied to tweet entity linking. In *Proceedings of the 53rd Annual Meeting of the Association for Compu-*

tational Linguistics and the 7th International Joint Conference on Natural Language Processing (Vol. 1: Long Papers, pp. 504–513). Beijing, China: Association for Computational Linguistics.
Yang, Z., Dhingra, B., Yuan, Y., Hu, J., Cohen, W. W., & Salakhutdinov, R. (2016). Words or characters? Fine-grained gating for reading comprehension. *CoRR*, abs/1611.01724.
Yih, W.-t., Chang, M.-W., He, X., & Gao, J. (2015). Semantic parsing via staged query graph generation: Question answering with knowledge base. In *ACL-IJCNLP*.
Yih, W.-t., He, X., & Meek, C. (2014). Semantic parsing for single-relation question answering. In *Proceedings of the 52nd Annual Meeting of the Association for Computational Linguistics* (Vol. 2: Short Papers, pp. 643–648). Baltimore, MD: Association for Computational Linguistics.
Yih, W.-t., Richardson, M., Meek, C., Chang, M.-W., & Suh, J. (2016). The value of semantic parse labeling for knowledge base question answering. In *Proceedings of the 54th Annual Meeting of the Association for Computational Linguistics* (Vol. 2: Short Papers, pp. 201–206). Berlin, Germany: Association for Computational Linguistics.
Yu, Y., Zhang, W., Hasan, K. S., Yu, M., Xiang, B., & Zhou, B. (2016). End-to-end reading comprehension with dynamic answer chunk ranking. *CoRR*, abs/1610.09996.
Zeng, D., Liu, K., Lai, S., Zhou, G., & Zhao, J. (2014). Relation classification via convolutional deep neural network. In *Proceedings of COLING 2014, the 25th International Conference on Computational Linguistics: Technical Papers* (pp. 2335–2344). Dublin, Ireland: Dublin City University and Association for Computational Linguistics.
Zhang, S., Feng, Y., Huang, S., Xu, K., Han, Z., & Zhao, D. (2015). Semantic interpretation of superlative expressions via structured knowledge bases. In *Proceedings of the 53rd Annual Meeting of the Association for Computational Linguistics and the 7th International Joint Conference on Natural Language Processing* (Vol. 2: Short Papers, pp. 225–230). Beijing, China: Association for Computational Linguistics.

Chapter 8
Deep Learning in Sentiment Analysis

Duyu Tang and Meishan Zhang

Abstract Sentiment analysis (also known as opinion mining) is an active research area in natural language processing. The task aims at identifying, extracting, and organizing sentiments from user-generated texts in social networks, blogs, or product reviews. Over the past two decades, many studies in the literature exploit machine learning approaches to solve sentiment analysis tasks from different perspectives. Since the performance of a machine learner heavily depends on the choices of data representation, many studies devote to building powerful feature extractor with domain expertise and careful engineering. Recently, deep learning approaches emerge as powerful computational models that discover intricate semantic representations of texts automatically from data without feature engineering. These approaches have improved the state of the art in many sentiment analysis tasks, including sentiment classification, opinion extraction, fine-grained sentiment analysis, etc. In this paper, we give an overview of the successful deep learning approaches sentiment analysis tasks at different levels.

8.1 Introduction

Sentiment analysis (also known as opinion mining) is a field that automatically analyzes people's opinions, sentiments, emotions from user-generated texts (Pang et al. 2008; Liu 2012). Sentiment analysis is a very active research area in natural language processing (Manning et al. 1999; Jurafsky 2000), and is also widely studied in data mining, web mining,and social media analytics as sentiments are key influencers of

---

D. Tang
Microsoft Research Asia, Beijing, China
e-mail: dutang@microsoft.com

M. Zhang (✉)
Heilongjiang University, Harbin, Heilongjiang, China
e-mail: mszhang@hlju.edu.cn

L. Deng and Y. Liu (eds.), *Deep Learning in Natural Language Processing*, https://doi.org/10.1007/978-981-10-5209-5_8

Table 8.1 An example that illustrates the definition of sentiment

Target	Sentiment	Holder	Time
iPhone	Positive	Alice	June 4, 2015
Touch screen	Positive	Alice	June 4, 2015
Price	Negative	Alice	June 4, 2015

human behaviors. With the rapid growth of social media such as Twitter,[1] Facebook[2], and review sites such as IMDB,[3] Amazon,[4] Yelp,[5] sentiment analysis draws growing attention from both the research and industry communities (Table 8.1).

According to the definition from (Liu 2012), sentiment (or an opinion) is represented as a quintuple e, a, s, h, t, where e is the name of an entity, a is the aspect of e, s is the sentiment on aspect a of entity e, h is the opinion holder, and t is the time when the opinion is expressed by h. In this definition, a sentiment s can be a positive, negative, or neutral sentiment, or a numeric rating score expressing the strength/intensity of the sentiment (e.g., 1–5 stars) in review sites like Yelp and Amazon. The entity can be a product, service, topic organization, or event (Hu and Liu 2004; Deng and Wiebe 2015).

Let us use an example to explain the definition of "sentiment". Supposing a user named Alice posted a review "I bought an iPhone a few days ago. It is such a nice phone. The touch screen is really cool. However, the price is a little high." at June 4, 2015. Three sentiment quintuples are involved in this example, as shown in Table 8.1.

Based on the definition of "sentiment", sentiment analysis aims at discovering all the sentiment quintuples in a document. Sentiment analysis tasks are derived from the five components of the sentiment quintuple. For example, document/sentence-level sentiment classification (Pang et al. 2002; Turney 2002) targets at the third component (sentiment such as positive, negative, and neutral) while ignoring the other aspects. Fine-grained opinion extraction focuses on the first four components of the quintuple. Target-dependent sentiment classification focuses on the second and the third aspects.

Over the past two decades, machine learning-driven methods have dominated most sentiment analysis tasks. Since feature representation greatly affects the performance of a machine learner (LeCun et al. 2015; Goodfellow et al. 2016), a lot of studies in the literature focus on effective features in hand with domain expertise and careful engineering. But this can be avoided by representation learning algorithms, which automatically discover discriminative and explanatory text representations from data. Deep learning is a kind of representation learning approach, which learns multiple levels of representation with nonlinear neural networks, each of which transforms

[1] https://twitter.com/.

[2] https://www.facebook.com.

[3] http://www.imdb.com/.

[4] https://www.amazon.com/.

[5] https://www.yelp.com/.

the representation at one level into a representation at a higher and more abstract level. The learned representations can be naturally used as features and applied to detection or classification tasks. In this chapter, we introduce successful deep learning algorithms for sentiment analysis. The notation of "deep learning" in this chapter stands for the use of neural network approaches to learning continuous and real-valued text representation/feature automatically from data.

We organize this chapter as follows. Since word is the basic computational unit of natural language, we first describe the methods to learn continuous word representation, also called word embedding. These word embeddings can be used as inputs to subsequent sentiment analysis tasks. We describe semantic compositional methods that compute representations of longer expressions (e.g., sentence or document) for sentence/document-level sentiment classification task (Socher et al. 2013; Li et al. 2015; Kalchbrenner et al. 2014), followed by neural sequential models for fine-grained opinion extraction. We finally conclude this paper and provide some future directions.

8.2 Sentiment-Specific Word Embedding

Word representation aims at representing aspects of word meaning. For example, the representation of "cellphone" may capture the facts that cellphones are electronic products, that they include battery and screen, that they can be used to chat with others, and so on. A straightforward way is to encode a word as a one-hot vector. It has the same length as the size of the vocabulary, and only one dimension is 1, with all others being 0. However, the one-hot word representation only encodes the indices of words in a vocabulary, while failing to capture rich relational structure of the lexicon.

One common approach to discover the similarities between words is to learn word clusters (Brown et al. 1992; Baker and McCallum 1998). Each word is associated with a discrete class, and words in the same class are similar in some respect. This leads to a one-hot representation over a smaller vocabulary size. Instead of characterizing the similarity with a discrete variable based on clustering results which correspond to a soft or hard partition of the set of words, many researchers target at learning a continuous and real-valued vector for each word, also known as word embedding. Existing embedding learning algorithms are typically based on the distributional hypothesis (Harris 1954), which states that words in similar contexts have similar meanings. Towards this goal, many matrix factorization methods can be viewed as modeling word representations. For example, Latent Semantic Indexing (LSI) (Deerwester et al. 1990) can be regarded as learning a linear embedding with a reconstruction objective, which uses a matrix of "term–document" co-occurrence statistics, e.g., each row stands for a word or term and each column corresponds to an individual document in the corpus. Hyperspace Analogue to Language (Lund and Burgess 1996) utilizes a matrix of term–term co-occurrence statistics, where both rows and columns correspond to words and the entries stand for the number of times a given word occurs in the context of another word. Hellinger PCA (Lebret et al. 2013) is also investigated to learn word embeddings over "term–term" co-occurrence statistics. Since standard matrix factorization methods do not incorporate task-specific information, it is not clear whether they are useful enough for a target goal. Supervised Semantic Indexing (Bai et al. 2010) tackles this problem and takes the supervised information of a specific task (e.g. information retrieval) into consideration. They learn the embedding model from click-through data with a margin ranking loss. DSSM (Huang et al. 2013; Shen et al. 2014) also could be considered as learning task-specific text embeddings with weak supervision in information retrieval.

A pioneering work that explores neural network approaches is given by (Bengio et al. 2003), which introduces a neural probabilistic language model that learns simultaneously a continuous representation for words and a probability function for word sequences based on these word representations. Given a word and its preceding context words, the algorithm first maps all these words to continuous vectors with a shared lookup table. Afterward, word vectors are fed to a feed-forward neural network with softmax as output layer to predict the conditional probability of next word. The parameters of neural network and lookup table are jointly estimated with backpropagation. Following Bengio et al. (2003)'s work, several approaches are proposed to speed-up training processing or capturing richer semantic information. Bengio et al. (2003) introduce a neural architecture by concatenating the vectors of context words and current word, and use importance sampling to effectively optimize the model with observed "positive sample" and sampled "negative samples". Morin and Bengio (2005) develop hierarchical softmax to decompose the conditional probability with a hierarchical binary tree. Mnih and Hinton (2007) introduce a log-bilinear language model. Collobert and Weston (2008) train word embeddings with a ranking-type hinge loss function by replacing the middle word within a window with a randomly selected one. Mikolov et al. (2013a, b) introduce continuous bag-of-words (CBOW) and continuous skip-gram, and release the popular word2vec[6] toolkit. The CBOW model predicts the current word based on the embeddings of its context words, and the skip-gram model predicts surrounding words given the embedding of current word. Mnih and Kavukcuoglu (2013) accelerate the word embedding learning procedure with Noise Contrastive Estimation (Gutmann and Hyvärinen 2012). There are also many algorithms developed for capturing richer semantic information, including global document information (Huang et al. 2012), word morphemes (Qiu et al. 2014), dependency-based contexts (Levy and Goldberg 2014), word–word co-occurrence (Levy and Goldberg 2014), sense of ambiguous words (Li and Jurafsky 2015), semantic lexical information in WordNet (Faruqui et al. 2014), hierarchical relations between words (Yogatama et al. 2015).

The aforementioned neural network algorithms typically only use the contexts of words to learn word embeddings. As a result, the words with similar contexts but opposite sentiment polarity like "good" and "bad" are mapped into close vectors in the embedding space. This is meaningful for some tasks such as POS tagging as the two words have similar usages and grammatical roles, but this is problematic

[6]https://code.google.com/p/word2vec/.

for sentiment analysis as "good" and "bad" have the opposite sentiment polarity. In order to learn word embeddings tailored for sentiment analysis tasks, some studies encode sentiment of texts in continuous word representation. Maas et al. (2011) introduce a probabilistic topic model by inferring the polarity of a sentence based on the embedding of each word it contains. Labutov and Lipson (2013) re-embed an existing word embedding with logistic regression by regarding sentiment supervision of sentences as a regularization item. Tang et al. (2014) extend the C&W model and develop three neural networks to learn sentiment-specific word embedding from tweets. Tang et al. (2014) use the tweets that contain positive and negative emoticons as training data. The positive and negative emoticon signals are regarded as weak sentiment supervision.

We describe two sentiment-specific approaches that incorporate sentiment of sentences to learn word embeddings. The model of Tang et al. (2016c) extends the context-based model of Collobert and Weston (2008), and the model of Tang et al. (2016a) extends the context based model of Mikolov et al. (2013b). We describe the relationships between these models.

The basic idea of the context-based model (Collobert and Weston 2008) is to assign a real word-context pair (w_i, h_i) a higher score than an artificial noise (w^n, h_i) by a margin. The model is learned to minimize the following hinge loss function, where T is the training corpora:

$$loss = \sum_{(w_i,h_i)\in T} max(0, 1 - f_\theta(w_i, h_i) + f_\theta(w^n, h_i)). \tag{8.1}$$

The scoring function $f_\theta(w, h)$ is achieved with a feed forward neural network. Its input is the concatenation of the current word w_i and context words h_i, and the output is a linear layer with only one node which stands for the compatibility between w and h. During training, an artificial noise w^n is randomly selected from the vocabulary.

The basic idea of sentiment-specific approach of Tang et al. (2014) is that if the gold sentiment polarity of a word sequence is positive, the predicted positive score should be higher than the negative score. Similarly, if the gold sentiment polarity of a word sequence is negative, its positive score should be smaller than the negative score. For example, if a word sequence is associated with two scores $[f_{pos}^{rank}, f_{neg}^{rank}]$, then the values of $[0.7, 0.1]$ can be interpreted as a positive case because the positive score 0.7 is greater than the negative score 0.1. By that analogy, the result with $[-0.2, 0.6]$ indicates a negative polarity. The neural network-based ranking model is given in Fig. 8.1b, which shares some similarities with (Collobert and Weston 2008). As is shown, the ranking model is a feed-forward neural network consisting of four layers ($lookup \rightarrow linear \rightarrow hTanh \rightarrow linear$). Let us denote the output vector of ranking model as f^{rank}, where $C = 2$ for binary positive and negative classification. The margin ranking loss function for model training is described as below.

$$loss = \sum_{t}^{T} max(0, 1 - \delta_s(t) f_0^{rank}(t) + \delta_s(t) f_1^{rank}(t)) \tag{8.2}$$

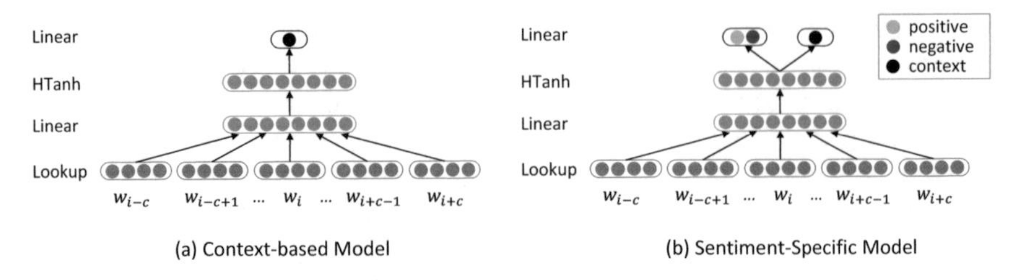

Fig. 8.1 An extension on ranking-based model for learning sentiment-specific word embeddings

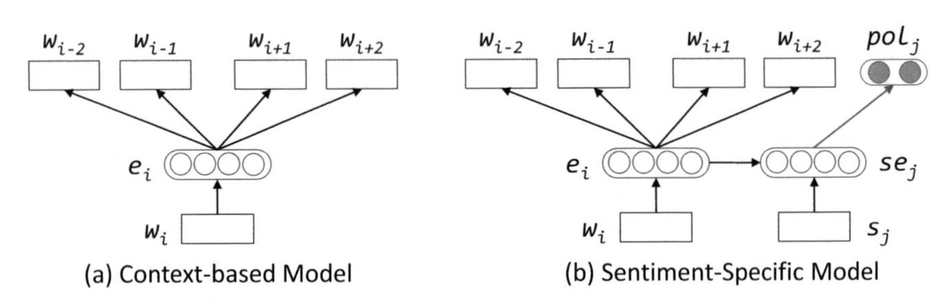

Fig. 8.2 An extension on skip-gram for learning sentiment-specific word embeddings

where T is the training corpus, f_0^{rank} is the predicted positive score, f_1^{rank} is the predicted negative score, $\delta_s(t)$ is an indicator function which reflects the gold sentiment polarity (positive or negative) of a sentence.

$$\delta_s(t) = \begin{cases} 1 & \text{if } f^g(t) = [1, 0] \\ -1 & \text{if } f^g(t) = [0, 1] \end{cases} \tag{8.3}$$

Holding a similar idea, an extension of skip-gram (Mikolov et al. 2013b) is developed to learn sentiment-specific word embeddings. Given a word w_i, skip-gram maps it into its continuous representation e_i, and utilizes e_i to predict the context words of w_i, namely w_{i-2}, w_{i-1}, w_{i+1}, w_{i+2}, et al. The objective of skip-gram is to maximize the average log probability:

$$f_{SG} = \frac{1}{T}\sum_{i=1}^{T}\sum_{-c\leq j\leq c, j\neq 0} log\ p(w_{i+j}|e_i), \tag{8.4}$$

where T is the occurrence of each phrase in the corpus, c is the window size, e_i is the embedding of the current phrase w_i, w_{i+j} is the context words of w_i, $p(w_{i+j}|e_i)$ is calculated with hierarchical softmax. The basic *softmax* unit is calculated as $softmax_i = exp(z_i)/\sum_k exp(z_k)$.

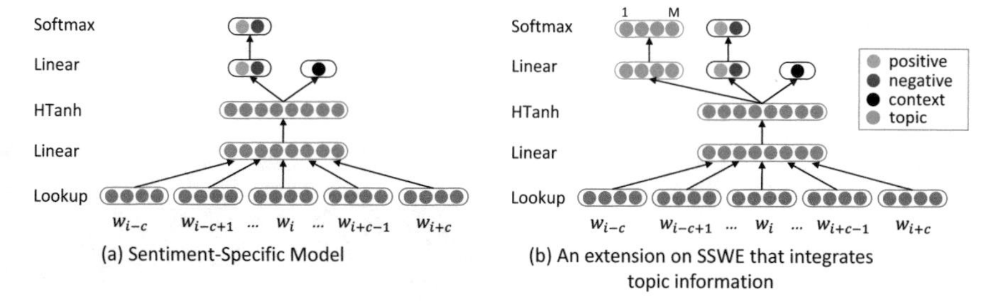

Fig. 8.3 Different ways to learn sentiment-specific word embeddings (**a**), and to incorporate topic information of texts (**b**)

The sentiment-specific model is given in Fig. 8.2b. Given a triple $\langle w_i, s_j, pol_j \rangle$ as input, where w_i is a phrase contained in the sentence s_j whose gold sentiment polarity is pol_j, the training objective is to not only utilize the embedding of w_i to predict its context words, but also to use the sentence representation se_j to predict the gold sentiment polarity of s_j, namely pol_j. The sentence vector is calculated by averaging the embeddings of words contained in a sentence. The objective is to maximize the weighted average loss function as given below.

$$f = \alpha \cdot \frac{1}{T} \sum_{i=1}^{T} \sum_{-c \le j \le c, j \ne 0} log\ p(w_{i+j}|e_i) + (1-\alpha) \cdot \frac{1}{S} \sum_{j=1}^{S} log\ p(pol_j|se_j), \tag{8.5}$$

where S is the occurrence of each sentence in the corpus, α weights the context, and the sentiment parts, $\sum_k pol_{jk} = 1$. For binary classification between positive and negative, the distribution of [0, 1] is for positive and [0, 1] is for negative.

There are different ways to guide the embedding learning process with sentiment information of texts. For example, the model of Tang et al. (2014) extends the ranking model of Collobert and Weston (2008) and use the hidden vector of text span to predict the sentiment label. Ren et al. (2016b) extend SSWE and further predicts the topic distribution of text based on input n-grams. These two approaches are given in Fig. 8.3.

8.3 Sentence-Level Sentiment Classification

Sentence-level sentiment analysis focuses on classifying the sentiment polarities of a given sentence. Typically, for one sentence $w_1 w_2 \dots w_n$, we divide its polarities into two ($\pm$) or three ($\pm/0$) categories, where + denotes positive, - denotes negative, and 0 denotes neutral. The task is a representative sentence classification problem.

Under the neural network setting, sentence-level sentiment analysis can be modeled as a two-phase framework, one being a sentence representation module by using

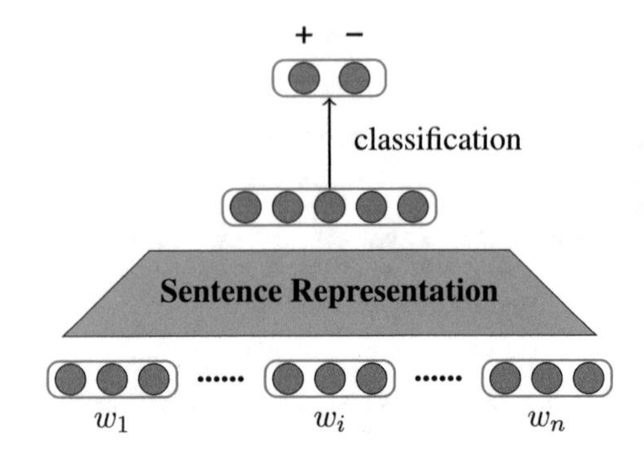

Fig. 8.4 Framework of sentiment classification

sophisticated neural structures, and the other being a simple classification module which can be resolved by a softmax operation. Figure 8.4 shows the overall framework.

Basically, with word embeddings for each sentential word, one can use pooling strategies to obtain a simple representation for a sentence, A pooling operation is able to summary salient features from a sequential input with variable length. Formally, we can use the equation $\mathbf{h} = \sum_{i=1}^{n} a_i \mathbf{x}_i$ to define popular pooling functions. For example, the widely adopted average (avg), max, and min pooling operations can be formalized as follows:

$$a_i^{avg} = \frac{1}{n}, \quad a_{ij}^{min} = \begin{cases} 1, & \text{if } i = \text{argmin}_k \mathbf{x}_{kj} \\ 0, & \text{otherwise,} \end{cases}, \quad a_{ij}^{max} = \begin{cases} 1, & \text{if } i = \text{argmax}_k \mathbf{x}_{kj} \\ 0, & \text{otherwise.} \end{cases} \tag{8.6}$$

Tang et al. (2014) exploit the three pooling methods to verify their proposed sentiment-encoded word embeddings, The method is just one simple example to represent sentences. In fact, recent advances on sentence representation for sentence classification are far beyond it. A number of sophisticated neural network structures have been proposed in the literature. As a whole, we summarize the related work by four categories: (1) convolutional neural networks, (2) recurrent neural networks, (3) recursive neural networks, (4) enhanced sentence representation by auxiliary resources. We introduce these works in the following subsections, respectively.

8.3.1 Convolutional Neural Networks

In the pooling neural network, we are only able to use word-level features. When the order of words changes in a sentence, the sentence representation result remains unchanged. In traditional statistical models, n-gram word features are adopted in order to alleviate the issue, showing improved performances. For neural network models, a convolution layer can be exploited to achieve a similar effect.

Formally, a convolution layer performs nonlinear transformations by traversing a sequential input with a fixed-size local filter. Give an input sequence $\mathbf{x}_1\mathbf{x}_2\ldots\mathbf{x}_n$, assuming that the size of local filter is K, then we can obtain a sequential output of $\mathbf{h}_1\mathbf{h}_2\ldots\mathbf{h}_{n-K+1}$:

$$\mathbf{h}_i = f\left(\sum_{k=1}^{K} W_k \mathbf{x}_{i+K-k}\right),$$

where f is an activation function such as $\tanh(\cdot)$ and $\text{sigmoid}(\cdot)$. When $K = 3$ and $\mathbf{x}_i$ is the input word embedding, the resulting $\mathbf{h}_i$ is a nonlinear combination of $\mathbf{x}_i$, $\mathbf{x}_{i+1}$, and $\mathbf{x}_{i+2}$, similar to the mixed unigram, bigram, and trigram features, which concatenate the surface forms of the corresponding words in a hard way.

Typically, convolutional neural network (CNN) is a certain network that integrates a convolution layer and a pooling layer together, as shown in Fig. 8.5, which has been widely studied for sentence-level sentiment classification. An initial attempt by directly applying of a standard CNN is introduced by Collobert et al. (2011). The study obtains the final sentence representation by using a convolutional layer over a sequence of input word embeddings, and using a further max pooling over the resulting hidden vectors.

Kalchbrenner et al. (2014) extend the basic CNN model for better sentence representation by two aspects. On the one hand, they use dynamic k-max pooling, where top-k values are reserved during pooling instead of only one value for each dimension in the simple max pooling. The value k is defined according to sentence length dynamically. On the other hand, they enlarge the layer number of CNN, using multilayer CNN structures, motivated by the intuition that deeper neural networks can encode more sophisticated features. Figure 8.6 shows the framework of multilayer CNNs.

Several CNN variations have been studied to better represent sentences. One most representative work is the nonlinear, nonconsecutive convolution operator proposed by Lei et al. (2015), as shown in Fig. 8.7. The operator aims to extract all n-word combinations through tensor algebra, no matter whether the words are consecutive. The process is conducted recursively, first one word, then two-word and further

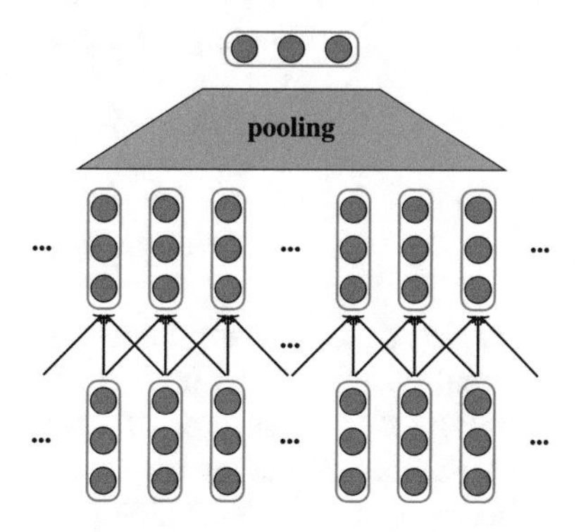

Fig. 8.5 Framework of CNN

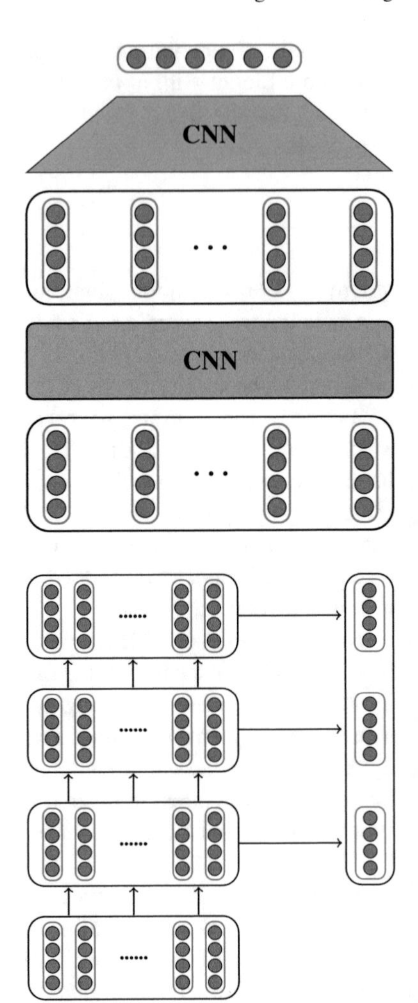

Fig. 8.6 Multilayer CNNs

Fig. 8.7 Nonlinear, nonconsecutive convolution

three-word combinations, respectively. They extract all unigram, bigram, and trigram features by the following formulas:

$$\begin{aligned}
\mathbf{f}_i^1 &= P\mathbf{x}_i \\
\mathbf{f}_i^2 &= s_{i-1}^1 \odot Q\mathbf{x}_i \ \text{ where } \ \mathbf{s}_i^1 = \lambda \mathbf{s}_{i-1}^1 + \mathbf{f}_i^1 \\
\mathbf{f}_i^3 &= s_{i-1}^2 \odot R\mathbf{x}_i \ \text{ where } \ \mathbf{s}_i^2 = \lambda \mathbf{s}_{i-1}^2 + \mathbf{f}_i^2,
\end{aligned}$$

where P, Q, and R are model parameters, λ is a hyper-parameter, and $\odot$ denote element-wise product. Finally, they make compositions of the three kinds of features, forming the representation of a sentence.

A number of studies have focused their attention on the exploration of heterogeneous input word embeddings. For example, Kim (2014) studies three different

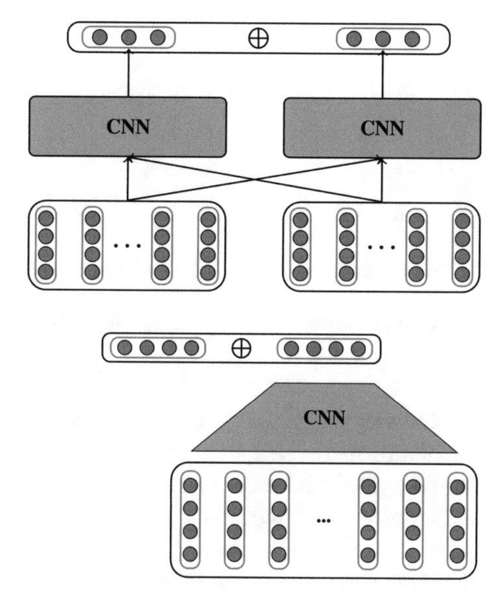

Fig. 8.8 Multichannel CNNs

Fig. 8.9 Enhanced word representations with character features

methods of using word embedding. The author concerns two different embeddings, a randomly initialized embedding and a pretrained embedding, considering the effect of dynamic fine-tuning over these embeddings. Finally, it combines the two kinds of embeddings and proposes the multichannel CNNs based on heterogeneous word embeddings, as shown in Fig. 8.8. The work is extended by Yin and Schütze (2015), who use several different word embeddings by multichannel multilayer CNNs. And, in addition, they exploit extensive pretraining techniques for the model weight initialization. However, a simpler version of it is presented by Zhang et al. (2016d), which meanwhile shows better performances.

Another extension of word embeddings is to enhance word representation by character-level features. The neural network to build word representations based on input character sequences is in spirit similar to that of sentence representations from input word sequences. Thus, we can also apply a standard CNN structure over the character embedding sequences to derive word representations. dos Santos and Gatti (2014) study the effect of such an extension. The resulting character-level word representations are concatenated with the original word embeddings, shown in Fig. 8.9, thus can enhance the final word representations for sentence encoding.

8.3.2 Recurrent Neural Networks

The CNN structure uses a fixed-size of word window to capture the local composition features around a given position, achieving promising results. However, it ignores the long-distance dependency features that reflect syntactic and semantic information, which are particularly important in understanding natural language sentences. These

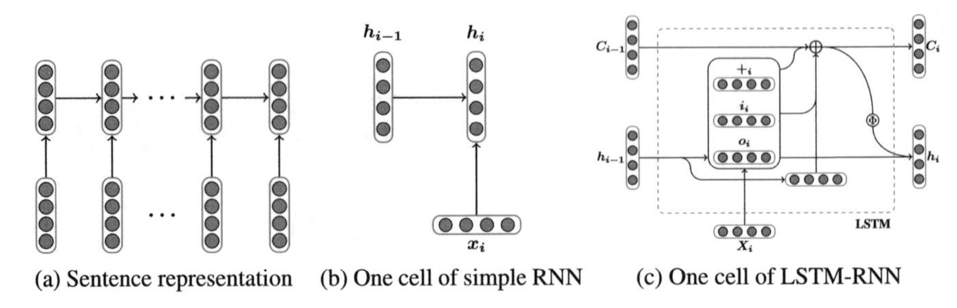

(a) Sentence representation (b) One cell of simple RNN (c) One cell of LSTM-RNN

Fig. 8.10 Sentence representation by using RNN

dependency-based features are addressed by recurrent neural network (RNN) under the neural setting, achieving great success. Formally, a standard RNN computes the output hidden vectors sequentially by $\mathbf{h}_i = f(W\mathbf{x}_i + U\mathbf{h}_{i-1} + \mathbf{b})$, where $\mathbf{x}_i$ denotes the input vector. According to the equation, we can see that the current output $\mathbf{h}_i$ relies not only on the current input $\mathbf{x}_i$, but also on the previous hidden output $\mathbf{h}_{i-1}$. In this manner, the current hidden output can have connections with previous input and output vectors without bound.

Wang et al. (2015) propose the first work of using long short-term memory (LSTM) neural networks for tweet sentiment analysis. Figure 8.10 shows the sentence representation method by using RNN, as well as the internal structures of standard and LSTM-RNN. First they apply a standard RNN over an input embedding sequence $\mathbf{x}_1\mathbf{x}_2\ldots\mathbf{x}_n$, and exploit the last hidden output $\mathbf{h}_n$ as the final representation of one sentence. Then the authors suggest a substitution by using LSTM-RNN structure, since standard RNNs may suffer the gradient explosion and diminish problems, while LSTM is much better by using three gates and a memory cell to connect input and output vectors. Formally, LSTM can be computed by

$$\begin{aligned}
\mathbf{i}_i &= \sigma(W_1\mathbf{x}_i + U_1\mathbf{h}_{i-1} + \mathbf{b}_1) \\
\mathbf{f}_i &= \sigma(W_2\mathbf{x}_i + U_2\mathbf{h}_{i-1} + \mathbf{b}_2) \\
\tilde{\mathbf{c}}_i &= \tanh(W_3\mathbf{x}_i + U_3\mathbf{h}_{i-1} + \mathbf{b}_3) \\
\mathbf{c}_i &= \mathbf{f}_i \odot \mathbf{c}_{i-1} + \mathbf{i}_i \odot \tilde{\mathbf{c}}_i \\
\mathbf{o}_i &= \sigma(W_4\mathbf{x}_i + U_4\mathbf{h}_{i-1} + \mathbf{b}_4) \\
\mathbf{h}_i &= \mathbf{o}_i \odot \tanh(\mathbf{c}_i),
\end{aligned}$$

where $W, U, \mathbf{b}$ are model parameters and σ denotes the sigmoid function.

Further, Teng et al. (2016) extend their work by two points. Figure 8.11 shows their framework. First, they exploit bidirectional LSTM instead, rather than a single left-to-right LSTM. The bidirectional can represent a sentence more comprehensively, where the hidden output of each point can have connections with both previous and future words. Second, they model sentence-level sentiment classification as a structural learning problem, predicting polarities for all sentiment words in a sentence and accumulating together as the evidence to determine the sentential polarity. By

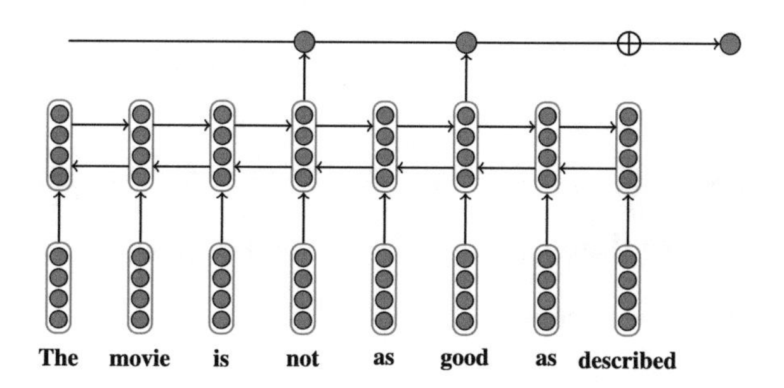

Fig. 8.11 The framework of Teng et al. (2016)

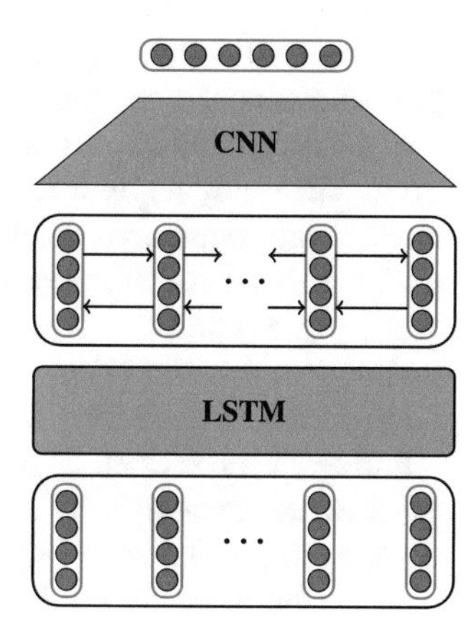

Fig. 8.12 A combination of RNN and CNN

the second extension, their model can effectively integrate the sentiment lexicons, which has been widely used in traditional statistical models.

CNN and RNN model natural language sentences in totally different ways. For example, CNN can better capture local window-based compositions, while RNN is efficient in learning implicit long-distance dependencies. Thus, one natural idea is to combine them together, taking advantages of both neural structures. Zhang et al. (2016c) propose a dependency-sensitive CNN model, which combines a LSTM and a CNN, making a CNN network structure being able to capture long- distance word dependencies as well. Concretely, first they construct a left-to-right LSTM on the input word embeddings, and then a CNN is built on the hidden outputs of the LSTM. Thus the final model can make full use of both local window-based features and global dependency-sensitive features. Figure 8.12 shows the framework of their combination model.

8.3.3 Recursive Neural Networks

Recursive neural network is recently proposed to model tree structural inputs, which are produced by explicit syntactic parsers. Socher et al. (2012) present a recursive matrix-vector neural network to compose two leaf nodes, resulting in

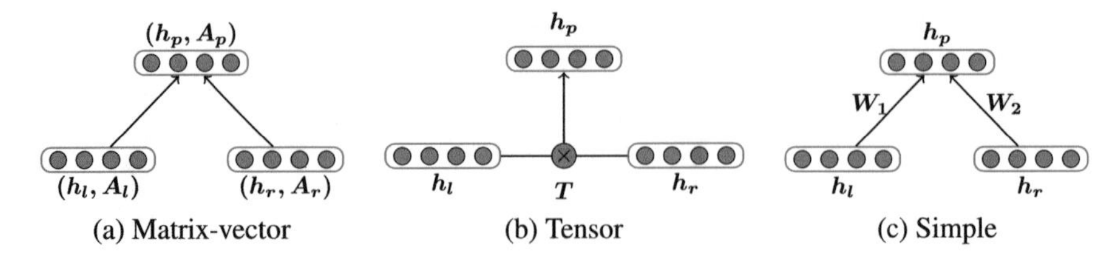

Fig. 8.13 Recursive neural network

the representation of the parent node. By this way, the sentence representation is constructed recursively from bottom to up. They first preprocess the input constituent trees, converting them into a binarized tree, where each parent node has two leaf nodes. Then they apply a recursive neural network over the binary tree by using matrix-vector operations. Formally, they represent each node by a hidden vector $\mathbf{h}$ and a matrix A. As shown in Fig. 8.13a, given the representations of the two child nodes, $(\mathbf{h}_l, A_l)$ and $(\mathbf{h}_r, A_r)$, respectively, the representation of the parent node is computed as follows: (1) $\mathbf{h}_p = f(A_r\mathbf{h}_l, A_l\mathbf{h}_r)$ and (2) $A_p = g(A_l, A_r)$, where $f(\cdot)$ and $g(\cdot)$ are transformation functions with model parameters.

Further, Socher et al. (2013) adopt low-rank tensor operations to substitute the matrix-vector recursion, by using $\mathbf{h}_p = f(\mathbf{h}_l T \mathbf{h}_r)$ to compute the representation of parent nodes, as shown in Fig. 8.13b, where T denotes a tensor. The model achieves better performances due to the tensor composition, which is intuitively simple than matrix-vector operation and has much less number of model parameters. In addition, they define the sentiment polarities over the non-root nodes of syntactic trees, thus can better capture the transition of sentiments from phrases to sentences.

The line of work is extended with three different directions. First, several work tries to find stronger composition operations for tree composition. For example, a number of works simply use $\mathbf{h}_p = f(W_1\mathbf{h}_l, W_2\mathbf{h}_r)$ to compose the leaf nodes, as shown in Fig. 8.13c. The method is much simpler, but suffers from the problem of gradient explosion or diminish, making the parameter learning extremely difficult. Motivated by the work of LSTM-RNN, several studies propose the LSTM adaption for recursive neural network. The representative work includes (Tai et al. 2015) and (Zhu et al. 2015), both of which show the effectiveness of LSTM over tree structures.

Second, sentence representation-based recursive neural network can be strengthened by using multichannel compositions. Dong et al. (2014b) study the effectiveness of such an enhancement. They apply C homogeneous compositions, arriving at C output hidden vectors, which are further used to represent the parent node by using an attention integration. Figure 8.14 shows the framework of their neural network. They apply the method on simple recursive neural networks, achieving consistent better performances on several benchmark datasets.

The third direction is to investigate recursive neural network by using deeper neural network structures, similar to the work of multilayer CNN. Briefly speaking, as the first layer, recursive neural network is applied over the input word embeddings. When all output hidden vectors are ready, the same recursive neural network can be

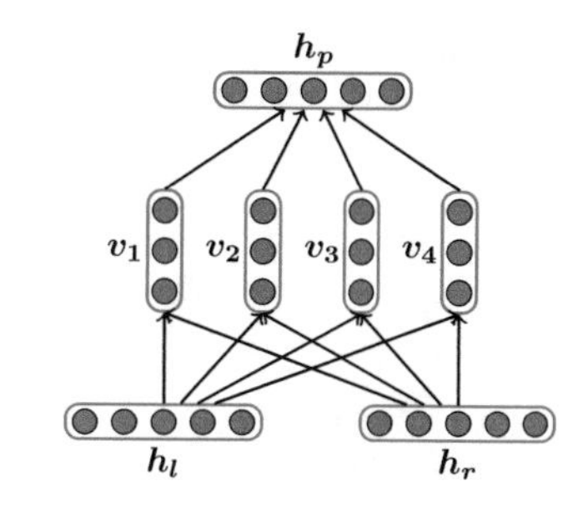

Fig. 8.14 Recursive neural network with multi-compositions

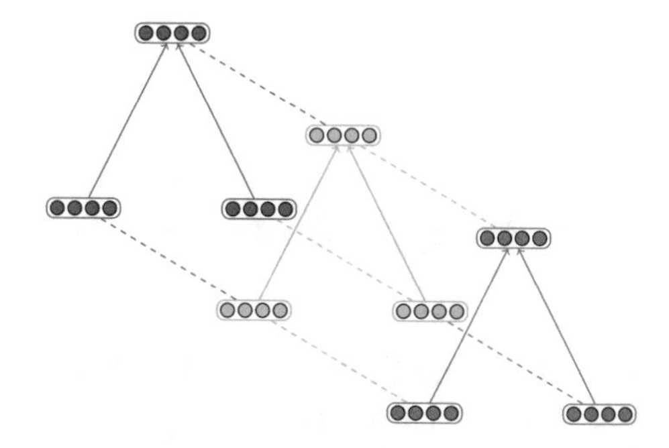

Fig. 8.15 Multilayer recursive neural network

applied by once again. The method is empirically studied by Irsoy and Cardie (2014a). Figure 8.15 shows their framework by using a three-layer recursive neural network. The experimental results demonstrate that deeper recursive neural network can bring better performances than a single-layer recursive neural network.

The above studies all construct recursive neural network over well-formed binary syntactic trees, which is seldom satisfied. Thus, they require certain preprocessing to convert original syntactic structures into binarized ones, which may be problematic without expert supervision. Recently, several studies propose to model trees with unbounded leaf nodes directly. For example, Mou et al. (2015) and Ma et al. (2015) both present a pooling operation based on the child nodes to compose variable length of inputs. Teng and Zhang (2016) perform the pooling process considering the left and right children. In addition, they suggest bidirectional LSTM recursive neural network, considering the top-to-down recursive operation, which is similar with the bidirectional LSTM-RNN.

It is worth to notice that, several works consider sentence representation by using recursive neural network without syntactic tree structures. These work suggest pseudo tree structures based on raw sentence inputs. For example, Zhao et al. (2015) construct a pseudo- directed acyclic graph in order to apply recursive neural network, as shown in Fig. 8.16. In addition, Chen et al. (2015) use a simpler method as shown in Fig. 8.17 to build a tree structure for a sentence automatically. Both the works achieve competitive performances for sentence-level sentiment analysis.

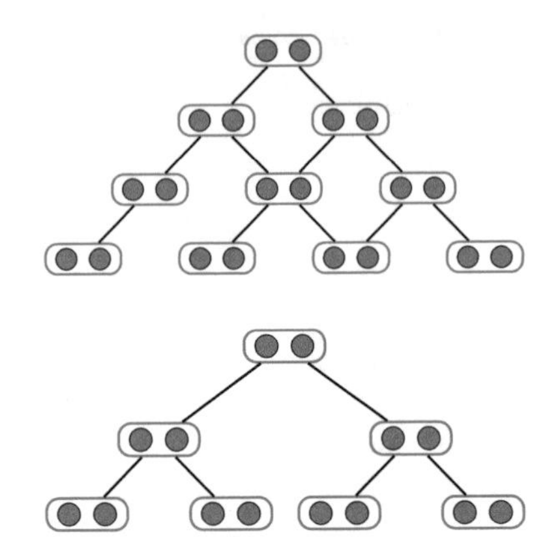

Fig. 8.16 Pseudo-directed acyclic graph of Zhao et al. (2015)

Fig. 8.17 Pseudo binary tree structure of Chen et al. (2015)

8.3.4 Integration of External Resources

The above subsections concern various neural structures for sentence representation, with the information from the source input sentences only, including words, parsing trees. Recently, another line of important work is to enhance sentence representation by integration with external resources. The major resources can be divided into three categories, including the large-scale raw corpus to pretrain supervised model parameters, external human-annotated or automatically extracted sentiment lexicons, and the background knowledge under a certain setting, for example, Twitter sentiment classification.

The exploration of large-scale corpus to enhance sentence representation has been investigated by a number of studies. Among these studies, the sequence autoencoder model proposed by Hill et al. (2016) are most representative. Figure 8.18 shows an example for the model, which first represents sentences by LSTM-RNN encoder, and then tries to generate the original sentential word step by step, thus model parameters are learned by this supervision, which are further used as external information for sentence representation. In particular, Gan et al. (2016) suggest a CNN encoder instead, aiming to solve the low-efficiency problem in LSTM-RNN.

External sentiment lexicons have been largely investigated in the statistical models, while there remains relatively little work under the neural setting, although there has been much work on automatically constructing sentiment lexicons. There are two exceptions. Teng et al. (2016) incorporate context-sensitive lexicon features in a LSTM-RNN neural network, treating sentence-level sentiment scores as a weighted sum of prior sentiment scores of negation words and sentiment words. Qian et al. (2017) go further, investigating the sentiment shifting effect of sentiment, negation, and intensity word, proposing a linguistically regularized LSTM model for sentence-level sentiment analysis.

There are several studies to investigate other information for sentence-level sentiment analysis under certain settings. In the Twitter sentiment classification, we can use several contextual information, including the tweet author's history tweets,

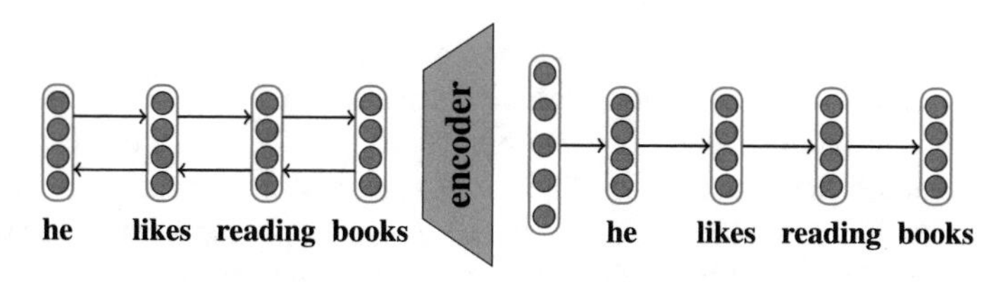

Fig. 8.18 Autoencoder by LSTM-RNN

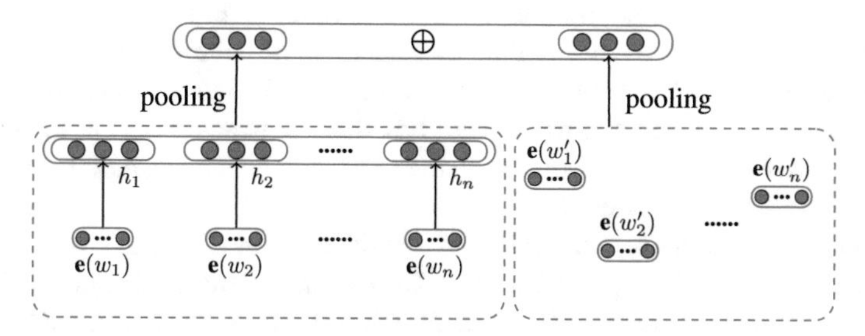

Fig. 8.19 Sentiment classification with contextual features

the conversational tweets surrounding the tweet, and the topic-related tweets. These information can be all severed as background information, which is intuitively helpful to decide the sentiment of a tweet. Ren et al. (2016a) exploit these related information in a neural network model by an additional contextual part, as shown in Fig. 8.19, to enhance sentiment analysis in Twitter. For the source input sentences, they apply a CNN to represent it, while for the contextual part, they apply a simple pooling neural network over a set of salient contextual words. Recently, Mishra et al. (2017) suggest an integration of cognitive features from gaze data to enhance sentence-level sentiment analysis, which is achieved by using an additional CNN structure to model the gaze features.

8.4 Document-Level Sentiment Classification

Document-level sentiment classification aims at identifying the sentiment label of a document (Pang et al. 2002; Turney 2002). The sentiment labels could be two categories such as *thumbs up* and *thumbs down* (Pang et al. 2002) or multiple categories such as the 1–5 stars on review sites (Pang and Lee 2005).[7]

In the literature, existing sentiment classification approaches could be grouped into two directions: lexicon- based approach and corpus-based approach. Lexicon-based approaches (Turney 2002; Taboada et al. 2011) mostly use a dictionary of

[7]In practice, it is time consuming to obtain the document- level sentiment labels via human annotation. Researchers typically leverage the review documents from IMDB, Amazon, and Yelp, and regard the associated rating stars as the sentiment labels.

sentiment words with their associated sentiment polarity, and incorporate negation and intensification to compute the sentiment polarity for each document. A representative lexicon-based method is given by (Turney 2002), which consists of three steps. Phrases are first extracted, if their POS tags conform to the predefined patterns. Afterward, the sentiment polarity of each extracted phrase is estimated through pointwise mutual information (PMI), which measures the degree of statistical dependence between two terms. In Turney's work, the PMI score is calculated by feeding queries to a search engine and collecting the number of hits. Finally, he averages the polarity of all phrases in a review as its sentiment polarity. Ding et al. (2008) apply negation words like "not", "never", "cannot", and contrary words like "but" to enhance the performance of lexicon-based method. Taboada et al. (2011) integrate intensifications and negation words with the sentiment lexicons annotated with their polarities and sentiment strengths.

Corpus-based methods treat sentiment classification as a special case of text categorization problem (Pang et al. 2002). They mostly build a sentiment classifier from documents with annotated sentiment polarity. The sentiment supervision can be manually annotated, or automatically collected by sentiment signals like emoticons in tweets or human ratings in reviews. Pang et al. (2002) pioneer to treat the sentiment classification of reviews as a special case of text categorization problem and first investigate machine learning methods. They employ Naive Bayes, Maximum Entropy, and Support Vector Machines (SVM) with a diverse set of features. In their experiments, the best performance is achieved by SVM with bag-of-words features. Following Pang et al.'s work, many studies focus on designing or learning effective features to obtain a better classification performance. On movie and product reviews, Wang and Manning (2012) present NBSVM, which trade-off between Naive Bayes and NB-feature enhanced SVM. Paltoglou and Thelwall (2010) learn feature weights by investigating variants weighting functions from Information Retrieval, such as tf.idf and its BM25 variants. Nakagawa et al. (2010) utilize dependency trees, polarity-shifting rules and conditional random fields with hidden variables to compute the document feature.

The intuition of developing neural network approach is that feature engineering is typically labor intensive. Neural network approaches instead have the ability to discover explanatory factors from the data and make the learning algorithms less dependent on extensive feature engineering. Bespalov et al. (2011) represent each word as a vector (embedding), and then get the vectors for phrases with temporal convolutional network. The document embedding is calculated by averaging the phrase vectors. Le and Mikolov (2014) extend the standard skip-gram and CBOW models Mikolov et al. (2013b) to learn the embeddings for sentences and documents. They represent each document by a dense vector which is trained to predict words in the document. Specifically, the PV-DM model extends the skip-gram model by averaging/concatenating the document vector with context vectors to predict the middle word. The models of Denil et al. (2014); Tang et al. (2015a); Bhatia et al. (2015); Yang et al. (2016); Zhang et al. (2016c) have the same intuition. They model the embedding of sentences from the words, and then use sentence vectors to compose the document vector. Specifically, Denil et al. (2014) use the same convolutional

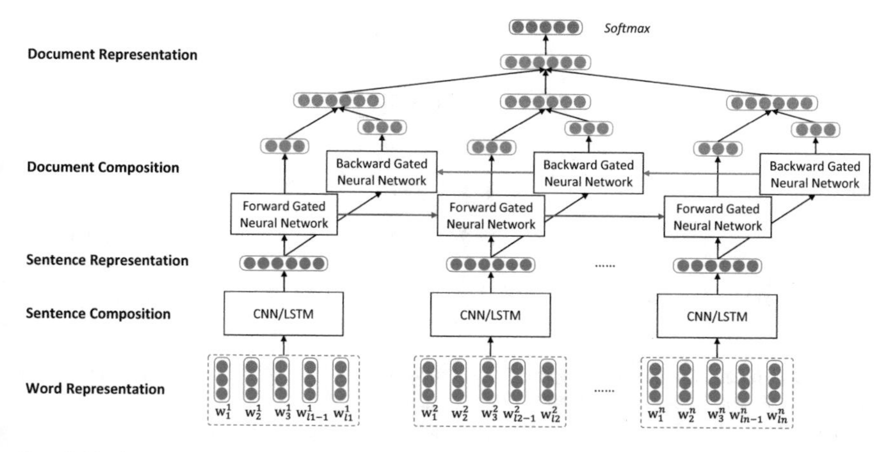

Fig. 8.20 A neural network architecture for document-level sentiment classification (Tang et al. 2015a).

neural network as the sentence modeling component and the document modeling component. Tang et al. (2015a) use convolutional neural network to calculate the sentence vector, and then use bidirectional gated recurrent neural network to calculate the document embedding. The model is given in Fig. 8.20. Bhatia et al. (2015) calculate document vector based on the structure obtained from the RST parse. Zhang et al. (2016c) calculate sentence vector with recurrent neural network, and then use convolutional network to calculate the document vector. Yang et al. (2016) use two attention layers to get the sentence vectors, and the document vector, respectively. In order to calculate the weights of different words from a sentence and the weights of different sentences of a document, they use two "context" vectors, which are jointly learned in the training process. Joulin et al. (2016) introduces a simple and efficient approach, which averages the word representations into a text representation, and then feeds the results to a linear classifier. Johnson and Zhang (2014, 2015, 2016) develop convolutional neural networks that take one-hot word vector as input and represent a document with the meanings of different regions. The aforementioned studies regard word as the basic computational unit, and compose the document vector based on word representation. Zhang et al. (2015b) and Conneau et al. (2016) use characters as the basic computational units, and explore convolutional architectures to calculate the document vector. The vocabulary for characters is dramatically smaller than the standard vocabulary of words. In Zhang et al. (2015b), the alphabet consists of 70 characters, including 26 English letters, 10 digits, 33 other characters, and the new line character. The model of Zhang et al. (2015b) has 6 convolution layers, and the model of Conneau et al. (2016) consists of 29 layers.

There also exist studies that explore side information such as individual preferences of users or overall qualities of products to improve document-level sentiment classification. For example, Tang et al. (2015b) incorporate user-sentiment consistency and user-text consistency to an existing convolutional neural network. In the

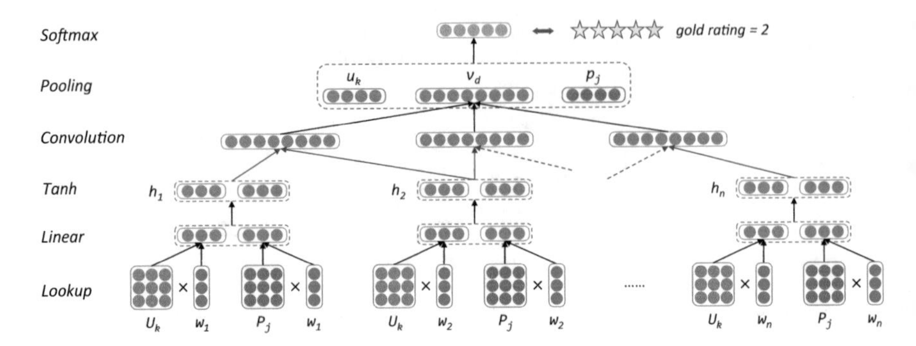

Fig. 8.21 The neural network approach that incorporates user and product information for document- level sentiment classification (Tang et al. 2015b).

user-text consistency, each user is represented as a matrix to modify the meaning of a word. In the user-sentiment consistency, each user is encoded as a vector, which is directly concatenated with the document vector and regarded as a part of the features for sentiment classification. The model is given in Fig. 8.21. Chen et al. (2016) make an extension and develop attention models to take into account the importance of words.

8.5 Fine-Grained Sentiment Analysis

In this section, we introduce the recent advances in fine-grained sentiment analysis using deep learning. Different from sentence/document-level sentiment classification, fine-grained sentiment analysis involves a number of tasks, most of which have their own characteristics. Thus, these tasks are modeled differently, carefully considering their special application settings. Here, we introduce five different topics of fine-grained sentiment analysis, including opinion mining, targeted sentiment analysis, aspect-level sentiment analysis, stance detection, and sarcasm detection.

8.5.1 Opinion Mining

Opinion mining has been a hot topic in the NLP community, which aims to extract structured opinions from user- generated reviews. Figure 8.22 shows several examples of opinion mining. Typically, the task involves two subtasks. First opinion entities such as holders, targets, and expressions are identified, and second we build relations over these entities, for example, the IS-ABOUT relation which specifies the target of a certain opinion expression, and the IS-FROM relation which links an opinion

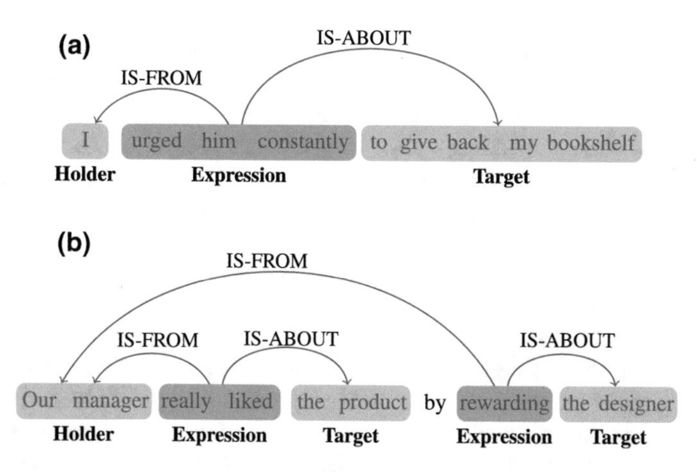

Fig. 8.22 Examples of opinion mining

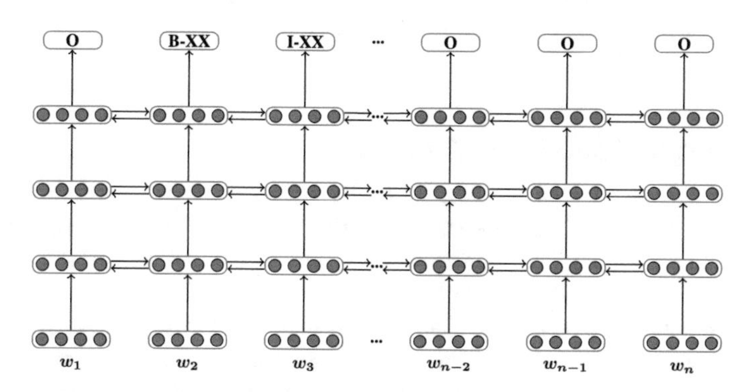

Fig. 8.23 A three-layer Bi-LSTM model for opinion entity detection

expression with its holder. In addition, the classification of sentiment polarities is an important task as well.

Opinion mining is a typical structural learning problem, which has been studied extensively by using traditional statistical models with human-designed discrete features. While recently, motivated by the great success of deep learning models on other NLP tasks, especially on sentiment analysis, neural network-based models have received grown attentions on the task as well. In the below, we describe several representative studies of this task by using neural networks.

The early work of neural network models focuses on the detection of opinion entities, treating the task as a sequence labeling problem to recognize boundaries of opinion entities. Irsoy and Cardie (2014b) investigate the RNN structure for the task. They apply the Elman-type RNNs, studying the effectiveness bidirectional RNN, and observing the influence of the RNN depth, as shown in Fig. 8.23. Their results show that bidirectional RNN can obtain better performances, and a three-layer bidirectional RNN can achieve the best performance.

A similar work is proposed by Liu et al. (2015). They make a comprehensive investigation of RNN variations, including Elman-type RNN, Jordan-type RNN, and LSTM. They study the bidirectionality as well. In addition, they compare three kinds of input word embeddings. They compare these neural network models with discrete models, and make a combination of the two different types of features. Their experiments show that the LSTM neural network combining with discrete features can achieve the best performance.

The above two studies do not involve the identification of the relation between opinion entities. Most recently, Katiyar and Cardie (2016) propose the first neural network that exploits LSTM to jointly perform entity recognition and opinion relation classification. They treat the two subtasks by a multitask learning paradigm, introducing sentence-level training considering both entity boundaries and their relations, based on a shared multilayer bidirectional LSTM. In particular, they define two sequences to denote the distance to their left and right entities of certain relations, respectively. Experimental results on benchmark MPQA datasets show that their neural model achieve the top-performing results.

8.5.2 *Targeted Sentiment Analysis*

Targeted sentiment analysis studies the sentiment polarity toward a certain entity in one sentence. Figure 8.24 shows several examples for the task, where $\{+, -, 0\}$ denote the positive, negative, and neutral sentiment, respectively.

The first neural network model for targeted-dependent sentiment analysis is proposed by Dong et al. (2014a). The model is adapted from their previous work of Dong et al. (2014b), which we have introduced in the sentence-level sentiment analysis. Similarly, they build recursive neural networks from a binarized dependency tree structure, by using multi- compositions from the child nodes. However, this work is different in that they convert the dependency tree according to the input target, making the headword of the target as the root in the resulting tree, not the original head word of the input sentence. Figure 8.25 shows the composition methods and the resulting dependency tree structure, where "phone" is the target.

The above work highly relies on the input dependency parsing trees, which are produced by automatic syntactic parsers. The trees can have errors, thus suffering from the error propagation problem. To avoid the problem, recent studies suggest

I like [**this washing machine**]$_+$! Really convenient and easy to use !
Disgust food of [**the school canteen**]$_-$! I admire myself for eating in the canteen for four years !
Love [**La La Land**]$_+$ most ! Much better than Beauty and the Beast .
I have no interest in playing [**basketball**]$_-$ and also never watch any live of it .
I do not know [**Ryan Gosling**]$_0$, so I cannot answer any questions in your survey .

Fig. 8.24 Targeted sentiment analysis

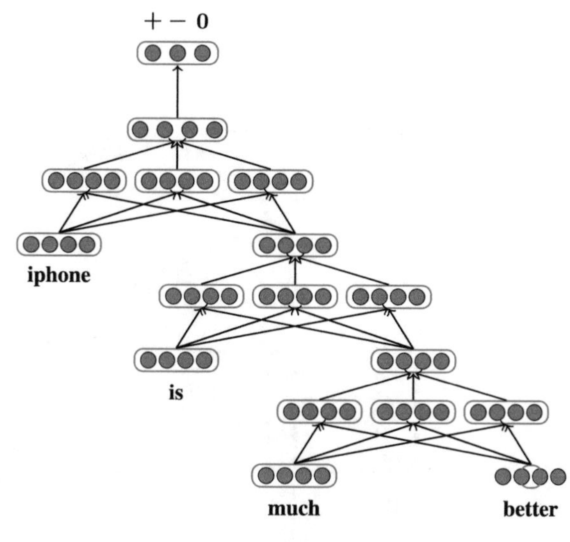

Fig. 8.25 The framework of Dong et al. (2014a)

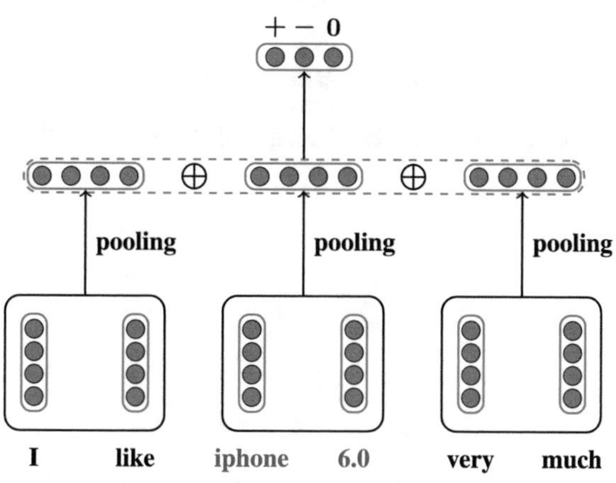

Fig. 8.26 The framework of Vo and Zhang (2015)

conducting targeted sentiment analysis with only raw sentence inputs. Vo and Zhang (2015) exploit various pooling strategies to extract a number of neural features for the task. They first divide the input sentence into three segments by a given target, and then apply different pooling functions over the three segments together with the whole sentence, as shown in Fig. 8.26. The resulting neural features are concatenated for further sentiment polarity prediction.

Recently, several works investigate the effectiveness of RNN for the task, which has brought promising performances in other sentiment analysis tasks. Zhang et al. (2016b) propose to use gated RNN to enhance the representation of sentential words. By using RNN, the resulting representations can capture context-sensitive information, as shown in Fig. 8.27. Further, Tang et al. (2016a) exploit LSTM-RNN as one basic neural layer to encode the input sequential words. Figure 8.28 shows the framework of their work. Both the works have achieved state-of-the-art performances in targeted sentiment analysis.

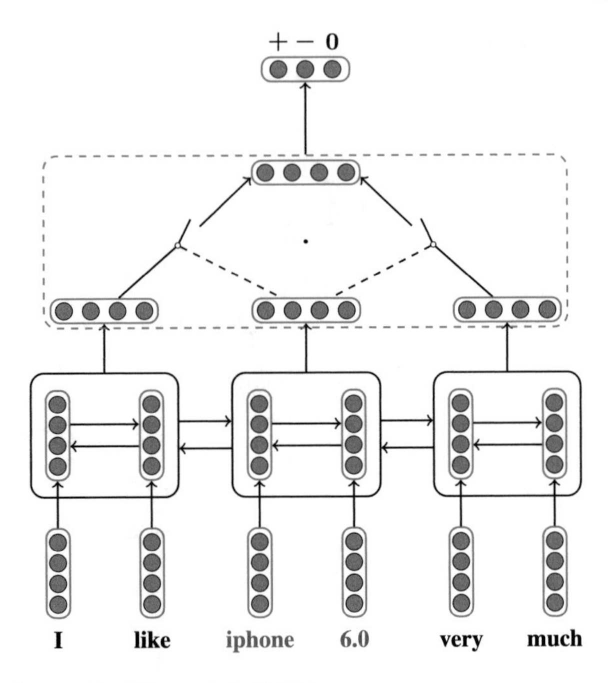

Fig. 8.27 The framework of Zhang et al. (2016b)

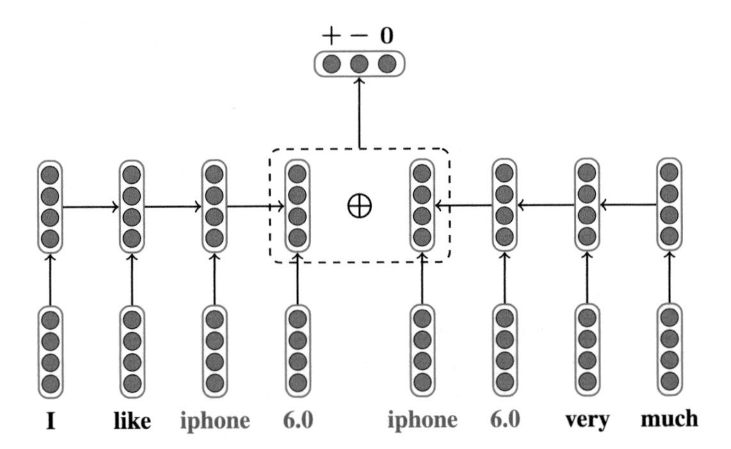

Fig. 8.28 The framework of Tang et al. (2016a)

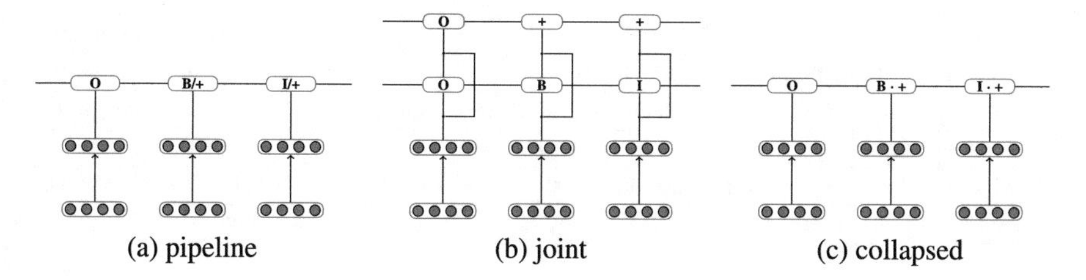

Fig. 8.29 Open domain-targeted sentiment analysis

Besides the use of RNN, Zhang et al. (2016b) present a gated neural network to compose the features of the left, right contexts by target supervision, as shown in Fig. 8.27. The main motivation behind is that the context-neural features should not be equally treated by simply pooling. The task should carefully consider the target as well in order to choose effective features. Liu and Zhang (2017) improve the gated mechanism further, by applying an attention strategy. With the attention, their model achieves the top performances on two benchmark datasets.

Previous work demonstrated that boundaries of the input target is important for the inferring of its sentiment polarities. They assume that well-posed targets are already given, which is not always a real scenario. For example, if we want to determine the sentiment polarities of open targets, it is required to recognize the these targets in advance. Zhang et al. (2015a) study the open domain-targeted sentiment analysis by using neural networks. They investigate the problem under various settings, including pipeline, joint, and collapsed frameworks. Figure 8.29 shows the three frameworks. In addition, they combine the neural and traditional discrete features in a single model, finding that better performances can be obtained consistently under the three settings.

8.5.3 Aspect-Level Sentiment Analysis

Aspect-level sentiment analysis aims to classify the sentiment polarities in a sentence for an aspect. An aspect is one attribute of a target, over which human can express their opinions. Figure 8.30 shows several examples of the task. Usually, the task is aimed to analyze user comments for a certain product, e.g., a hotel, an electronics, or a movie. Products may have a number of aspects. For example, the aspects of a hotel include environment, price, and service, and users usually post a review to express their opinions over certain aspects. Different from targeted sentiment analysis, aspects can be enumerated when the product is given, and the aspect may not be expressed regularly in one review in some cases.

Initially, the task is modeled as a sentence classification problem, thus we can exploit the same method as the sentence-level sentiment classification, expect that the categories are different. Typically, assuming that a product has N aspects which

Sentence	Aspect	Polarity
The screen of the laptop is nice. I like it very much .	screen	positive
It is a choice as a whole, although the owner is not as friendly.	service	negative
The phone is not bad, especially for its strong battery.	battery	positive
I likė the movie very much, in particular the story touches me greatly.	screenwriter	positive
I need to change my laptop now, since the key U does work.	keyboard	negative

Fig. 8.30 Aspect-level sentiment analysis

are predefined by expert, the aspect-level sentiment classification is actually a $3N$-classification problem, since each aspect can have three sentiment polarities: positive, negative, and neutral. Lakkaraju et al. (2014) propose a recursive neural network model-based matrix-vector composition for the task, which is similar to Socher et al. (2012) that performs sentence-level sentiment classification.

In later work, the task has been simplified by assuming that aspect has been given in an input sentence, thus it is equivalent to the aforementioned targeted sentiment analysis. Nguyen and Shirai (2015) propose a phrase-based recursive neural network model to the aspect-level sentiment analysis, where the input phrase structure trees are converted from dependency structures along with the input aspects. Tang et al. (2016b) apply a deep memory neural network under the same setting, without using syntactic trees. Their model achieves state-of-the-art performances, and meanwhile is highly efficient in speed in comparison with the neural models that exploit LSTM structures. Figure 8.31 shows their three-layer deep memory neural network. The final features for classification are extracted by attentions with aspect supervision.

In real scenarios, one aspect of a certain product can have several different expressions. Taking the laptop as an example, we can express the aspect screen by display, resolution, and look, which are closely related to screen. If we can group similar aspect phrases into one aspect, the results of aspect-level sentiment analysis are more helpful for further application. Xiong et al. (2016) propose the first neural network model for aspect phrase grouping. They learn representations of aspect phrase by simple multilayer feed-forward neural networks, extracting neural features with attention composition. The model parameters are trained by distant supervision with automatic training examples. Figure 8.32 shown their framework. He et al. (2017) exploit an unsupervised auto-encoder framework for aspect extraction, which can learn the scale of aspect words automatically by attention mechanism.

8.5.4 *Stance Detection*

The goal of stance detection is to recognize the attitude of one sentence toward a certain topic. Generally, the topic is specified for the task as one input, and the other input is the sentence that needs to be classified. Input sentences may not have explicit relations with the given topic. which makes the task rather different with

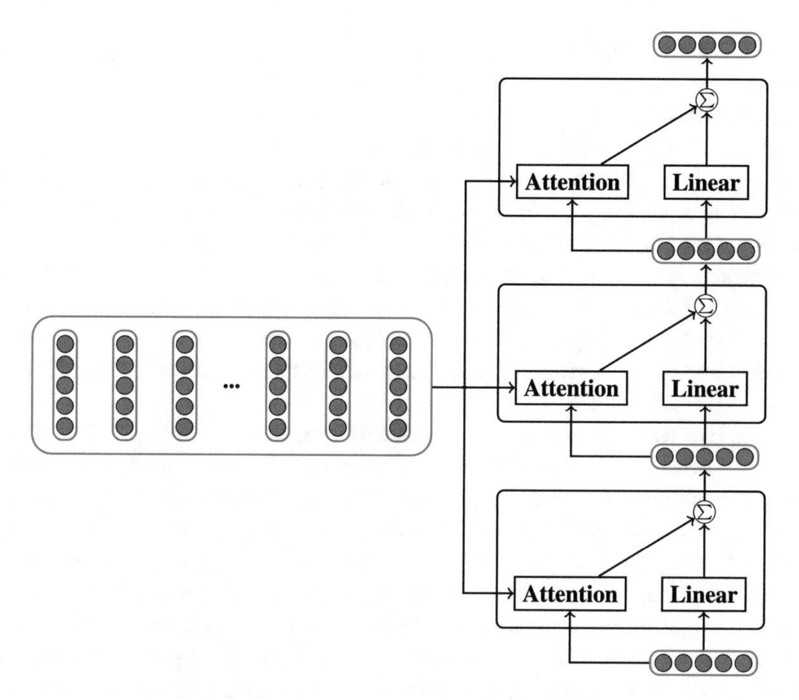

Fig. 8.31 The framework of Tang et al. (2016a)

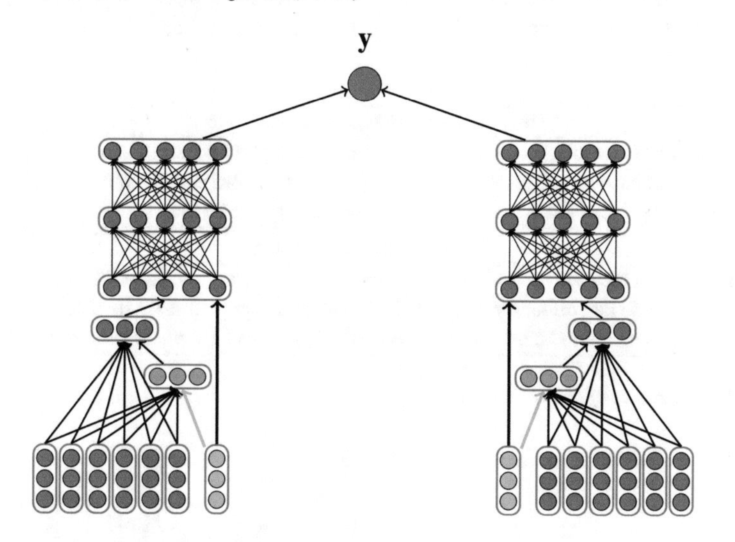

Fig. 8.32 The framework of Xiong et al. (2016)

target/aspect-level sentiment analysis, Thus stance detection is extremely difficult. Figure 8.33 shows several examples of the task.

Early work trains independent classifiers for each topic. Thus, the task is treated as a simple 3-way classification problem. For example, Vijayaraghavan et al. (2016) exploit a multilayer CNN model for the task. They integrate both word and character embeddings as inputs in order to solve the unknown words. In the SemEval 2016 task 6 of stance detection, the model of Zarrella and Marsh (2016) achieved the top performance, which builds a neural network based on LSTM-RNN, who has strong capabilities of learning syntactic and semantic features. In addition, motivated by the spirit of transfer learning, they learn the model parameters by the priori knowledge from hashtags in the Twitter, because the raw input sentences of the SemEval task are crawled from Twitter.

The above work models stance classification of different topics independently, which has two main drawbacks. On the one hand, it is not as practical to annotate training examples for each topic, in order to classify the attitudes of a sentence for future topics. On the other hand, several topics may have close relations, for example, "Hillary Clinton" and "Donald Trump" while training the classifiers independently is unable of using this information. Augenstein et al. (2016) propose the first model to train a single model no matter the input topics as a whole, using LSTM neural networks. They model the input sentence and topic jointly, by using the resulting representation of the topics as the input for LSTM over the sentences. Figure 8.34 shows the framework of their method. Their model achieves significantly better performances than the individual classifiers of previous work.

Topic: Climate Change is a Real Concern	
Academy of Science talk Tech solutions for climate change with Barry Brook.	Favor
This just in, an ocean wave just broke an inch further on the beach than normal!	Against
I love this Pope. I don't care what religion you are, this guy is awesome.	NULL
Topic: Feminist Movement	
Because women are seen as "soft," and "emotional" in the eyes of male politicians.	Favor
If the confederate flag offends you, good. Stop making things politically correct.	Against
People say I'm young to be into politics. Honestly, I just stand for what I believe in.	NULL

Fig. 8.33 Examples of stance detection

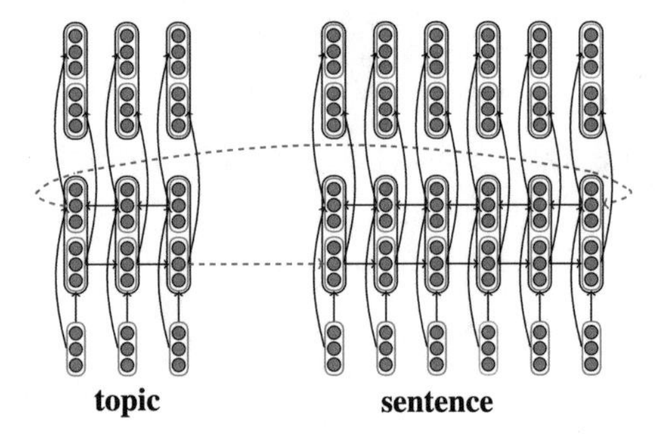

Fig. 8.34 Conditional LSTM for stance detection

8.5.5 *Sarcasm Recognition*

In this section, we discuss a special language phenomenon that has close connections with sentiment analysis, namely sarcasm or irony. This phenomenon usually makes change of a sentence's literal meaning, and greatly influence the sentiment expressed by the sentence. Figure 8.35 shows several examples.

Typically, sarcasm detection is modeled as a binary classification problem, which is similar with sentence-level sentiment analysis is essential. The major difference between the two tasks lies in their goals. Ghosh and Veale (2016) study various neural network models for the task in detail, including CNN, LSTM, and deep feed-forward neural networks. They present several different neural models, and investigate their effectiveness empirically. The experimental results show that a combination of these neural networks can bring the best performances. The final model is composed by a two-layer CNN, a two-layer LSTM and another one feed-forward layer, as shown in Fig. 8.36.

For sarcasm detection in social media such as Twitter, author-based information is one kind of useful features. Zhang et al. (2016a) propose a contextualized neural model for Twitter sarcasm recognition. Concretely, they extract a set of salient words from the tweet authors' historical posts, using these words to represent the tweet author. Their proposed neural network model consists two parts, as shown in Fig. 8.37, one being a gated RNN to represent sentences, and the other being a simple pooling neural network to represent tweet author.

Sometimes idiots just brighten my day , incredible !!
I love waking up in the dark and coming home in the dark.
Now I know where you get your manners from.
The only bad thing about weed is getting caught with it #makessense
My life is so exciting......., I just can not believe what have happened.
Glad my dryer has ruined two of my camis now

Fig. 8.35 Sarcasm examples

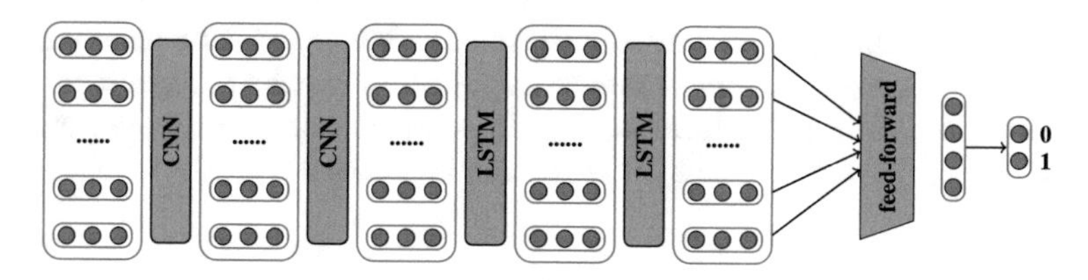

Fig. 8.36 The framework of Ghosh and Veale (2016)

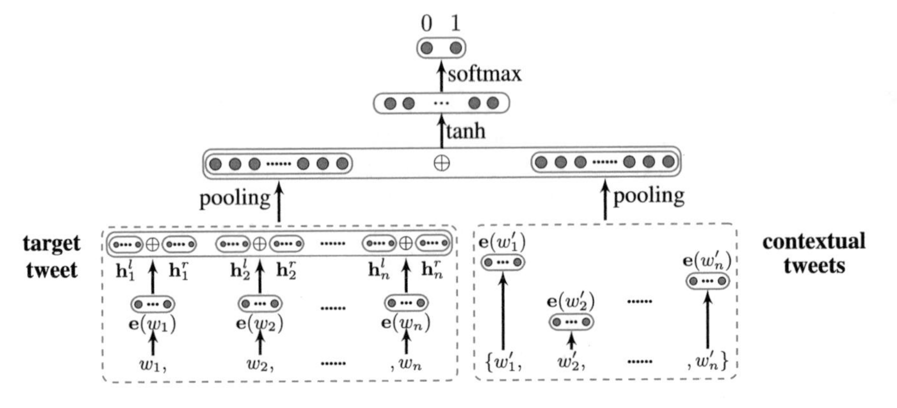

Fig. 8.37 The framework of Zhang et al. (2016a).

8.6 Summary

In this chapter, we give an overview on the recent success of neural network approaches in sentiment analysis. We first describe how to integrate sentiment information of texts to learn sentiment-specific word embeddings. Then, we describe sentiment classification of sentences and documents, both of which require semantic composition of texts. We then present how to develop neural network models to deal with fine-grained tasks.

Despite deep learning approaches have achieved promising performances on sentiment analysis tasks in recent years, there are some potential directions to further improve this area. The first direction is explainable sentiment analysis. The current deep learning models are accurate yet unexplainable. Leveraging knowledge from cognitive science, common sense knowledge, or extracted knowledge from text corpus might be a potential direction to improve this area. The second direction is learning a robust model for a new domain. The performance of a deep learning model depends on the amount and the quality of the training data. Therefore, how to learn a robust sentiment analyzer for a domain with little/no annotated corpus is very challenging yet important for real application. The third direction is how to understand the emotion. Majority of existing studies focus on opinion expressions, targets, and holders. Recently, new attributes have been suggested to better understand the emotion, such as opinion causes and stances. Pushing forward this area requires powerful models and large corpora. The fourth direction is fine-grained sentiment analysis, which receives increasing interests recently. Improving this area requires larger training corpus.

References

Augenstein, I., Rocktäschel, T., Vlachos, A., & Bontcheva, K. (2016). Stance detection with bidirectional conditional encoding. In *EMNLP2016* (pp. 876–885).

Bai, B., Weston, J., Grangier, D., Collobert, R., Sadamasa, K., Qi, Y., et al. (2010). Learning to rank with (a lot of) word features. *Information Retrieval, 13*(3), 291–314.

Baker, L. D. & McCallum, A. K. (1998). Distributional clustering of words for text classification. In *Proceedings of the 21st Annual International ACM SIGIR Conference on Research and Development in Information Retrieval* (pp. 96–103). ACM.
Bengio, Y., Ducharme, R., Vincent, P., & Jauvin, C. (2003). A neural probabilistic language model. *Journal of Machine Learning Research, 3*(Feb), 1137–1155.
Bespalov, D., Bai, B., Qi, Y., & Shokoufandeh, A. (2011). Sentiment classification based on supervised latent n-gram analysis. In *Proceedings of the 20th ACM International Conference on Information and Knowledge Management* (pp. 375–382). ACM.
Bhatia, P., Ji, Y., & Eisenstein, J. (2015). Better document-level sentiment analysis from rst discourse parsing. arXiv:1509.01599.
Brown, P. F., Desouza, P. V., Mercer, R. L., Pietra, V. J. D., & Lai, J. C. (1992). Class-based n-gram models of natural language. *Computational Linguistics, 18*(4), 467–479.
Chen, X., Qiu, X., Zhu, C., Wu, S., & Huang, X. (2015). Sentence modeling with gated recursive neural network. In *Proceedings of the 2015 Conference on Empirical Methods in Natural Language Processing* (pp. 793–798). Lisbon, Portugal: Association for Computational Linguistics.
Chen, H., Sun, M., Tu, C., Lin, Y., & Liu, Z. (2016). Neural sentiment classification with user and product attention. In *Proceedings of EMNLP*.
Collobert, R. & Weston, J. (2008). A unified architecture for natural language processing: Deep neural networks with multitask learning. In *Proceedings of the 25th International Conference on Machine Learning* (pp. 160–167). ACM.
Collobert, R., Weston, J., Bottou, L., Karlen, M., Kavukcuoglu, K., & Kuksa, P. (2011). Natural language processing (almost) from scratch. *Journal of Machine Learning Research, 12*(Aug), 2493–2537.
Conneau, A., Schwenk, H., Barrault, L., & Lecun, Y. (2016). Very deep convolutional networks for natural language processing. arXiv:1606.01781.
Deerwester, S., Dumais, S. T., Furnas, G. W., Landauer, T. K., & Harshman, R. (1990). Indexing by latent semantic analysis. *Journal of the American Society for Information Science, 41*(6), 391.
Deng, L. & Wiebe, J. (2015). MPQA 3.0: An entity/event-level sentiment corpus. In *HLT-NAACL* (pp. 1323–1328).
Denil, M., Demiraj, A., Kalchbrenner, N., Blunsom, P., & de Freitas, N. (2014). Modelling, visualising and summarising documents with a single convolutional neural network. arXiv:1406.3830.
Ding, X., Liu, B., & Yu, P. S. (2008). A holistic lexicon-based approach to opinion mining. In *Proceedings of the 2008 International Conference on Web Search and Data Mining* (pp. 231–240). ACM.
Dong, L., Wei, F., Tan, C., Tang, D., Zhou, M., & Xu, K. (2014a). Adaptive recursive neural network for target-dependent twitter sentiment classification. In *ACL* (pp. 49–54).
Dong, L., Wei, F., Zhou, M., & Xu, K. (2014b). Adaptive multi-compositionality for recursive neural models with applications to sentiment analysis. In *AAAI* (pp. 1537–1543).
dos Santos, C. & Gatti, M. (2014). Deep convolutional neural networks for sentiment analysis of short texts. In *Proceedings of COLING 2014, The 25th International Conference on Computational Linguistics: Technical Papers* (pp. 69–78). Dublin, Ireland: Dublin City University and Association for Computational Linguistics.
Faruqui, M., Dodge, J., Jauhar, S. K., Dyer, C., Hovy, E., & Smith, N. A. (2014). Retrofitting word vectors to semantic lexicons. arXiv:1411.4166.
Gan, Z., Pu, Y., Henao, R., Li, C., He, X., & Carin, L. (2016). Unsupervised learning of sentence representations using convolutional neural networks. arXiv:1611.07897.
Ghosh, A., & Veale, D. T. (2016). Fracking sarcasm using neural network. In *Proceedings of the 7th Workshop on Computational Approaches to Subjectivity, Sentiment and Social Media Analysis* (pp. 161–169).
Goodfellow, I., Bengio, Y., & Courville, A. (2016). *Deep learning*. Cambridge: MIT Press.
Gutmann, M. U., & Hyvärinen, A. (2012). Noise-contrastive estimation of unnormalized statistical models, with applications to natural image statistics. *Journal of Machine Learning Research,13*(Feb), 307–361.

Harris, Z. S. (1954). *Distributional structure. Word, 10*(2–3), 146–162.
He, R., Lee, W. S., Ng, H. T., & Dahlmeier, D. (2017). An unsupervised neural attention model for aspect extraction. In *Proceedings of the 55th ACL* (pp. 388–397). Vancouver, Canada: Association for Computational Linguistics.
Hill, F., Cho, K., & Korhonen, A. (2016). Learning distributed representations of sentences from unlabelled data. In *NAACL* (pp. 1367–1377).
Hu, M. & Liu, B. (2004). Mining and summarizing customer reviews. In *Proceedings of the Tenth ACM SIGKDD International Conference on Knowledge Discovery and Data Mining* (pp. 168–177). ACM.
Huang, P.-S., He, X., Gao, J., Deng, L., Acero, A., & Heck, L. (2013). Learning deep structured semantic models for web search using clickthrough data. In *Proceedings of the 22nd ACM International Conference on Information and Knowledge Management* (pp. 2333–2338). ACM.
Huang, E. H., Socher, R., Manning, C. D., & Ng, A. Y. (2012). Improving word representations via global context and multiple word prototypes. In *Proceedings of the 50th Annual Meeting of the Association for Computational Linguistics: Long Papers-Volume 1* (pp. 873–882). Association for Computational Linguistics.
Irsoy, O. & Cardie, C. (2014a). Deep recursive neural networks for compositionality in language. In *Advances in neural information processing systems* (pp. 2096–2104).
Irsoy, O. & Cardie, C. (2014b). Opinion mining with deep recurrent neural networks. In *Proceedings of the 2014 EMNLP* (pp. 720–728).
Johnson, R. & Zhang, T. (2014). Effective use of word order for text categorization with convolutional neural networks. arXiv:1412.1058.
Johnson, R. & Zhang, T. (2015). Semi-supervised convolutional neural networks for text categorization via region embedding. In *Advances in neural information processing systems* (pp. 919–927).
Johnson, R. & Zhang, T. (2016). Supervised and semi-supervised text categorization using LSTM for region embeddings. arXiv:1602.02373.
Joulin, A., Grave, E., Bojanowski, P., & Mikolov, T. (2016). Bag of tricks for efficient text classification. arXiv:1607.01759.
Jurafsky, D. (2000). *Speech and language processing*. New Delhi: Pearson Education India.
Kalchbrenner, N., Grefenstette, E., & Blunsom, P. (2014). A convolutional neural network for modelling sentences. In *Proceedings of the 52nd Annual Meeting of the Association for Computational Linguistics (Volume 1: Long Papers)* (pp. 655–665), Baltimore, Maryland: Association for Computational Linguistics.
Katiyar, A. & Cardie, C. (2016). Investigating LSTMs for joint extraction of opinion entities and relations. In *Proceedings of the 54th ACL* (pp. 919–929).
Kim, Y. (2014). Convolutional neural networks for sentence classification. In *Proceedings of the 2014 Conference on Empirical Methods in Natural Language Processing (EMNLP)* (pp. 1746–1751). Doha, Qatar: Association for Computational Linguistics.
Labutov, I., & Lipson, H. (2013). Re-embedding words. In *ACL* (Vol. 2, pp. 489–493).
Lakkaraju, H., Socher, R., & Manning, C. (2014). Aspect specific sentiment analysis using hierarchical deep learning. In *NIPS Workshop on Deep Learning and Representation Learning*.
Le, Q. V. & Mikolov, T. (2014). Distributed representations of sentences and documents. In *ICML* (Vol. 14, pp. 1188–1196).
Lebret, R., Legrand, J., & Collobert, R. (2013). *Is deep learning really necessary for word embeddings?*. Idiap: Technical Report.
LeCun, Y., Bengio, Y., & Hinton, G. (2015). Deep learning. *Nature, 521*(7553), 436–444.
Lei, T., Barzilay, R., & Jaakkola, T. (2015). Molding CNNs for text: Non-linear, non-consecutive convolutions. In *Proceedings of the 2015 Conference on Empirical Methods in Natural Language Processing* (pp. 1565–1575). Lisbon, Portugal: Association for Computational Linguistics.
Levy, O. & Goldberg, Y. (2014). Dependency-based word embeddings. In *ACL*, (Vol. 2, pp. 302–308). Citeseer.

Li, J. & Jurafsky, D. (2015). Do multi-sense embeddings improve natural language understanding? arXiv:1506.01070.

Li, J., Luong, M.-T., Jurafsky, D., & Hovy, E. (2015). When are tree structures necessary for deep learning of representations? arXiv:1503.00185.

Liu, J. & Zhang, Y. (2017). Attention modeling for targeted sentiment. In *Proceedings of EACL* (pp. 572–577).

Liu, P., Joty, S., & Meng, H. (2015). Fine-grained opinion mining with recurrent neural networks and word embeddings. In *Proceedings of the 2015 EMNLP* (pp. 1433–1443).

Liu, B. (2012). Sentiment analysis and opinion mining. *Synthesis Lectures on Human Language Technologies*, *5*(1), 1–167.

Lund, K., & Burgess, C. (1996). Producing high-dimensional semantic spaces from lexical co-occurrence. *Behavior Research Methods, Instruments, and Computers*, *28*(2), 203–208.

Ma, M., Huang, L., Zhou, B., & Xiang, B. (2015). Dependency-based convolutional neural networks for sentence embedding. In *Proceedings of the 53rd Annual Meeting of the Association for Computational Linguistics and the 7th International Joint Conference on Natural Language Processing (Volume 2: Short Papers)* (pp. 174–179), Beijing, China: Association for Computational Linguistics.

Maas, A. L., Daly, R. E., Pham, P. T., Huang, D., Ng, A. Y., & Potts, C. (2011). Learning word vectors for sentiment analysis. In *Proceedings of the 49th Annual Meeting of the Association for Computational Linguistics: Human Language Technologies-Volume 1* (pp. 142–150). Association for Computational Linguistics.

Manning, C. D., Schütze, H., et al. (1999). *Foundations of Statistical Natural Language Processing* (Vol. 999). Cambridge: MIT Press.

Mikolov, T., Chen, K., Corrado, G., & Dean, J. (2013a). Efficient estimation of word representations in vector space. arXiv:1301.3781.

Mikolov, T., Sutskever, I., Chen, K., Corrado, G. S., & Dean, J. (2013b). Distributed representations of words and phrases and their compositionality. In *Advances in Neural Information Processing Systems* (pp. 3111–3119).

Mishra, A., Dey, K., & Bhattacharyya, P. (2017). Learning cognitive features from gaze data for sentiment and sarcasm classification using convolutional neural network. In *Proceedings of the 55th ACL* (pp. 377–387). Vancouver, Canada: Association for Computational Linguistics.

Mnih, A. & Hinton, G. (2007). Three new graphical models for statistical language modelling. In *Proceedings of the 24th International Conference on Machine Learning* (pp. 641–648). ACM.

Mnih, A. & Kavukcuoglu, K. (2013). Learning word embeddings efficiently with noise-contrastive estimation. In *Advances in neural information processing systems* (pp. 2265–2273).

Morin, F. & Bengio, Y. (2005). Hierarchical probabilistic neural network language model. In *Aistats* (Vol. 5, pp. 246–252). Citeseer.

Mou, L., Peng, H., Li, G., Xu, Y., Zhang, L., & Jin, Z. (2015). Discriminative neural sentence modeling by tree-based convolution. In *Proceedings of the 2015 Conference on Empirical Methods in Natural Language Processing* (pp. 2315–2325). Lisbon, Portugal: Association for Computational Linguistics.

Nakagawa, T., Inui, K., & Kurohashi, S. (2010). Dependency tree-based sentiment classification using CRFs with hidden variables. In *Human Language Technologies: The 2010 Annual Conference of the North American Chapter of the Association for Computational Linguistics* (pp. 786–794). Association for Computational Linguistics.

Nguyen, T. H. & Shirai, K. (2015). PhraseRNN: Phrase recursive neural network for aspect-based sentiment analysis. In *EMNLP* (pp. 2509–2514).

Paltoglou, G. & Thelwall, M. (2010). A study of information retrieval weighting schemes for sentiment analysis. In *Proceedings of the 48th Annual Meeting of the Association for Computational Linguistics* (pp. 1386–1395). Association for Computational Linguistics.

Pang, B., & Lee, L. (2005). Seeing stars: Exploiting class relationships for sentiment categorization with respect to rating scales. In *Proceedings of the 43rd Annual Meeting on Association for Computational Linguistics* (pp. 115–124). Association for Computational Linguistics.

Pang, B., Lee, L., & Vaithyanathan, S. (2002). Thumbs up?: Sentiment classification using machine learning techniques. In *Proceedings of the ACL-02 Conference on Empirical Methods in Natural Language Processing-Volume 10* (pp. 79–86). Association for Computational Linguistics.

Pang, B., Lee, L., et al. (2008). Opinion mining and sentiment analysis. Foundations and trends®. *Information Retrieval*, *2*(1–2), 1–135.

Qian, Q., Huang, M., Lei, J., & Zhu, X. (2017). Linguistically regularized LSTM for sentiment classification. In *Proceedings of the 55th ACL* (pp. 1679–1689). Vancouver, Canada: Association for Computational Linguistics.

Qiu, S., Cui, Q., Bian, J., Gao, B., & Liu, T.-Y. (2014). Co-learning of word representations and morpheme representations. In *COLING* (pp. 141–150).

Ren, Y., Zhang, Y., Zhang, M., & Ji, D. (2016a). Context-sensitive twitter sentiment classification using neural network. In *AAAI* (pp. 215–221).

Ren, Y., Zhang, Y., Zhang, M., & Ji, D. (2016b). Improving twitter sentiment classification using topic-enriched multi-prototype word embeddings. In *AAAI* (pp. 3038–3044).

Shen, Y., He, X., Gao, J., Deng, L., & Mesnil, G. (2014). Learning semantic representations using convolutional neural networks for web search. In *Proceedings of the 23rd International Conference on World Wide Web* (pp. 373–374). ACM.

Socher, R., Huval, B., Manning, C. D., & Ng, A. Y. (2012). Semantic compositionality through recursive matrix-vector spaces. In *Proceedings of the 2012 Joint Conference on Empirical Methods in Natural Language Processing and Computational Natural Language Learning* (pp. 1201–1211). Jeju Island, Korea: Association for Computational Linguistics.

Socher, R., Perelygin, A., Wu, J., Chuang, J., Manning, C. D., Ng, A., & Potts, C. (2013). Recursive deep models for semantic compositionality over a sentiment treebank. In *Proceedings of the 2013 Conference on Empirical Methods in Natural Language Processing* (pp. 1631–1642). Seattle, Washington, USA: Association for Computational Linguistics.

Taboada, M., Brooke, J., Tofiloski, M., Voll, K., & Stede, M. (2011). Lexicon-based methods for sentiment analysis. *Computational Linguistics*, *37*(2), 267–307.

Tai, K. S., Socher, R., & Manning, C. D. (2015). Improved semantic representations from tree-structured long short-term memory networks. In *Proceedings of the 53rd Annual Meeting of the Association for Computational Linguistics and the 7th International Joint Conference on Natural Language Processing (Volume 1: Long Papers)* (pp. 1556–1566). Beijing, China: Association for Computational Linguistics.

Tang, D., Qin, B., & Liu, T. (2015a). Document modeling with gated recurrent neural network for sentiment classification. In *EMNLP* (pp. 1422–1432).

Tang, D., Qin, B., & Liu, T. (2015b). Learning semantic representations of users and products for document level sentiment classification. In *ACL* (Vol. 1, pp. 1014–1023).

Tang, D., Qin, B., & Liu, T. (2016a). Aspect level sentiment classification with deep memory network. In *Proceedings of the 2016 EMNLP* (pp. 214–224).

Tang, D., Qin, B., Feng, X., & Liu, T. (2016b). Effective LSTMs for target-dependent sentiment classification. In *Proceedings of COLING, 2016* (pp. 3298–3307).

Tang, D., Wei, F., Yang, N., Zhou, M., Liu, T., & Qin, B. (2014). Learning sentiment-specific word embedding for twitter sentiment classification. In *Proceedings of the 52nd Annual Meeting of the Association for Computational Linguistics (Volume 1: Long Papers)* (pp. 1555–1565). Baltimore, Maryland: Association for Computational Linguistics.

Tang, D., Wei, F., Qin, B., Yang, N., Liu, T., & Zhou, M. (2016c). Sentiment embeddings with applications to sentiment analysis. *IEEE Transactions on Knowledge and Data Engineering*, *28*(2), 496–509.

Teng, Z., & Zhang, Y. (2016). Bidirectional tree-structured lstm with head lexicalization. arXiv:1611.06788.

Teng, Z., Vo, D. T., & Zhang, Y. (2016). Context-sensitive lexicon features for neural sentiment analysis. In *Proceedings of the 2016 Conference on Empirical Methods in Natural Language Processing* (pp. 1629–1638). Austin, Texas: Association for Computational Linguistics.

Turney, P. D. (2002). Thumbs up or thumbs down?: Semantic orientation applied to unsupervised classification of reviews. In *Proceedings of the 40th Annual Meeting on Association for Computational Linguistics* (pp. 417–424). Association for Computational Linguistics.

Vijayaraghavan, P., Sysoev, I., Vosoughi, S., & Roy, D. (2016). Deepstance at semeval-2016 task 6: Detecting stance in tweets using character and word-level CNNs. In *SemEval-2016* (pp. 413–419).

Vo, D.-T. & Zhang, Y. (2015). Target-dependent twitter sentiment classification with rich automatic features. In *Proceedings of the IJCAI* (pp. 1347–1353).

Wang, S. & Manning, C. D. (2012). Baselines and bigrams: Simple, good sentiment and topic classification. In *Proceedings of the 50th Annual Meeting of the Association for Computational Linguistics: Short Papers-Volume 2* (pp. 90–94). Association for Computational Linguistics.

Wang, X., Liu, Y., Sun, C., Wang, B., & Wang, X. (2015). Predicting polarities of tweets by composing word embeddings with long short-term memory. In *Proceedings of the 53rd Annual Meeting of the Association for Computational Linguistics and the 7th International Joint Conference on Natural Language Processing (Volume 1: Long Papers)* (pp. 1343–1353), Beijing, China: Association for Computational Linguistics.

Xiong, S., Zhang, Y., Ji, D., & Lou, Y. (2016). Distance metric learning for aspect phrase grouping. In *Proceedings of COLING, 2016* (pp. 2492–2502).

Yang, Z., Yang, D., Dyer, C., He, X., Smola, A., & Hovy, E. (2016). Hierarchical attention networks for document classification. In *Proceedings of NAACL-HLT* (pp. 1480–1489).

Yin, W. & Schütze, H. (2015). Multichannel variable-size convolution for sentence classification. In *Proceedings of the Nineteenth Conference on Computational Natural Language Learning* (pp. 204–214). Beijing, China: Association for Computational Linguistics.

Yogatama, D., Faruqui, M., Dyer, C., & Smith, N. A. (2015). Learning word representations with hierarchical sparse coding. In *ICML* (pp. 87–96).

Zarrella, G. & Marsh, A. (2016). Mitre at semeval-2016 task 6: Transfer learning for stance detection. In *SemEval-2016* (pp. 458–463).

Zhang, R., Lee, H., & Radev, D. R. (2016c). Dependency sensitive convolutional neural networks for modeling sentences and documents. In *Proceedings of the 2016 NAACL* (pp. 1512–1521). San Diego, California: Association for Computational Linguistics.

Zhang, Y., Roller, S., & Wallace, B. C. (2016d). MGNC-CNN: A simple approach to exploiting multiple word embeddings for sentence classification. In *Proceedings of the 2016 NAACL* (pp. 1522–1527). San Diego, California: Association for Computational Linguistics.

Zhang, M., Zhang, Y., & Fu, G. (2016a). Tweet sarcasm detection using deep neural network. In *Proceedings of COLING 2016, The 26th International Conference on Computational Linguistics: Technical Papers* (pp. 2449–2460). Osaka, Japan: The COLING 2016 Organizing Committee.

Zhang, M., Zhang, Y., & Vo, D.-T. (2015a). Neural networks for open domain targeted sentiment. In *Proceedings of the 2015 Conference on EMNLP*.

Zhang, M., Zhang, Y., & Vo, D.-T. (2016b). Gated neural networks for targeted sentiment analysis. In *AAAI* (pp. 3087–3093).

Zhang, X., Zhao, J., & LeCun, Y. (2015b). Character-level convolutional networks for text classification. In *Advances in neural information processing systems* (pp. 649–657).

Zhao, H., Lu, Z., & Poupart, P. (2015). Self-adaptive hierarchical sentence model. arXiv:1504.05070.

Zhu, X.-D., Sobhani, P., & Guo, H. (2015). Long short-term memory over recursive structures. In *ICML* (pp. 1604–1612).

Chapter 9
Deep Learning in Social Computing

Xin Zhao and Chenliang Li

Abstract The goal of social computing is to devise computational systems to learn mechanisms and principles to explain and understand the behaviors of each individual and collective teams, communities, and organizations. The unprecedented online data in social media provides a fruitful resource for this purpose. However, traditional techniques have a hard time in handling the complex and heterogeneous nature of social media for social computing. Fortunately, the recent revival and success of deep learning brings new opportunities and solutions to address these challenges. This chapter introduces the recent progress of deep learning on social computing in three aspects, namely user-generated content, social connections, and recommendation, which have covered most of the core elements and applications in social computing. Our focus lies in the discussions on how to adapt deep learning techniques to mainstream social computing tasks.

9.1 Introduction to Social Computing

The essence of human behaviors is *profoundly social*, which is reflected by various kinds of human activities in their social life. For example, people communicate with their families, purchase products from business retailers, and watch movies with friends. With these social activities, everyone is remarkably influenced by and affect other people around us and beyond (Homans 1974). Social behaviors are not the product of the development of our modern society or technical advances, but a critical building block of human society. Back to the Stone Age, individuals gather together to form the tribes, which can be considered as a kind of community. Within a tribe, people share their experiences about the world and make exchange with other people within or outside the tribe (Sahlins 2017). Through successive generations, social

X. Zhao (✉)
Renmin University of China, Beijing, China
e-mail: batmanfly@gmail.com

C. Li
Wuhan University, Wuhan, China
e-mail: lichenliang.whu@gmail.com

L. Deng and Y. Liu (eds.), *Deep Learning in Natural Language Processing*, https://doi.org/10.1007/978-981-10-5209-5_9

regulations and conventions regarding the individuals, organizations, and societies are then developed to guide their behaviors.

In recent years, the rapid growth of Internet technology leads to the prosperity of numerous online social media services, which not only refer to popular social networks, like Facebook, Twitter, or Sina Weibo, but also relate to any online services that are powered by some Internet technology with social features. Online social media has greatly changed or affected the way that people live. It is time to think over how to model users' social behaviors and improve online social services. That is the topic that *social computing* focuses in the era of social media. *Social computing* is defined to be systems that support the gathering, representation, processing, use, and dissemination of information that is distributed across social collectives such as teams, communities, organizations, and markets (Wang et al. 2007; Parameswaran and Whinston 2007). Moreover, the information is not "anonymous" but is significant precisely because it is linked to people, who are in turn linked to other people (Schuler 1994). In other words, social computing is the discipline of understanding the activities of individuals in a social context.

The advent of different online social media services brings about unprecedented information explosion in human history. Compared to traditional websites that restrict the users to be only information consumers, online social media enables the users to produce information via diverse interactions with information items, such as Wikipedia and Open Directory Project (ODP) for collaborative knowledge building; Delicious, BibSonomy, and CiteULike for collaboratively tagging documents; Digg for evaluating web content, Facebook, and Twitter; Weibo for information sharing and commenting among friends; Netflix and IMDB for evaluating movies; YouTube for sharing videos, Yahoo! Answers; Quora for knowledge sharing, etc.

A major feature of online social media is that users are highly connected via various linking mechanisms (Kaplan and Haenlein 2010). Due to the elaborate design of online social networks, there exist multiple types of social connections between users. Take Twitter as an instance. On Twitter, there are three major types of social links between two users: (1) following, a user has added another user in her friend list; (2) retweeting, a user has forwarded a tweet from another user; and (3) mentioning, a user has included another user in her own tweet. Rich user connections significantly enhance the collaborative, interactive environment of online social media. The connections are also likely to convey topic semantics or interest similarities to some extent (Weng et al. 2010). For example, two users may edit the same Wikipedia article because both are interested in some common topic.

Besides content and connection, another important issue is how to satisfy users' complicated, diverse, and varying information needs on information resources. Following the convention in recommender system (Adomavicius and Tuzhilin 2005), we refer to an information resource as an *item* on social media, which can be a tweet, a movie, a song, a product, etc. Most social media platforms provide their own recommender systems to facilitate the access of information by users. The recommendation scenario can be understood as a process of social interaction between users and items. A user can provide either explicit or implicit feedbacks to the items during the interaction process. These feedback information encode important evidence to infer users' interests or needs over the items.

Given the explosive content, rich social connections, and complicated information needs, social computing is strongly tied to user behaviors and user interests in social media. The ultimate goal of social computing is to devise computational systems to learn mechanisms and principles (or called knowledge/intelligence) to explain and understand the behaviors of each individual and collective teams, communities, and organizations (Wang et al. 2007). To achieve this goal, three fundamental aspects underlying the success of social computing are highlighted here:

- Deep semantic understanding of user-generated text content. People participate in online social media to write or share real-time posts, rate and leave opinions for products and services, tag web pages, and so on. One critical step for social computing is to enable semantic information extraction and understanding from the user-generated text content in an automatic way. Moreover, given the flexible mechanism of online social media services, social text can be presented in diverse forms and with new features. Hence, an effective modeling approach is desired to be developed, which could help us identify the needles from a huge pile of haystacks efficiently and precisely.
- Effective representation learning for social connections. The rich social connections enable us to study and analyze user relations in a large social context. Online social networks are complex in nature. A key technique toward network or link analysis is to develop an effective network representation learning approach. The solution to network representation should be general to characterize multiple types of user links, and support a series of computation tasks such as community detection, influence maximization, expert finding, and other tasks. Moreover, it is important to combine knowledge from different perspectives by mining various explicit and implicit relations.
- Accurate recommendation with information resources. Recommender systems play an important role in online social media. Making recommendations or suggestions to users are able to increase their degree of engagement for websites. It will help reduce the efforts of a user in looking for interested items. Social media brings new challenges to traditional recommender systems by incorporating more social context information. The interaction between users and items has become more complex, and multiple kinds of feedback information are available for consideration. To develop an accurate recommender system, these new features from social media platforms should be considered.

Traditional techniques from natural language process (NLP), information retrieval (IR) (Manning et al. 2008), and machine learning (ML) (Alpaydin 2014) can be utilized in social computing to some extent. However, these techniques have difficulty in solving the challenges raised by social media. First, traditional data representations are usually based on the one-hot sparse representation. The high dimensionality of one-hot representations makes it difficult to discover the underlying knowledge/relations from sparse data and process large-scale data efficiently. Moreover, traditional data representations are powerless to capture deep semantics of social media data, e.g., the commonly used "bag-of-words" (BOW) scheme can not well capture the polysemy and synonym that exist naturally in human languages or activities. Second, traditional data models may not be capable of characterizing the complex nature of social media data. For example, matrix factorization essentially is a linear factorization model which is not able to capture nonlinear data characteristics. Although nonlinear models achieve more capacity in data modeling, they are usually either shallow models or difficult to learn, which cannot effectively solve complicated tasks on social media. Third, traditional techniques are not flexibly extended to online social media, since online social media brings new data features or challenges to social computing. For example, the user-generated content is a fruitful resource and fast channel for understanding the trends and opinions expressed by the users, and many social media platforms have added the news spreading mechanism, such as *retweet* on Twitter (Kwak et al. 2010). Traditional techniques may not be easy to adapt to these new features in social media.

Fortunately, the recent revival and success of deep learning brings us new opportunities and solutions to address these difficulties that traditional techniques are faced with in social computing. Research in deep learning makes better data representations by using distributed representations and is able to learn these representations from large-scale unlabeled data (Mikolov et al. 2013). Deep learning tries to build more powerful data models using flexible deep nonlinear structures, which is loosely based on interpretation of information processing and communication patterns in a nervous system. Deep learning algorithms transform their inputs through more layers than shallow learning algorithms, and the capacity of neural networks has been discussed and proved in the universal approximation theorem (Hornik 1991). Another important feature of deep learning is that it is usually designed and trained in an end-to-end way, which substantially reduces the accumulated discrepancy from multiple separate model components. Besides powerful data modeling capacity, deep learning is a fast-growing field, and new architectures, variants, or algorithms appear every few weeks, which provide us a flexible way to model new data types or features (e.g., sequence data models and tree-structured data models).

Based on the above discussions, in this chapter, the deep learning is the major approach to social computing. As being introduced before, three aspects are mainly considered here, namely user-generated content analysis, social connections, and recommendation. Focusing on these three aspects, the major progress made on social computing with deep learning will be reviewed. These three parts will be introduced in Sects. 9.2, 9.3, and 9.4 respectively. Finally, the conclusion of this chapter is made in Sect. 9.5.

9.2 Modeling User-Generated Content with Deep Learning

The major resource for social computing is the user-generated content across different social media services (Cortizo et al. 2012). Because each individual can share the stories, social events, and opinions without many constraints to the public in a real-time manner, a user in social media works as a social sensor, recording the timely information happened around her. In this sense, the user-generated content contains a wealth of timely information and beyond. The UGC has been widely recognized as an avenue for opinion extraction, expert finding, user profiling, user intent understanding, and so on. For example, a government officer of the Social Security department is likely to be made aware whenever a rumor outbreak is taking place. A fated and complicated task inside these semantic tasks is to derive a semantic representation for the information generated by social media users. Among different kinds of UGC, text is a dominating resource form. Hence, many discussions will be made about how to model user-generated text content in this section.[1]

Effective learning of word semantics is now feasible and practical with recent developments in neural network techniques, which have contributed improvements in many NLP tasks. Specifically, neural network language models (Mikolov et al. 2013) learn word embeddings (or dense word representations) with the aim of fully retaining the contextual information for each word, including both semantic and syntactic relations. Moreover, most task-driven neural networks are devised to learn the embedding representations for words, documents, users, and many metadata information. In this section, we first briefly review the traditional semantic representation models, and then introduce the shallow embedding techniques such as CBOW and skip-gram models, and deep neural network models such as convolutional neural networks (CNN) and recurrent neural networks (RNN) that are task-driven. Finally, an introduction about attentive mechanism for text-based neural network techniques is given. As the emphasis through the entire section, the discussions about how to adapt deep learning techniques to specific social tasks will be provided.

9.2.1 Traditional Semantic Representation Approaches

The conventional approaches to represent documents and words are one-hot vector representations and bag-of-words (BOW) scheme (Manning et al. 2008). In one-hot vector representation, a word w is represented as a sparse $|V|$-dimensional vector $\mathbf{x}_w$, where every element in $\mathbf{x}_w$ is zero and only the element corresponding to word w is 1, and $|V|$ is the size of the vocabulary V. For example, assume the vocabulary $V = \{$"*I*", "*like*", "*apple*"$\}$ contains three distinct words. By sorting words in alphabetical order, the one-hot vector representation $\mathbf{x}_{apple}$ is then represented as $[1, 0, 0]$. With the one-hot representation, a document can then be represented as a weighted sum of $\mathbf{x}_w$ for all the words contained in the document as follows:

$$\mathbf{x}_d = \sum_{w \in d} f(w, d)\mathbf{x}_w, \tag{9.1}$$

[1]Other data types such as images and videos are not considered, which are beyond the scope of this chapter.

where $f(w, d)$ is the weighting function for word w in context of document d. The widely used term weighting function is TF-IDF scheme, which takes term frequency and inverse document frequency into account. Although this BOW representation leads to promising performance for regular document retrieval, it results in much poorer performance in many social media-related IR/NLP tasks, because of the short and error-prone nature of the user-generated content. For example, words "car" and "automobile" share the same semantic meaning and syntactic function. However, by using one-hot sparse representation mentioned above, the cosine similarity based measure gives a score of 0 for them. It is desired to derive a dense vector representation such that the syntactic and semantic relations for a pair of words can be well captured.

9.2.2 Semantic Representation with Shallow Embedding

Distributed representations have been successfully applied in many NLP and IR tasks. Two popular models to learn word embeddings are Continuous Bag-of-Words (CBOW) and continuous skip-gram models (Mikolov et al. 2013). To learn the representation for text units of different lengths (e.g., sentences, paragraph, and documents) instead of single word, Le et al. propose two paragraph vector models, which can derive the dense representation for sentences, paragraphs, and documents (Le and Mikolov 2014). These models can be called as *shallow embedding approaches* since they only involve a hidden layer. Because they can project text units of different lengths and metadata information into the same hidden representation space, it is flexible to apply these techniques and the variants in many semantic applications (e.g., microblog recommendation). In what follows, the introduction about several representative shallow embedding techniques will be provided as well as the discussion about how to adapt them to characterize additional social features besides textual semantics.

The main idea of CBOW is to predict the target word with surrounding context words. For convenient, the surrounding words are symmetric (so as in skip-gram), i.e., a window with size m is predefined and the task is to predict the target word w_c with a sequence of words $(w_{c-m},\ldots,w_{c-1},w_{c+1},\ldots,w_{c+m})$, where w_i denotes the word at position c. In the input layer, every word is represented by a one-hot sparse vector, i.e., every word is represented as a $\mathbb{R}^{|V|*1}$ vector where $|V|$ is the size of vocabulary. Then, an input word matrix $\mathbf{V} \in \mathbb{R}^{n\times|V|}$ is defined such that the ith column of $\mathbf{V}$ is the n-dimensional embedded vector for word w_i. From the input layer to the hidden layer, each hidden embedding vector $\mathbf{v}_i$ of word w_i is calculated by multiplying matrix $\mathbf{V}$ by $\mathbf{x}_i$, i.e., $\mathbf{v}_i = \mathbf{V}\mathbf{x}_i$. Recall word vector $\mathbf{x}_i$ is a one-hot vector, so this multiplication essentially performs a lookup operation (i.e., select the corresponding column in $\mathbf{V}$ as output). Then, the embeddings of input word in a window are averaged to form vector $\hat{\mathbf{v}}$, i.e., $\hat{\mathbf{v}} = (\mathbf{v}_{c-m} + \mathbf{v}_{c-m+1} + \ldots + \mathbf{v}_{c+m})/2m$. To enable prediction, another output word matrix $\mathbf{U} \in \mathbb{R}^{|V|\times n}$ is defined such that the jth row of $\mathbf{U}$ is an n-dimensional embedded vector for word w_j. The likelihood score vector $\mathbf{z}$ is then calculated by multiplying $\mathbf{U}$ by $\hat{\mathbf{v}}$, i.e., $\mathbf{z} = \mathbf{U}\hat{\mathbf{v}}$. Then, the softmax function takes $\mathbf{z}$ as the input to get the output $\hat{\mathbf{y}}$ (note $\hat{\mathbf{y}}$ is a probabilistic distribution vector).

The main idea of skip-gram model is opposite to CBOW model, i.e., surrounding context words are predicted by using the target words. There are two main differences between skip-gram and CBOW models. The first is that only one-word vector is taken in the input layer, instead of the context words in CBOW model. The second is that $2 \cdot m$ context words are predicted in the output layer separately. Following CBOW model, a similar optimization approach can be adopted. As a follow-up work, Le et al. propose two Paragraph Vector (PV) models to learn the distributed representations for text units of different lengths (e.g., sentences, paragraphs, and documents) (Le and Mikolov 2014). The main idea is to take the paragraph as an extra word. The embedding vector associated with the paragraph is considered as paragraph vector. There are two distinct models in PV framework. One is Distributed Memory Model of Paragraph Vector (PV-DM), and the other is the Distributed Bag-of-Words version of Paragraph Vector (PV-DBOW). In PV-DM, the main idea is similar with CBOW model, in a paragraph (or sentence, document), paragraph vector and context word vectors are averaged or concatenated to predict the next words (not central words). In PV-DBOW, context words are ignored in the input, and paragraph vectors are used alone to predict words randomly sampled from the paragraph (they are in same window). Although both models can learn the vector representation for a paragraph, but as pointed out by the authors, PV-DM is consistently better than PV-DBOW. Moreover, by concatenating the vectors learnt from both models, further improvement is observed in terms of document classification tasks.

Microblogging services, as a real-time information sharing platform, have attracted a huge number of users across different domains. Specifically, many researchers also publish or share the academic advances in microblogging sites with their comments and emotions. Identifying the expertise and research interest of these users could enable us to recommend the relevant microblogs to them. The effective scholarly microblog recommendation accuracy could enable the researcher to easily follow the recent progress of the field of interest. To devise a personalized scholarly microblog recommendation approach, Yu et al. propose two User2Vec models to learn the user embedding as well as text/word embedding jointly (Yu et al. 2016). Then, the recommendation is accomplished by calculating the similarity between a user's vector and a scholarly microblog text's vector. These two User2Vec models were built on the basis of PV-DM model. In User2Vec#1 model (shown in Fig. 9.1a), the upper architecture is the same as PV-DM. However, in User2Vec#1 model, the paragraph vector is also estimated by using the average of the embedding vectors of the relevant users. Here, the author of the microblog and the users that have forwarded the microblog are considered as relevant users.

In User2Vec#1, every user is mapped to a vector represented by a column in matrix U, in addition to the microblog text matrix D and the word matrix W. Given a microblog text $d_i, w_i, w_2, \ldots, w_T$, our goal is to predict both w_{T+1} and microblog token d_i. We define all users related to d_i as $u_{i1}, u_{i2}, \ldots, u_{jh}$. The objective function that should be maximized is as follows:

$$J = \frac{1}{T} \sum_{t} [\log p(w_t | d_i, w_{t-k}, \ldots, w_{t+k}) + \log p(d_i | u_{i1}, \ldots, u_{ih})]. \tag{9.2}$$

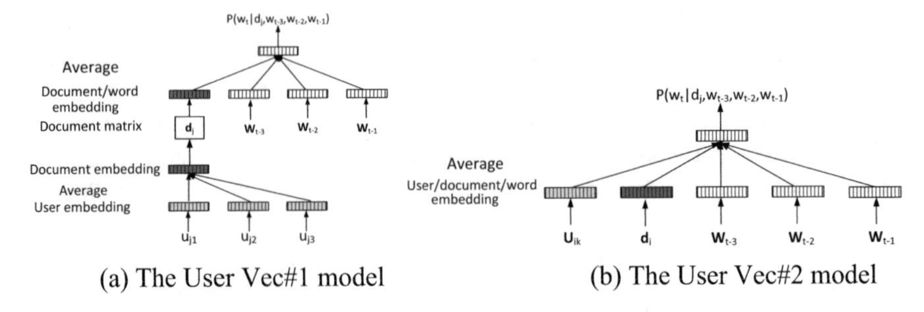

Fig. 9.1 The network diagrams for user embeddings (Yu et al. 2016)

In User2Vec#2 model (shown in Fig. 9.1b), user vectors and text/word vectors are in the same layer; these vectors are averaged to make prediction for next word. In this framework, user embeddings are learned as the contexts of documents as follows:

$$J = \frac{1}{T}\left(\sum_{t} \log p(w_t | d_i, w_{t-k}, \ldots, w_{t+k}, u_{i1}, \ldots, u_{ih})\right). \tag{9.3}$$

It has been shown that User2Vec#2 model achieves better performance because it learns user embeddings directly from word embeddings.

9.2.3 Semantic Representation with Deep Neural Networks

9.2.3.1 Learning Representations by Recurrent Neural Networks

Many types of social content have presented as a sequential semantic structure. For example, a social comment is a sequence of words in nature. Similarly, a conversation made in social media between users is a sequence of sentences. Exploiting this first-order sequential structure would lead us to better understand the social contexts. A standard Recurrent Neural Network (RNN) processes an arbitrary sequence of data by recurrently applying the transition function over the current input vector and last hidden state vector. The output of the transition function is the current hidden state vector. Given a sequence of words $d = \{w_1, w_2, \ldots, w_t\}$, the hidden state vector $\mathbf{h}_t$ at position t is computed by RNN as follows:

$$\mathbf{h}_t = \sigma(\mathbf{W}\mathbf{q}_t + \mathbf{C}\mathbf{h}_{t-1}), \tag{9.4}$$

where $\mathbf{q}_t$ is the embedding of word w_t at position t, $\mathbf{W}$ is the transition matrix from the input embedding to the hidden state, $\mathbf{C}$ is the state-to-state recurrent weight matrix, σ is the transition function and is often implemented by sigmoid, tanh, or

ReLU function. The hidden state vector $\mathbf{h}_t$ is expected to capture the hidden semantic features for the sequence $\{w_1, w_2, \ldots, w_t\}$.

Although the RNN structure can process sequential input, the gradient becomes smaller and smaller until it diminishes completely when the length of input is large. This is the *gradient vanishing* problem. A simple solution is to magnify the values of weight matrices. However, the strategy could turn out to cause the *gradient exploding* problem. Both of the problems will prevent RNN from learning the distant dependencies within the longer sequence appropriately. To address this problem, Long Short-Term Memory (LSTM) and Gated Recurrent Unit (GRU) are proposed with the gating mechanism to control the information flow. In recent years, recurrent neural networks have experienced great success in many fields. There are lots of works utilizing the aforementioned structures such as language modeling, image captioning, speech recognition, machine translation, computer-composed music, click prediction, etc. Since RNNs and their variants can be used to model a text unit of variable length, they have been widely investigated to derive the task-specific representation for the text units. We will introduce two representative works that address the tasks of rumor detection and automatic conversation-response modeling.

In the modern information age, rumors can cause public panic and social unrest. For example, a rumor about "salt can protect radiation" triggered the rush of salt tide. At the early stage, detecting rumors is through manual verification; however, the effect is very limited and has long debunking delay. Many existing works which employ machine learning methods rely on hand-crafted features and is time-consuming. Several RNN-based models have been proposed to detect rumors in (Ma et al. 2016). Given an event and a set of relevant tweets $\{(m_i, t_i)\}$ where m_i is a specific tweet and t_i is the corresponding publish time. First, the incoming streams of tweets are converted into continuous variable-length time series, and then RNN-based models are used to classify rumors. Three models are proposed in (Ma et al. 2016) to address this task. The corresponding architectures are illustrated in Fig. 9.2.

- tan h-RNN. It is a basic RNN structure whose input is the TF-IDF values of the vocabulary terms in the time interval. The hidden unit is computed as

$$\mathbf{h}_t = \tan h(\mathbf{U}\mathbf{x}_t + \mathbf{W}\mathbf{h}_{t-1} + \mathbf{b}) \tag{9.5}$$

$$\mathbf{o}_t = \mathbf{V}\mathbf{h}_t + \mathbf{c}, \tag{9.6}$$

  and then *softmax* operator will be employed to classify rumors and non-rumors. The goal is to minimize the squared error between probability distribution of the prediction and ground truth.
- Single-layer LSTM/GRU. An embedding layer is added in this model to transform the TF-IDF weights into embeddings, and the basic RNN unit is replaced with LSTM/GRU unit in order to capture long-distance dependencies, which is important in rumor detection.
- Multilayer GRU. The authors further extend the second GRU-based model by stacking another GRU layer. The higher level GRU layer is expected to capture more abstract features for the prediction.

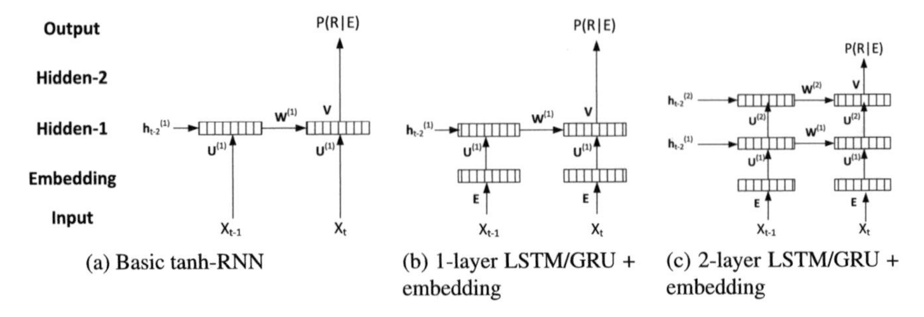

(a) Basic tanh-RNN (b) 1-layer LSTM/GRU + embedding (c) 2-layer LSTM/GRU + embedding

Fig. 9.2 The RNN-based rumor detection models (Ma et al. 2016)

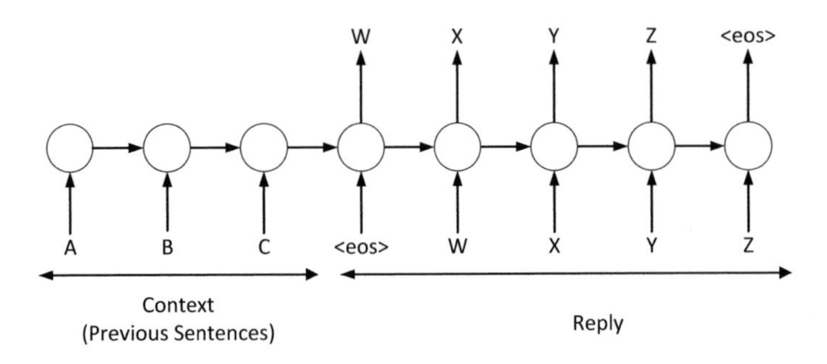

Fig. 9.3 Using the *seq2seq* framework for modeling conversations (Vinyals and Le 2015)

The corresponding architectures are illustrated in Fig. 9.2. All the models are trained using backpropagation to compute the derivatives of the loss and then update their parameters. The experimental results show that the significantly better performance is obtained by RNN-based models, compared with the existing state-of-the-art alternatives.

Building the intelligent conversation system is an important task in natural language processing and artificial intelligence. Most of the existing works focus on the development of task-oriented conversation systems. Although these works have achieved promising performance for some specific tasks in some limited domain, however, building an open domain conversation system that enables general-purpose conversation with human beings is still challenging. The recurrent processing manner of RNN models shed light on the further advance of open domain conversation, because of its ability to model variable-length text. Vinyals and Le propose a neural conversation model by modeling the word sequence with LSTM model (Vinyals and Le 2015).

This sequence-to-sequence model takes in a sequence of tokens as input, and produces an output sequence by generating each token recurrently (shown in Fig. 9.3). During training, the golden response in the form of a sequence of tokens is passed to the model, and backpropagation is utilized to update the parameters through the cross-entropy loss function. And during inference, when to predict a token, the previous prediction is passed as input to predict the current output. The proposed model makes some modifications by predicting the next sentence given the previous sentence rather than the token. For example, the task is to predict "WXYZ" given "ABC", the sentence vector of the input is the hidden state after processing the symbol "<eos>" which indicates the end symbol of a sentence. The model predicts the tokens in next sentence one by one given the last hidden state. This neural network architecture requires very little feature engineering or specific domain knowledge while retains state-of-the-art performance.

9.2.3.2 Learning Representations by Convolutional Neural Networks

Besides its prevalence in the domain of computer vision, Convolutional Neural Network (CNN) has also been applied in social computing. For example, the #TagSpace model has been proposed to address the hashtag prediction task (Weston et al. 2014). By projecting the words, textual post, and hashtags into the same vector space, #TagSpace is able to calculate the relevance score between a hashtag and a post using the inner product between their embeddings.

Figure 9.4 presents the framework of *#TagSpace*. Unlike image pixels in computer vision, the inputs to most NLP tasks are words or sentences. So the authors first convert each word of an input document into d-dimensional embedding vector by using the word lookup table, resulting in a matrix of size $l_d \times d$, where l_d is the document length. This operation incorporates a matrix of $N \times d$ parameters, called the lookup-table layer, where N is the vocabulary size. After that, a convolutional operation is applied to the $l_d \times d$ matrix. Specifically, the authors construct H filter matrices of size $K \times d$ and slide each filter matrix over the original input matrix from position 1 to l_d, where K is the sliding window size. To account for words at two boundaries of a document, the both ends are padded with special vectors, so that we

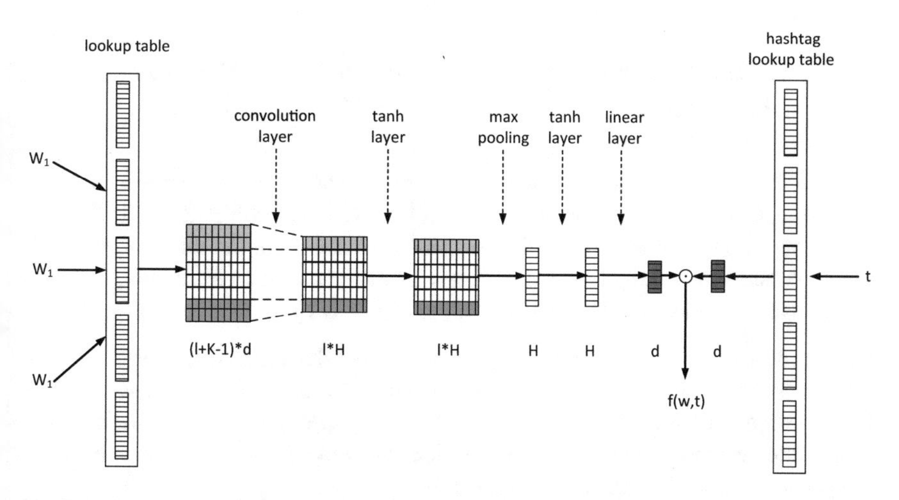

Fig. 9.4 The architecture of the TagSpace model (Weston et al. 2014)

can apply the filter that can be applied to bordering elements of input matrix. After convolutional step, we use nonlinearity activation function such as tanh function for each element in the $l_d \times H$ matrix. And then, we apply max pooling operation over the $l_d \times H$ matrix to extract a fixed-size (H-dimensional) global vector, which contains the features of the input document. It is noted that the d-dimensional global vector obtained from the CNN is independent of the length of the document. At last, the tanh nonlinear activation function and a full connected linear layer of size $H \times d$ are employed. Consequently, a single document is converted into a d-dimensional vector, representing the entire content in the original embedding space.

Similarly, a candidate hashtag can be represented by a d-dimensional embedding vector using a lookup table. In this way, textual post and hashtags have been represented by d-dimensional vectors, respectively, in the same embedding space. The inner product is adopted to calculate the semantic relatedness between document w and hashtag t:

$$f(w, t) = \mathbf{e}_{conv}(w)^{\top} \cdot \mathbf{e}_{lt}(t), \tag{9.7}$$

where $\mathbf{e}_{conv}(w)$ is the embedding of the document calculated by the CNN, and $\mathbf{e}_{lt}(t)$ is the embedding of the candidate hashtag t using lookup table. We can rank all the candidate hashtags according to the scores $f(w, t)$. The larger the score is, the more relevance the hashtag and the post are.

To train the #TagSpace model, the pairwise hinge loss is used as the objective function:

$$\mathscr{L} = \max\{0, m - f(w, t^{+}) + f(w, t^{-})\}, \tag{9.8}$$

where t^{+} is a positive tag, t^{-} is a negative example sampled from training set, and m is the predefined margin. The lookup-table layers are initialized with the pretrained embeddings to expedite the convergence.

9.2.4 Enhancing Semantic Representation with Attention Mechanism

In this subsection, we will discuss how to apply the attention mechanism to model social text. Originating from the field of computer vision (Mnih et al. 2014; Xu et al. 2015), attention mechanism enables the model to select important information to attend to based on the input and what it has produced so far. In NLP field, attention mechanism is used to enhance text modeling typically by

- handling long input sequences (e.g., sentences or documents) and ensuring that the output can acquire useful information as much as possible (Luong et al. 2015).
- alleviating the order variation and discrepancy problem in some tasks (e.g., machine translation and text summarization) by producing soft alignment between input and output (Bahdanau et al. 2014).

Distributed representation models, such as skip-gram and CBOW, have been shown to be effective in capturing word semantic relations. However, they are incapable of capturing the syntactical relations between words because they do not consider the word order. To tackle this issue, a simple extension which adds the attention mechanism into CBOW has been proposed (Ling et al. 2015). The intuition behind this model is that the prediction of a word mainly depends on certain words and their positions within the context. For instance, in the sentence of *"We won the game!"*. The prediction of the word *"game"* is mainly based on both the syntactic relation from the word *"the"*, since it is always followed by nouns, and on the semantic relation from the word *"won"*. The word *"We"* contributes very little to the prediction of *"game"*. In this case, assigning different weights to words at different positions in a fixed-length context is necessary for word prediction.

In this model, each word $w \in V$ at position i is assigned with an attention score $a_i(w)$:

$$a_i(w) = \frac{\exp(k_{w,i} + s_i)}{\sum_{j \in [-b,b]-\{0\}} \exp(k_{w,j} + s_j)}, \tag{9.9}$$

where $k_{w,i}$ denotes the importance of the word w at position i, s_i is an offset of position i within the context window, and b is the window size. After the attention calculation, the context vector $\mathbf{c}$ is calculated as follows:

$$\mathbf{c} = \sum_{i \in [-b,b]-\{0\}} a_i(w_i)\mathbf{v_i}, \tag{9.10}$$

where $\mathbf{v_i}$ denotes the embedding of word i. With CBOW, a weighted sum of the individual word embeddings is taken in Eq. 9.10 instead of simply taking the average of word embeddings within the context. Finally, the model predicts the target word by maximizing the following probability:

$$p(v_0|w_{[-b,b]-\{0\}}) = \frac{\exp(\mathbf{u}_0^T\mathbf{c})}{\sum_{w \in V} \exp(\mathbf{u}_w^T\mathbf{c})}. \tag{9.11}$$

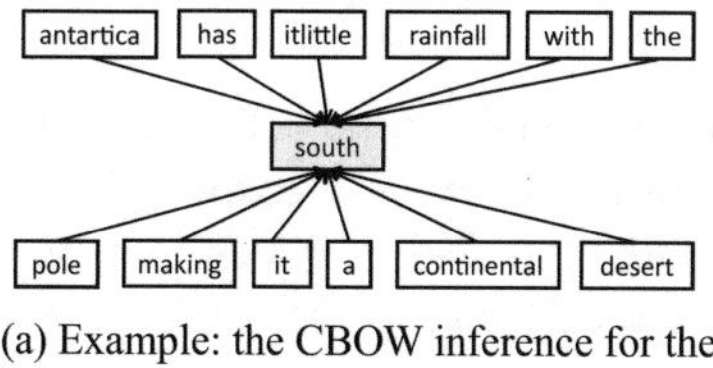

(a) Example: the CBOW inference for the prediction of word "south"

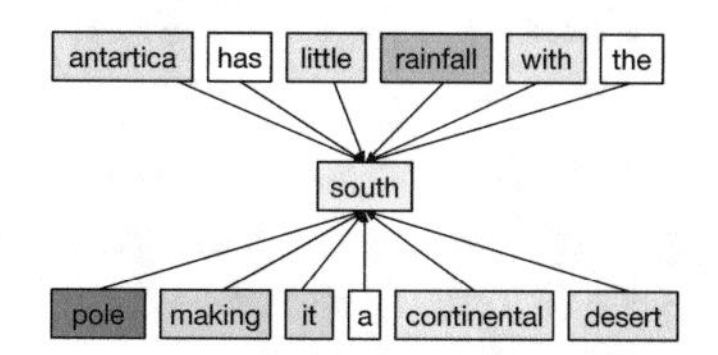

(b) Example: the attention-based CBOW inference for the prediction of word "south"

Fig. 9.5 The comparison between attention-based CBOW and CBOW models (Ling et al. 2015)

Figure 9.5a, b shows the different prediction mechanisms of CBOW and attention-based CBOW. In CBOW, all context words contribute evenly to the prediction of target word *south*, including function words. While, in attention-based CBOW, darker cells indicate higher weights (ref. Eq. 9.9) for predicting target word *south*. The experimental results in (Ling et al. 2015) have demonstrated that word embeddings learnt by the attention-based CBOW retain better syntactical relations between words. Attention mechanism has been widely used in different tasks. For example, several works have adopted attention mechanism for hashtag recommendation and obtained the state-of-the-art recommendation performance (Gong and Zhang 2016; Zhang et al. 2017).

9.3 Modeling Social Connections with Deep Learning

9.3.1 Social Connections on Social Media

As already discussed in Sect. 9.1, a major feature of online social media platforms is that they provide rich social connections. Social networking sites typically utilize explicit or implicit linking mechanisms to enhance the interactions or connections between users. User links can either unidirectional or bidirectional. For example, on Twitter, a user can follow another user unilaterally. As a comparison, on Facebook, the user link is constructed in a bidirectional way. Typically, these user links indicate friendship or interest similarity (Weng et al. 2010). In some cases, the links can also explicitly relate to trust information (e.g., EPinion[2]). In addition to explicit links, the implicit relations are prevalent on social media. For example, a user can forward (*a.k.a.,* retweet) one's tweet without following her. Such implicit links are also important to consider in conveying useful semantic information (Welch et al. 2011; Zhao et al. 2013, 2015; Wang et al. 2014).

9.3.2 A Network Representation Learning Approach to Modeling Social Connections

In this section, we discuss how to model user links in a general perspective. With the revival of deep learning in recent years, network representation learning has become a hot research topic (Perozzi et al. 2014; Tang et al. 2015), which aims to embed vertices into a low-dimensional space, and the derived representations are usually called *node embeddings*.

[2]http://www.epinionglobal.com/.

Table 9.1 Categorization of network embedding models

Classes		Models
Shallow	Neighborhood	DEEPWALK (Perozzi et al. 2014), NODE2VEC (Grover and Leskovec 2016)
	Proximity	LINE (Tang et al. 2015), GRAREP (Cao et al. 2015)
	Heterogeneous	HINE (Huang and Mamoulis 2017), ESIM (Shang et al. 2016)
Deep	Neighborhood	GRUWALK (Li et al. 2016)
	Proximity	SDNE (Wang et al. 2016a), GRAREP (Cao et al. 2015)
	Heterogeneous	HNE (Chang et al. 2015)

Formally, let $\mathscr{G} = (\mathscr{V}, \mathscr{E}, \mathbf{W})$ denote a general social network representation, where $\mathscr{V}$ is the vertex set, $\mathscr{E}$ is the edge set, and $\mathbf{W}$ is the weight matrix for edges. If there exists an edge from vertex u to vertex v, then $(u, v) \in \mathscr{E}$. Let $w_{u,v}$ denote the weight of the edge from u to v. Both unidirectional or bidirectional, weighted or unweighted networks can be modeled in this definition.[3] Network representation learning aims to generate a d-dimensional latent representation $\mathbf{e}_v \in \mathbb{R}^d$ for each vertex $v \in \mathscr{V}$. Usually, the dimensionality (i.e., d) varies from 50 to several hundreds. In Table 9.1, we introduce the methods for network representation learning in two categories: (1) shallow embedding based methods; and (2) deep neural networks based methods. The first category refers to the models which derive distributed representations with a shallow neural architecture[4]. As a comparison, the second category utilizes standard neural network models for learning network representation.

The learned representations in existing studies (Perozzi et al. 2014) are mainly utilized for network reconstruction or node classification, but the approach can be easily extended to solve some specific tasks (Chen and Sun 2016). Here, our focus is the general network representation learning, while other types of information are ignored here, such as text data (Yang et al. 2015). Especially, knowledge graphs can be considered as a specific type of heterogeneous networks, and many studies from natural language processing (Xie et al. 2016; Guo et al. 2016) are related to network representation learning. In this section, we will only focus on the existing studies for social network analysis.

[3] In our case, a vertex corresponds to a user, and the graph corresponds to the user network. Unless specified, we will use "network" for short instead of "user network", since our methods are general and can be essentially applied to any networks of other types.

[4] Strictly speaking, the embedding based models are not standard neural networks, such as WORD2VEC (Mikolov et al. 2013).

9.3.3 Shallow Embedding Based Models

9.3.3.1 Traditional Graph Embedding Models

In the early literature of machine learning and pattern recognition, an important task is dimensionality reduction and data representation. These methods take the dyadic data-feature matrix as the input, and each row in the data-feature matrix corresponds to a high-dimensional observation point. The essence of these early methods lies in the transformation of high-dimensional observations into low-dimensional representations via dimensionality reduction. Some well-known methods include IsoMap (Balasubramanian and Schwartz 2002), LLE (Roweis and Saul 2000), and Laplacian Eigenmaps (Belkin and Niyogi 2001). Typically, early studies largely borrow the ideas of *Principal Components Analysis (PCA)*, *Multidimensional Scaling*, *Graph Laplacian*, and *Manifold Learning*. Usually, these algorithms have high computational complexity, which is not easy to deploy on large-scale datasets. In recent years, matrix factorization technique is also applied to network embedding (Wang et al. 2011), which decomposes the network matrix (e.g., the adjacency matrix) into a product between two matrices.

9.3.3.2 Neighborhood-Based Embedding

The key idea of the neighborhood-based approach aims to model the relation between a target vertex and its neighborhood constructed by random walks using some strategy.

DeepWalk (Perozzi et al. 2014) is the first network embedding model which borrows the idea from word embedding. In word embedding (e.g., word2vec), the basic elements are sentences (or word sequences) and words, and the purpose is to learn the latent representations of words by characterizing the relations between a target word and its context information in local windows. Let w denote a word and $\mathbf{C}_w$ denote the contexts (i.e., the context words) of word w. Word embedding models essentially model the conditional probabilities of $P(w|\mathbf{C}_w)$ or $P(\mathbf{C}_w|w)$. DeepWalk considers *vertices* as words and *vertex sequences* as sentences. While, there are explicit vertices and links on a graph, but not vertex sequences. To solve this problem, DeepWalk first generates short random walks based on the graph structure. These walks can be considered as short sentences, and it estimates the likelihood of observing a specific vertex given the surrounding vertices visited in random walks. More formally, DeepWalk models the conditional probability of $P(\mathbf{N}_v|v)$, where $\mathbf{N}_v$ denotes the neighbors of vertex v in the generated random walks given the graph $\mathcal{G}$. The model is implemented by using the skip- gram architecture of word2vec and optimized by the hierarchical softmax algorithm. The elegance of DeepWalk lies in the connections between word sentences and vertex random walks.

Based on DeepWalk, an extended model node2vec (Grover and Leskovec 2016) has been proposed. It defines a flexible notion of a node's network neighborhood as the vertex set generated by a family of parameterized and biased random walks. The resulting algorithm is flexible in the control of random walks through two tunable parameters: return parameter p and in-out parameter q. Parameter p controls the likelihood of immediately revisiting a node in the walk, while parameter q allows the search to differentiate between inward and outward nodes. The two parameters p and q allow the search procedure to (approximately) interpolate between *Breadth-First Search* and *Depth-First Search*. Overall, NODE2VEC generalizes DEEPWALK with the parameterized control on the search of neighborhood.

9.3.3.3 Proximity-Based Embedding

The second class of embedding models aims to characterize the pairwise vertex similarities using the latent node representations. There can be multiple ways to measure pairwise vertex similarity on graphs. In particular, we will present embedding models based on the k-order ($k \geq 1$) similarity derived from the original graph.

LINE (Tang et al. 2015) defines an objective function that preserves both the first-order and second-order proximities, which aims to model arbitrary types of information networks and scale to millions of nodes. Specially, the first-order proximity characterizes the local structures reflected by the observed links in the networks. As a complement, the second-order proximity characterizes the indirect similarity between two vertices through the shared first-order neighborhood structures of the vertices. Both kinds of proximities are modeled by probability values, and subsequently Kullback–Leibler divergence is adopted to derive the objective function. LINE has given several important practical considerations on efficiency, including negative/edge sampling and alias table, which makes it efficient to scale to very large datasets.

GRAREP (Cao et al. 2015) is an embedding model which can capture k-order proximity when $k \geq 2$. The key idea is to estimate the proximity using the transition probabilities derived from higher order transition matrices. The work is based on an important property that SKIP- GRAM model with negative sampling is mathematically equivalent to matrix factorization over the (shifted) pointwise mutual information (PMI) co-occurrence matrix. Specially, GRAREP models the k-order transitions for all $k = 1, \ldots, K$, where K is a predefined parameter. For each k-order transition matrix, we can derive the corresponding node representations. The final representation is constructed by concatenating all the representations corresponding to each k-order representation. GRAREP extends LINE by modeling high-order similarity, and setting different representations for varying orders.

9.3.3.4 Community Enhanced Embedding

The above methods mainly focus on the local vertex links, while group or community structure has not been modeled. In this part, we discuss embedding with community or group structure information. The community structure characterizes the community member relations and considers the vertex relatedness in a wider range than local neighborhood.

GENE (Chen et al. 2016) is an embedding model which can incorporate community structure for network representation. The key idea is to model the community as a vertex. In this way, the community vertex is considered as the context for generating a specific vertex. A community vertex is modeled as the shared context for all the vertices from the corresponding community. GENE makes the analogy as follows: a community is treated as a document, while a vertex is treated as a word belonging to some document. GENE borrows the idea of DOC2VEC (Le and Mikolov 2014) in its two architectures, namely *Distributed Memory* (DM) and *Distributed Bag-of-Words* (DBOW). GENE combines both architectures and jointly models both neighboring users and group information.

Similar to early works on community detection (Wang et al. 2011), a Modularized Nonnegative Matrix Factorization (M-NMF) (Wang et al. 2017) has been proposed for learning the vertex representations and preserving the community structure. In M-NMF, it first applies the classic modularity-based method for community detection. Then, it builds the objective function which involves three factors, which corresponds to the factorization of similarity matrix, the factorization of the community membership matrix, and the community-preserving loss. The key to connect the first two factors lies in the shared vertex representations, and the community-preserving loss is defined based on the community membership matrix. In this way, a unified nonnegative matrix factorization approach jointly optimizes the above three factors.

9.3.3.5 Heterogeneous Network Embedding

Previously, vertex similarity is evaluated based on homogeneous networks. In practice, many information networks are heterogeneous. For example, in the scientific collections, different types of entities form a heterogeneous network, where there may be author, paper, and venue vertices. These heterogeneous networks describe the relations between objects (i.e., network vertices) of different types. To deal with them, a commonly adopted method is the meta-path based algorithm (Sun et al. 2011). A meta-path is a sequence of object types with edge types in between modeling a particular relationship. Next, we discuss how to apply the meta-path based algorithm to enhance the network embedding models for heterogeneous networks.

A straightforward method is to transform meta-path based information into similarities (Huang and Mamoulis 2017). In this way, we can build a meta-path based graph, where the edge weights are derived from the meta-path based similarities. Once the similarity matrix (i.e., the adjacency matrix in the graph) has been constructed, the problem becomes a standard network embedding task, and we can apply any existing network embedding models. For calculating the meta-path based similarities, truncated k-length paths are considered and a dynamic programming algorithm has been applied to efficiently calculate the similarities. After similarity calculation, the first-order loss function of LINE is adopted to learn the vertex representations.

A note is that LINE can characterize the edge weights by using a sampling-based method.

Instead of simply evaluating meta-path based similarities, ESIM (Shang et al. 2016) models the meta-path based similarity by incorporating path-specific embeddings. Given two vertices, their path-specific similarities can be composed into four parts: a path-specific constant, the inner product between vertex embeddings, and two inner products between a vertex embedding and the path embedding. Formally, the path-specific conditional probability from vertex v_1 to v_2 via the path type t can be given as

$$Pr(v_2|v_1, t) = \frac{\exp(f(v_1, v_2, t))}{\sum_{v' \in V} \exp(f(v_1, v', t))}, \tag{9.12}$$

where $f(v_1, v_2, t)$ is a score function measuring the importance of the path $v_1 \rightarrow_t v_2$ defined as $f(v_1, v_2, t) = \mu_t + \mathbf{e}_{v_1}^\top \cdot \mathbf{e}_t + \mathbf{e}_{v_2}^\top \cdot \mathbf{e}_t + \mathbf{e}_{v_1}^\top \cdot \mathbf{e}_{v_2}$. To learn the vertex and path embedding, ESIM further proposes two kinds of optimization methods, namely sequential and pairwise learning methods.

9.3.4 Deep Neural Network Based Models

Above, we have extensively discussed various embedding based models for learning latent vertex representations. All these works share the common point that they mainly rely on shallow embedding models for deriving the similarities. In some cases, link information in a network can be very complex, which may be difficult for shallow models to explain and generate. In this part, we turn to deep neural networks for more powerful modeling capability.

9.3.4.1 Deep Random Walk Based Models

The essence of DEEPWALK can be summarized in two points: first, transform graph structure into node sequences; second, learn node representations based on sequence embedding models. However, strictly speaking, WORD2VEC model is not a real sequence model: the context words are order insensitive. Indeed, we can apply any kind of sequence neural network models to learn node and sequence representations based on node sequences, e.g., the widely used recurrent neural network. To characterize long sequences, Gated Recurrent Unit (GRU) and Long Short-Term Memory (LSTM) are two well-known variants for improving basic RNNs. Li et al. (2016) have applied the bidirectional GRU to encode node sequences, which applies a forward GRU that reads the sequence from left to right, and a backward GRU from right to left. We call such a model GRUWALK. Similarly, other sequence neural networks can apply here for learning node representations.

9.3.4.2 Deep Proximity-Based Models

We consider two studies which model low-order and high-order proximity, respectively.

SDNE (Wang et al. 2016a) is the first study to characterize low-order similarity using deep neural networks. It emphasizes three important properties in network reconstruction, namely high nonlinearity, structure-preserving, and sparsity-resistant. In terms of methodology, SDNE can be loosely understood as a neuralized generation of LINE. To capture the nonlinear linking characteristics, a deep autoencoder model is adopted, which takes the neighborhood information (using the one-hot representation) of a vertex both as input and output. The autoencoder aims to reconstruct the input by first projecting it into a low-dimensional embedding (with several nonlinear layers) and then recovers the output from the embedding (with several nonlinear layers). The embeddings in the middlemost layer of the autoencoder model can be considered as the latent representations of vertices, usually called *code*. With these vertex codes, the first-order proximities are characterized via the graph-based regularization loss, which forces the codes of connected vertices to be similar. The autoencoder model implicitly characterizes the second-order proximities due to the fact that the model parameters are shared by all the vertices. In this way, vertices with similar neighborhood will have similar codes, since their neighborhood information will be fed into the same autoencoder model.

For capturing high-order proximity, the DNGR model (Cao et al. 2016) extends the GRAREP model (Cao et al. 2015) by using deep neural network models. DNGR first performs the random surfing, and then estimates the transition probabilities using random walk with restart. In the original DEEPWALK, random walks are generated without considering the effect of the start vertex. As a comparison, DNGR enhances the effect of the start vertex via the restart vector, and it tends to assign a larger probability to a vertex which is closer to the start vertex. The above random surfing model is adopted to estimate the PMI co-occurrence matrix for the network vertices. Unlike SKIP- GRAM with negative sampling, which directly factorizes the PMI matrix in a shallow way, DNGR tries to reconstruct the PMI matrix using stacked denoising autoencoder. By combing the above two steps, DNGR is supposed to generate better-quality random walks and enhance the capacity in characterizing the complex relations, which is expected to perform better on network embedding.

9.3.4.3 Deep Heterogeneous Information Network Fusion

Heterogeneous information networks usually contain different types of nodes and links, and it is more challenging to deriving effective representations for heterogeneous information. Chang et al. (2015) propose the HNE model to fuse heterogeneous information with different data types. The fusion approach is intuitive. For each data type, we first project the data points into a latent space with deep neural networks, so that the data characteristics for each local domain can be preserved. The model further makes an assumption that after a series of nonlinear transformations, the local data features from different domains can be mapped into a shared space. By preserving both the within-domain and cross-domain similarity, the final loss function jointly optimizes the data embedding via a deep architecture.

9.3.5 Applications of Network Embedding

In social computing, analyzing user connections is a fundamental and important step. The network embedding based approach can generate effective representations from social connection structures, which can be utilized in various downstream tasks. The above network embedding models provide a general network representation approach for various applications related to social network analysis, including network reconstruction, link prediction, node classification, node clustering, and visualization (Perozzi et al. 2014; Tang et al. 2015). In these tasks, network embedding serves as an automatic and unsupervised feature engineering procedure. More recently, some studies have also tried to develop task-driven network embedding models. For example, the network embedding approach has been extended by incorporating task-specific labeled information (Huang et al. 2017; Chen and Sun 2016).

9.4 Recommendation with Deep Learning

9.4.1 Recommendation on Social Media

On social media, recommendation is a ubiquitous task, which aims to match users' interests or needs with suitable information resources (i.e., items) (Adomavicius and Tuzhilin 2005; King et al. 2009). For example, a news portal website can recommend news or tweets to users with potential interests. The resource items are defined in a general way, which can be a news, a tweet, a friend, etc. In the recommendation task, a set of users $\mathscr{U}$ and a set of items $\mathscr{I}$ are the core elements.

- *Rating Prediction*: It aims to infer the preference degree of a user u on an item i given some context information C. Specially, let $r_{u,i}$ denote the rating of user u on item i. Rating prediction aims to infer the missing values for $r_{u,i}$.
- *Top-N Recommendation*: It aims to generate a recommendation ranklist of N items from $\mathscr{I}$ to a target user $u \in \mathscr{U}$ given some context information C.

The two tasks are highly correlated. In what follows, we mainly focus on the models themselves but will not discriminate between the two tasks unless specified. The introduced models are summarized in Table 9.2 by two approaches, namely shallow embedding based and deep neural network based.

Table 9.2 Categorization of deep learning recommendation models. "Integration" indicates the utilization of side information

Classes		Models
Shallow	Word embedding	PRODUCT2VEC (Zhao et al. 2016b), MC-TEM (Zhou et al. 2016), HRM (Wang et al. 2015b)
	Network embedding	NERM (Zhao et al. 2016a)
	Embedding regularization	COFACTOR (Liang et al. 2016)
Deep	Traditional	RBM (Salakhutdinov et al. 2007)
	Interaction (MLP)	NEUMF (He et al. 2017), NMF (He and Chua 2017)
	Interaction (Auoencoder)	CDAE (Wu et al. 2017a)
	Interaction (Sequence)	NADE (Zheng et al. 2016), NASA (Yang et al. 2017)
	Integration (Profile)	DUP (Covington et al. 2016), Wide and Deep (Cheng et al. 2016), RRN (Wu et al. 2017b), DeepCoNN (Zheng et al. 2017)
	Integration (Content)	SDAE (Wang et al. 2015a), DCMR (van den Oord et al. 2013)
	Integration (Knowledge)	CKE (Zhang et al. 2016)
	Integration (Cross-domain)	MV-DSSM (Elkahky et al. 2015)

9.4.2 Traditional Recommendation Algorithms

Various recommendation methods have been proposed for recommender systems in the past, including collaborative filtering methods (Su and Khoshgoftaar 2009), content-based methods (Lops et al. 2011), and hybrid methods (De Campos et al. 2010). Collaborative filtering methods build a model from a user's past behaviors as well as the decisions made by other similar users. Content-based methods extract a set of important features from an item in order to recommend other items with similar features. In the collaborative filtering approach, Matrix Factorization (MF) is widely adopted in various recommendation tasks (Koren et al. 2009). Different from traditional methods such as UserKNN and ItemKNN, MF can generate a latent factor for a user or an item, and the recommendation task can be solved by calculating the similarity between these latent vectors. A major merit of MF is that it can be flexibly modified to incorporate various kinds of contextual information for adapting to new task settings. The MF methods perform very well in practice and serve as competitive baselines in many tasks to the date.

9.4.3 *Shallow Embedding Based Models*

Shallow embedding based models largely borrow the idea of distributed representation learning, especially the works on word embedding (e.g., WORD2VEC). The basic idea is to map users, items, and related contextual information into a low-dimensional space. Furthermore, the recommendation task can be casted into a similarity measurement problem in the latent embedding space.

9.4.3.1 Recommendation as "Word" Embedding

The core idea of word embedding is that the semantics of a given word depends on its contextual words. The similar idea can be utilized to model item adoption sequences, where the items have shown sequential relatedness.

Zhao et al. (2016b) propose a straightforward application of word embedding models in recommender systems. In this work, product purchase records are first grouped by users, and then for each user the purchased products are sorted according to their timestamps chronologically. We make the following analogy: a product is considered as a word and the entire purchase sequence of a user is considered as a document. In this way, DOC2VEC can be applied to model product purchase sequences, called PRODUCT2VEC. The assumption made here is that the consecutive product purchases by a user are highly related in terms of the product semantics. Hence, we can infer the semantics of a product using its surrounding contexts in the purchase sequence. In (Zhou et al. 2016), the DOC2VEC model is only used for learning high-quality feature representations for both users and items. Subsequently, these features are further utilized in feature-based recommendation algorithms, i.e., LIBFM (Rendle 2012).

The PRODUCT2VEC model mainly captures the interactions between users and items. In some application scenarios, many kinds of contextual information can be utilized in the recommendation algorithms. In (Zhou et al. 2016), the DBOW architecture of DOC2VEC is extended to incorporate more contextual information, called *MC-TEM* model. The extension is relatively straightforward. It first discretizes the contextual information into discrete values, and each value will be associated with a unique embedding in the same latent space. To utilize various kinds of contexts, average pooling is used to combine multiple kinds of embeddings into a single context embedding. Although the approach is simple, it can be implemented very efficiently. Especially, all the contextual information has been modeled in the same latent space, and it is convenient to analyze the relations between different contextual information using simple similarity measurement on the embeddings (e.g., cosine similarity). A potential problem is that the contextual information itself may not be additive in terms of their latent representations, and using average pooling might lose information and hurt the performance in some cases.

The above methods treat the purchase record of a user as a whole sequence. Wang et al. (2015b) propose the HRM model which splits the purchase records into transactions, called *baskets*. It is essentially built on the DBOW architecture of DOC2VEC. The major difference lies in the generation of the items for the next basket, which is modeled in a hierarchical way. To generate an item, the contextual information consists of the user and the items purchased in the last transaction. Compared with (Zhou et al. 2016), HRM has a more clear and intuitive definition about the sequential contexts: only the purchased products in the last transaction are considered as the contexts for the current transaction. To aggregate the items from the last transaction, different pooling operations have been proposed, such as max pooling and average pooling.

In recommender systems, MF models decompose the observed rating or interaction matrix into user and item latent factors. Such an approach mainly characterizes the two-way user–item interactions. While, for embedding models, such as WORD2VEC, their advantage lies in capturing the local or sequential relatedness in the item sequences. Based on these considerations, the COFACTOR model (Liang et al. 2016) is proposed to combine the benefits of the two approaches into a unified model. Specially, the SKIP- GRAM model with negative sampling can be mathematically equivalent as the factorization of the (shifted) PMI co-occurrence matrix (Levy and Goldberg 2014). Based on this idea, the final model is built by incorporating the factorization of the user–item matrix and the regularization of the item–item PMI matrix. In this way, the global user–item preference relations and the local item–item relatedness have been jointly considered.

9.4.3.2 Recommendation as "Network" Embedding

Recommendation problems can be solved in different perspectives. As a perspective, the recommendation task can be casted as similarity evaluation on graphs, and adopt the graph-based algorithm for recommendation, such as SIMRANK (Jeh and Widom 2002). In Sect. 9.3, we have extensively discussed the studies on network embedding. If the recommendation problems can be formulated in a graph setting, it is possible to reuse the existing approaches from network embedding for recommendation. Specially, the NERM model (Zhao et al. 2016a) is proposed to transform the recommendation task into a task of embedding K-partite adoption network. A K-partite network consists of K types of entities in the recommender systems. Most recommendation settings can be characterized by a K-partite adoption network. Then, the network embedding is performed for the K-partite graphs by treating all the types of entities equivalently. The final recommendation task is solved by calculating the inner product between corresponding embeddings for users, items, and related contexts.

9.4.4 Deep Neural Network Based Models

9.4.4.1 Restricted Boltzmann Machines for Recommendation

The first study that applies deep learning for recommender systems can be dated back to the work in (Salakhutdinov et al. 2007), which describes a class of two-layer undirected graphical models that generalize Restricted Boltzmann Machines (RBM) to modeling rating data. The RBM model consists of two major parts, namely the binary hidden features and visible rating data (represented as one-hot vectors). A weight matrix connects both parts. Overall, the number of parameters in the weight matrix is large and the learning procedure is relatively difficult and slow. To reduce the number of parameters, a commonly used technology is to factorize the weight matrix into two small-sized matrices. Such a method is effective to reduce the number of parameters with little performance decreasing. As the first attempt, however, the RBM model does not give very promising results: only a slight performance improvement over the standard matrix factorization has been achieved.

9.4.4.2 Deep Learning Models for Interaction Characterization

Basically speaking, the recommendation task is mainly concerned about how we model the interactions between users and items. In what follows, we will discuss both non-sequential and sequential interaction-based models for recommendation.

Most existing traditional recommendation methods capture linear relations between the representations of users and items, which may not be effective to characterize complex user–item interactions. He et al. (2017) propose the NeuMF model by utilizing deep neural networks for learning arbitrary interaction function from data, which presents a general framework for collaborative filtering based on neural networks. In NeuMF, it first maps the one-hot representations of users and items into embeddings using a lookup-table layer. Then, it aggregates the user and item embeddings using some pooling operations, e.g., concatenation and element-wise product. In this way, each user–item interaction pair will be modeled as an embedding vector. The derived embedding vector will be subsequently fed into a Multilayer Perceptron (MLP) model, which is composed of a series of nonlinear transformation layers. The output of the MLP component will be directly tied with the loss function. The NeuMF essentially exploits the capability of deep neural networks on capturing complex data relations or characteristics. As a follow-up of NeuMF, the Neural Factorization Machine (NFM) has been proposed in (He and Chua 2017), which is a neuralized instantiation of linear Factorization Machine (Rendle 2012). NFM incorporates a bi-interaction layer to perform the bi-interaction pooling for the two embeddings corresponding to two features. The derived bi-interaction pooling vector will be transformed into the predicted rating value with an MLP component.

Instead of predicting the outcome for an individual item separately, the CDAE model (Wu et al. 2017a) treats the feedbacks of a user u on all the items as a vector $\mathbf{y}$. Its purpose is to build a mapping function which takes the corrupted input

$\tilde{\mathbf{y}}$ and reconstructs the real feedback vector $\mathbf{y}$. CDAE implements the corrupted self-mapping function using the Denoising Autoencoders (DAE) model with only a hidden layer. With a hidden layer, the model parameters needed to learn include the weight parameters $\mathbf{W}$ connecting the input with hidden layers, and the weight parameters $\mathbf{W}'$ connecting the hidden with output layers. Formally, we can have the following formulas:

$$\begin{aligned} \mathbf{z} &= g(\mathbf{W}^\top \cdot \tilde{\mathbf{y}} + \mathbf{b}), \\ \mathbf{y} &= h(\mathbf{W}'^\top \cdot \mathbf{z} + \mathbf{b}'), \end{aligned} \tag{9.13}$$

where $g(\cdot)$ and $h(\cdot)$ are mapping functions consisting of multiple nonlinear layers. The latent vector $\mathbf{z}$ is often called *code*. Note that the parameters of the DAE model are shared for all the users. Hence, only taking the feedbacks as the input may not be effective to characterize the personalization. CDAE makes an extension by incorporating the user-specific embedding $\mathbf{e}_u$ into the input layer. Formally, the hidden layer is derived using the following formula:

$$\mathbf{z} = g(\mathbf{W}^\top \cdot \tilde{\mathbf{y}} + \mathbf{e}_u + \mathbf{b}). \tag{9.14}$$

In this way, the derived code (i.e., $\mathbf{z}$) takes the users' preference into consideration for better personalization.

The interactions between users and items are essentially a sequential process, while the above models cannot characterize sequential user behaviors. Hence, a natural consideration is to apply sequential neural networks to model user behaviors for recommendation. In the literature, Recurrent Neural Networks (RNN) are an important class of sequential neural networks (Mikolov et al. 2010), in which they maintain an internal state of the network which allows it to exhibit dynamic temporal behavior. RNN-based models have been successfully applied in various domains, including natural language processing and speech processing. A major obstacle in applying RNN in dealing with long sequences is the vanishing gradient problem. To tackle this problem, two well-known unit models have been proposed, namely the Long Short-Term Memory unit (LSTM) (Hochreiter and Schmidhuber 1997) and Gated Recurrent Unit (GRU) (Chung et al. 2014). With the improved RNN models, it is relatively straightforward to apply them to recommender system, and it is possible to build either an overall or user-specific RNN model to characterize users' behavior sequences. In (Yang et al. 2017), an extended RNN model (called *NASA* model) is utilized for POI recommendation, in which both long- and short-term sequential contexts have been considered. Meanwhile, the user's preference has also been incorporated into the recommendation model. As another kind of sequential recommendation models, a neural autoregressive model (Zheng et al. 2016) has been proposed for rating prediction. It is built based on Restricted Boltzmann Machine (RBM) and the Neural Autoregressive Distribution Estimator (NADE). The main idea is to treat the rating records of a user as a sequence, and the rating for the current item is predicted conditioned on the previous ratings from the user. The parameters include item embeddings and weight parameters. Like the classic RBM model, the preference of a specific user is not explicitly modeled by an embedding vector, but is reflected in her rating records. They also propose two major improvement techniques by sharing the parameters across ratings and factorizing the large-scale weight matrix.

The traditional user profiling method is usually static, which cannot reflect dynamic nature of user interests. Wu et al. (2017b) propose the Recurrent Recommender Network (RRN) model to predict future behavioral traces by creating dynamic user and item profiles. The key idea is to model user and item states and characterize state transitions using recurrent neural networks. The final prediction is from a combined model with the results from both dynamic and static profiling models.

9.4.4.3 Deep Learning Models for Side-information Integration and Utilization

Above, deep neural networks are mainly utilized to enhance the modeling of the user–item interactions. The side information (*a.k.a.,* context information) is not considered in these models. In what follows, we discuss how to utilize deep learning for modeling auxiliary information.

In many recommendation scenarios, the content information from the item side can be leveraged to improve the recommendation performance. Indeed, this is the key idea of the classic content-based approach (Lops et al. 2011), which performs the recommendations based on the description of an item and builds the profile of the user's interests. For consistency, we call the descriptions of an item *content information*. A major difficulty in achieving this purpose is that the content information itself may not be in a form directly applicable to the recommendation task, even noisy or sparse in some cases. It is necessary to transform or map the content information in a suitable form, which can be effectively utilized by the recommender systems. Fortunately, deep learning has the excellent capability of characterizing or learning complex data characteristics. A solution will be integrating content information into recommender systems using deep learning models. As a representative work, Wang et al. (2015a) propose the CDL model, which utilizes the content information for improving the recommender systems. It characterizes the content information by using a Stacked Denoising Autoencoder Model (SDAE). The final item representation is derived by concatenating a bias vector with the middlemost-layer code learned from the SDAE model. The CDL model is a deep learning implementation of the previous collaborative topic regression model (CTR) (Wang and Blei 2011). The results reported in (Wang et al. 2015a) show that the performance of CDL is better than CTR in the given tasks. A direct extension of the CDL model is to improve the text modeling part. The CDL model makes the bag-of-words assumption by using the SDAE for modeling text. The following work (Wang et al. 2016b) further proposes a Collaborative Recurrent Autoencoder (CRAE) which is a Denoising Recurrent Autoencoder (DRAE) that models the generation of content sequences in the Collaborative Filtering (CF) setting. The major improvement lies in the sequential modeling of text information. The content-based approaches are more appealing when the interaction data is not sufficient, especially in a cold-start setting. The work in (van den Oord et al. 2013) presents a solution to cold-start music recommendation with deep content-based recommendation algorithms. For ease of understanding, we will slightly simplify the original model in (van den Oord et al. 2013). Specially, a standard matrix factorization approach to recommender systems can be formulated as below:

$$\min_{\mathbf{x}_\cdot, \mathbf{y}_\cdot} \sum_{u,i} (r_{u,i} - \mathbf{x}_u^\top \cdot \mathbf{y}_i) + \lambda\big(\sum_u \parallel \mathbf{x}_u \parallel^2 + \sum_i \parallel \mathbf{y}_i \parallel^2\big), \tag{9.15}$$

where the user–item matrix (with the ratings of $r_{u,i}$) is factorized into the product between a user latent vector (i.e., $\mathbf{x}_u$) and an item latent vector (i.e., $\mathbf{y}_i$). The latent vectors are actually the parameters of the matrix factorization model. However, in a cold-start setting, the goal is to recommend new items to users, and there is little interaction information to train the MF model. The basic idea in (van den Oord et al. 2013) is to first train the latent vectors with the existing interaction data of "old" items, and then build the mapping relations between latent vectors and content information. Formally, let $\mathbf{f}_i$ denote the extracted content information for item i, which can be transformed into the latent vector $\mathbf{y}_i$ via a deep learning model

$$\hat{\mathbf{y}}_i = g(\mathbf{f}_i), \tag{9.16}$$

where the mapping function $g(\cdot)$ can be learned by minimizing the differences between $\hat{\mathbf{y}}_i$ and $\mathbf{y}_i$. Once such a mapping model has been effectively learned, making predictions on a new item becomes simple since its latent vector can be inferred using the content information. We call the model *Deep Cold-start Music Recommendation (DCMR)*. In the above two models, deep learning is utilized to transform side information into a representation form that is ready in recommender systems.

In addition to content information, structural knowledge graph is another important kind of information to improve the performance of recommender systems. The items from the recommender systems can be also considered as the entities in knowledge graphs. Knowledge graphs provide an effective way to organize and index entities via the typed edge or relations.

To model the items in these two different views, the CKE model (Zhang et al. 2016) is proposed to first embed entities using the structural knowledge graph, and then utilize the derived structural item embedding for improving recommendation. For embedding entities in knowledge graphs, a Bayesian structural embedding model has been adopted. For embedding entities in recommender systems, a similar approach to the aforementioned CDL model was proposed by integrating multiple signals, including visual, textual, and the structural embeddings. The CKE model makes an important assumption that the embedding vectors extracted from knowledge graph, images, and text can be fused in an additive way directly. In CKE, both visual and textual features are extracted using the stacked autoencoders.

For recommender systems, a key task is user profiling, which aims to build an effective user model for accurate recommendation (Zhao et al. 2014, 2016c). User profiling has become a fundamental task in various social media platforms, not limited to recommender systems, since it is the first step to understand a user. Covington et al. (2016) propose a deep neural network architecture for building effective user profiling model, called *DUP* model. The idea is to combine various kinds of context information using deep learning, including search history, watch history, demographic, and geographic information. After a series of nonlinear transformations (i.e., ReLU activation function), the final prediction is modeled by a softmax function over the set of items. A note is that a two-stage recommendation method has been adopted in (Covington et al. 2016), namely candidate generation and item ranking. Both stages are implemented with the similar DUP model architecture. As a representative of profile-enhanced models, the *Wide and Deep* model (Cheng et al. 2016) has built the similar deep neural network architecture for recommendation. A major difference is that both original and deep features are utilized for final prediction, that is why it is called *Wide & Deep*. As another interesting work, Zheng et al. (2017) propose the Deep Cooperative Neural Networks (DeepCoNN) model, which aims to build user and item profiles using review text. It consists of two parallel neural networks, where one neural network learns user profiles using the reviews written by the user, and the other learns item profiles using the reviews written for the item. A shared layer further combines these two profiles (i.e., two embeddings) as the input of factorization machines.

In real world, a user usually engages in multiple recommendation services. For example, a user may have both news App and video App for reading news and watching movies, respectively. Intuitively, the user information from different domains will complement each other. It is possible to build a more comprehensive and accurate user profile if we can jointly leverage information from multiple domains. Hence, a multi-view recommender system is preferred to improve the recommendation performance. The MV-DSSM model (Elkahky et al. 2015) is proposed to address the multi-view recommendation task. Specially, it utilizes the single-view Deep Structured Semantic Model (DSSM) (Huang et al. 2013) as the component, which is originally proposed in the field of information retrieval. The basic structure of DSSM is composed of two separate DNN components: the first component is for modeling the queries, while the second component is for modeling the documents. After a series of nonlinear transformations, the DSSM model ties the final embeddings from two parts in a shared space. The loss function follows a typical pairwise ranking way. If we would like to directly apply the single-view DSSM for recommendation in multiple domains, a straightforward approach is to set up multiple isolated DSSM models in different domains. Each DSSM model will be learned separately using the information from the individual domain. However, such an approach ignores the sharing and complementing of user information in multiple domains. The idea of MV-DSSM is intuitive, and it only reserves a single DNN component for a user, but sets multiple DNN components for items in each domain. The single-user DNN component will be integrated with multiple domain-specific DNN item components for building a global recommendation model. In this way, the user information is shared in multiple domains, which enhances the cross-domain recommendation performance.

9.5 Summary

Social computing is a multidisciplinary research area, where social science and computational approaches may blend to answer important and challenging questions about user behavior through online social media platforms. It relates to various interesting tasks which aim to produce intelligent and interactive social media applications. For a complete review of social computing, we suggest the readers refer to the survey (King et al. 2009; Wang et al. 2007) and the classic textbook (Easley and Kleinberg 2010).

This chapter focuses on three important aspects of social computing, namely social content analysis, social connection modeling, and recommendation. The three aspects cover most of the core elements and applications in social computing. Specially, we take deep learning as the major approach to social computing, and mainly review the recent progress made in social computing with deep learning. Deep learning techniques that have been reviewed so far include both shallow embedding based and deep neural network based methods. Our discussions emphasize how to adapt existing deep learning techniques to social computing tasks.

Nowadays, the exploration of applying deep learning techniques to social computing is still in an early stage. There are still many challenges or difficulties to address in this direction. As a major challenge, compared with traditional NLP tasks, the input and output of social computing tasks are much more flexible and diverse, and even hard to be formally defined in some cases. It is important and meaningful to study how to effectively model the varying settings of different social computing tasks, in which multi-modality data fusion, noisy data reduction, and complicated output prediction are possible issues to address. We believe this direction will increasingly attract the attention from both research and industry communities. As a result, in the near future, the improved social media platforms will provide better service to users with the progress of machine intelligence.

References

Adomavicius, G., & Tuzhilin, A. (2005). Toward the next generation of recommender systems: A survey of the state-of-the-art and possible extensions. *IEEE Transactions on Knowledge and Data Engineering*, *17*(6), 734–749.

Alpaydin, E. (2014). *Introduction to machine learning*. Cambridge: MIT press.

Bahdanau, D., Cho, K., & Bengio, Y. (2014). Neural machine translation by jointly learning to align and translate. *CoRR*. arXiv:1409.0473.

Balasubramanian, M., & Schwartz, E. L. (2002). The isomap algorithm and topological stability. *Science*, *295*(5552), 7–7.

Belkin, M. & Niyogi, P. (2001). Laplacian eigenmaps and spectral techniques for embedding and clustering. In *NIPS* (pp. 585–591).
Cao, S., Lu, W., & Xu, Q. (2015). GraRep: Learning graph representations with global structural information (pp. 891–900).
Cao, S., Lu, W., & Xu, Q. (2016). Deep neural networks for learning graph representations (pp. 1145–1152).
Chang, S., Han, W., Tang, J., Qi, G., Aggarwal, C. C., & Huang, T. S. (2015). Heterogeneous network embedding via deep architectures (pp. 119–128).
Chen, T. & Sun, Y. (2016). Task-guided and path-augmented heterogeneous network embedding for author identification. arXiv:1612.02814.
Chen, J., Zhang, Q., & Huang, X. (2016). Incorporate group information to enhance network embedding (pp. 1901–1904).
Cheng, H.-T., Koc, L., Harmsen, J., Shaked, T., Chandra, T., Aradhye, H., et al. (2016). Wide and deep learning for recommender systems. In *Proceedings of the 1st Workshop on Deep Learning for Recommender Systems, DLRS 2016* (pp. 7–10).
Chung, J., Gulcehre, C., Cho, K., & Bengio, Y. (2014). Empirical evaluation of gated recurrent neural networks on sequence modeling. arXiv:1412.3555.
Cortizo, J. C., Carrero, F. M., Cantador, I., Troyano, J. A., & Rosso, P. (2012). Introduction to the special section on search and mining user-generated content. *ACM TIST*, *3*(4), 65:1–65:3.
Covington, P., Adams, J., & Sargin, E. (2016). Deep neural networks for youtube recommendations. In *Proceedings of the 10th ACM Conference on Recommender Systems, Boston, MA, USA, September 15–19, 2016* (pp. 191–198).
De Campos, L. M., Fernández-Luna, J. M., Huete, J. F., & Rueda-Morales, M. A. (2010). Combining content-based and collaborative recommendations: A hybrid approach based on bayesian networks. *International Journal of Approximate Reasoning*, *51*(7), 785–799.
Easley, D., & Kleinberg, J. (2010). *Networks, crowds, and markets: Reasoning about a highly connected world*. Cambridge: Cambridge University Press.
Elkahky, A. M., Song, Y., & He, X. (2015). A multi-view deep learning approach for cross domain user modeling in recommendation systems. In *Proceedings of the 24th International Conference on World Wide Web, WWW 2015, Florence, Italy, May 18–22, 2015* (pp. 278–288).
Gong, Y. & Zhang, Q. (2016). Hashtag recommendation using attention-based convolutional neural network. In *Proceedings of the Twenty-Fifth International Joint Conference on Artificial Intelligence, IJCAI 2016, New York, NY, USA, 9–15 July 2016* (pp. 2782–2788).
Grover, A. & Leskovec, J. (2016). node2vec: Scalable feature learning for networks (pp. 855–864).
Guo, S., Wang, Q., Wang, L., Wang, B., & Guo, L. (2016). Jointly embedding knowledge graphs and logical rules. In *Proceedings of the 2016 Conference on Empirical Methods in Natural Language Processing* (pp. 1488–1498).
He, X. & Chua, T.-S. (2017). Neural factorization machines for sparse predictive analytics. In *Proceedings of The 40th International ACM SIGIR Conference on Research and Development in Information Retrieval*.
He, X., Liao, L., Zhang, H., Nie, L., Hu, X., & Chua, T.-S. (2017). Neural collaborative filtering. In *Proceedings of the 26th International World Wide Web Conference*.
Hochreiter, S., & Schmidhuber, J. (1997). Long short-term memory. *Neural Computation*, *9*(8), 1735–1780.
Homans, G. C. (1974). *Social behavior: Its elementary forms*.
Hornik, K. (1991). Approximation capabilities of multilayer feedforward networks. *Neural Networks*, *4*(2), 251–257.
Huang, Z. & Mamoulis, N. (2017). Heterogeneous information network embedding for meta path based proximity. *CoRR*. arXiv:1701.05291.
Huang, P.-S., He, X., Gao, J., Deng, L., Acero, A., & Heck, L. (2013). Learning deep structured semantic models for web search using clickthrough data. In *Proceedings of the 22nd ACM International Conference on Information and Knowledge Management* (pp. 2333–2338). ACM.
Huang, X., Li, J., & Hu, X. (2017). Label informed attributed network embedding (pp. 731–739).

Jeh, G. & Widom, J. (2002). SimRank: A measure of structural-context similarity. In *Proceedings of the Eighth ACM SIGKDD International Conference on Knowledge Discovery and Data Mining* (pp. 538–543). ACM.

Kaplan, A. M., & Haenlein, M. (2010). Users of the world, unite! the challenges and opportunities of social media. *Business Horizons*, *53*(1), 59–68.

King, I., Li, J., & Chan, K. T. (2009). A brief survey of computational approaches in social computing. In *International Joint Conference on Neural Networks, 2009. IJCNN 2009* (pp. 1625–1632). IEEE.

Koren, Y., Bell, R., & Volinsky, C. (2009). Matrix factorization techniques for recommender systems. *Computer*, *42*(8), 4179.

Kwak, H., Lee, C., Park, H., & Moon, S. (2010). What is twitter, a social network or a news media? In *Proceedings of the 19th International Conference on World Wide Web, WWW '10* (pp. 591–600). New York, NY, USA: ACM.

Le, Q. V. & Mikolov, T. (2014). Distributed representations of sentences and documents. In *Proceedings of the 31th International Conference on Machine Learning, ICML 2014, Beijing, China, 21–26 June 2014* (pp. 1188–1196).

Levy, O. & Goldberg, Y. (2014). Neural word embedding as implicit matrix factorization. In *Advances in neural information processing systems* (pp. 2177–2185).

Li, C., Ma, J., Guo, X., & Mei, Q. (2016). DeepCas: An end-to-end predictor of information cascades. arXiv:1611.05373.

Liang, D., Altosaar, J., Charlin, L., & Blei, D. M. (2016). Factorization meets the item embedding: Regularizing matrix factorization with item co-occurrence. In *Proceedings of the 10th ACM Conference on Recommender Systems, Boston, MA, USA, September 15–19, 2016* (pp. 59–66).

Ling, W., Tsvetkov, Y., Amir, S., Fermandez, R., Dyer, C., Black, A. W., et al. (2015). Not all contexts are created equal: Better word representations with variable attention. In *Proceedings of the 2015 Conference on Empirical Methods in Natural Language Processing* (pp. 1367–1372).

Lops, P., De Gemmis, M., & Semeraro, G. (2011). Content-based recommender systems: State of the art and trends. *Recommender systems handbook* (pp. 73–105). Boston: Springer.

Luong, T., Pham, H., & Manning, C. D. (2015). Effective approaches to attention-based neural machine translation. In *Proceedings of the 2015 Conference on Empirical Methods in Natural Language Processing, EMNLP 2015, Lisbon, Portugal, September 17–21, 2015* (pp. 1412–1421).

Ma, J., Gao, W., Mitra, P., Kwon, S., Jansen, B. J., Wong, K., et al. (2016). Detecting rumors from microblogs with recurrent neural networks. In *Proceedings of the Twenty-Fifth International Joint Conference on Artificial Intelligence, IJCAI 2016, New York, NY, USA, 9–15 July 2016* (pp. 3818–3824).

Manning, C. D., Raghavan, P., & Schütze, H. (2008). *Introduction to information retrieval*. Cambridge: Cambridge University Press.

Mikolov, T., Karafiát, M., Burget, L., Cernockỳ, J., & Khudanpur, S. (2010). Recurrent neural network based language model. In *Interspeech* (Vol. 2, p. 3).

Mikolov, T., Sutskever, I., Chen, K., Corrado, G. S., & Dean, J. (2013). Distributed representations of words and phrases and their compositionality. In *Advances in neural information processing systems* (pp. 3111–3119).

Mnih, V., Heess, N., Graves, A., & Kavukcuoglu, K. (2014). Recurrent models of visual attention. In *Advances in Neural Information Processing Systems 27: Annual Conference on Neural Information Processing Systems December 8–13, 2014, Montreal, Quebec, Canada* (pp. 2204–2212).

Parameswaran, M., & Whinston, A. B. (2007). Social computing: An overview. *Communications of the Association for Information Systems*, *19*(1), 37.

Perozzi, B., Al-Rfou, R., & Skiena, S. (2014). Deepwalk: Online learning of social representations (pp. 701–710).

Rendle, S. (2012). Factorization machines with libFM. *ACM Transactions on Intelligent Systems and Technology (TIST)*, *3*(3), 57.

Roweis, S. T., & Saul, L. K. (2000). Nonlinear dimensionality reduction by locally linear embedding. *Science*, *290*(5500), 2323–2326.

Sahlins, M. (2017). *Stone age economics*. Routledge: Taylor & Francis.
Salakhutdinov, R., Mnih, A., & Hinton, G. (2007). Restricted Boltzmann machines for collaborative filtering. In *International Conference on Machine Learning* (pp. 791–798).
Schuler, D. (1994). Social computing. *Communications of the ACM, 37*(1), 28–108.
Shang, J., Qu, M., Liu, J., Kaplan, L. M., Han, J., & Peng, J. (2016). Meta-path guided embedding for similarity search in large-scale heterogeneous information networks. *CoRR*. arXiv:1610.09769.
Su, X., & Khoshgoftaar, T. M. (2009). A survey of collaborative filtering techniques. *Advances in Artificial Intelligence, 2009*, 4.
Sun, Y., Han, J., Yan, X., Yu, P. S., & Wu, T. (2011). PathSim: Meta path-based top-k similarity search in heterogeneous information networks. *Proceedings of the VLDB Endowment, 4*(11), 992–1003.
Tang, J., Qu, M., Wang, M., Zhang, M., Yan, J., & Mei, Q. (2015). LINE: Large-scale information network embedding (pp. 1067–1077).
van den Oord, A., Dieleman, S., & Schrauwen, B. (2013). Deep content-based music recommendation. In *Advances in Neural Information Processing Systems 26: 27th Annual Conference on Neural Information Processing Systems 2013. Proceedings of a Meeting Held, December 5–8, 2013, Lake Tahoe, Nevada, United States* (pp. 2643–2651).
Vinyals, O. & Le, Q. V. (2015). A neural conversational model. *CoRR*. arXiv:1506.05869.
Wang, C. & Blei, D. M. (2011). Collaborative topic modeling for recommending scientific articles. In *Proceedings of the 17th ACM SIGKDD International Conference on Knowledge Discovery and Data Mining* (pp. 448–456). ACM.
Wang, D., Cui, P., & Zhu, W. (2016a). Structural deep network embedding (pp. 1225–1234).
Wang, X., Cui, P., Wang, J., Pei, J., Zhu, W., & Yang, S. (2017). Community preserving network embedding (pp. 203–209).
Wang, P., Guo, J., Lan, Y., Xu, J., Wan, S., & Cheng, X. (2015b). Learning hierarchical representation model for nextbasket recommendation. In *International ACM SIGIR Conference on Research and Development in Information Retrieval* (pp. 403–412).
Wang, H., Shi, X., & Yeung, D. (2016b). Collaborative recurrent autoencoder: Recommend while learning to fill in the blanks. In *Advances in Neural Information Processing Systems 29: Annual Conference on Neural Information Processing Systems 2016, December 5–10, 2016, Barcelona, Spain* (pp. 415–423).
Wang, H., Wang, N., & Yeung, D. (2015a). Collaborative deep learning for recommender systems. In *Proceedings of the 21th ACM SIGKDD International Conference on Knowledge Discovery and Data Mining, Sydney, NSW, Australia, August 10–13, 2015* (pp. 1235–1244).
Wang, J., Zhao, W. X., He, Y., & Li, X. (2014). Infer user interests via link structure regularization. *ACM TIST, 5*(2), 23:1–23:22.
Wang, F.-Y., Carley, K. M., Zeng, D., & Mao, W. (2007). Social computing: From social informatics to social intelligence. *IEEE Intelligent Systems, 22*(2), 79–83.
Wang, F., Li, T., Wang, X., Zhu, S., & Ding, C. (2011). Community discovery using nonnegative matrix factorization. *Data Mining and Knowledge Discovery, 22*(3), 493–521.
Welch, M. J., Schonfeld, U., He, D., & Cho, J. (2011). Topical semantics of twitter links. In *Proceedings of the Fourth ACM International Conference on Web Search and Data Mining* (pp. 327–336). ACM.
Weng, J., Lim, E., Jiang, J., & He, Q. (2010). Twitterrank: Finding topic-sensitive influential twitterers. In *Proceedings of the Third International Conference on Web Search and Web Data Mining, WSDM 2010, New York, NY, USA, February 4–6, 2010* (pp. 261–270).
Weston, J., Chopra, S., & Adams, K. (2014). #tagspace: Semantic embeddings from hashtags. In *Proceedings of the 2014 Conference on Empirical Methods in Natural Language Processing, EMNLP 2014, October 25–29, 2014, Doha, Qatar, A meeting of SIGDAT, A Special Interest Group of the ACL* (pp. 1822–1827).
Wu, C., Ahmed, A., Beutel, A., Smola, A. J., & Jing, H. (2017a). Recurrent recommender networks. In *Proceedings of the Tenth ACM International Conference on Web Search and Data Mining, WSDM 2017, Cambridge, United Kingdom, February 6–10, 2017* (pp. 495–503).

Wu, C., Ahmed, A., Beutel, A., Smola, A. J., & Jing, H. (2017b). Recurrent recommender networks. In *Proceedings of the Tenth ACM International Conference on Web Search and Data Mining, WSDM 2017, Cambridge, United Kingdom, February 6–10, 2017* (pp. 495–503).

Xie, R., Liu, Z., Jia, J., Luan, H., & Sun, M. (2016). Representation learning of knowledge graphs with entity descriptions. In *AAAI* (pp. 2659–2665).

Xu, K., Ba, J., Kiros, R., Cho, K., Courville, A. C., Salakhutdinov, R., et al. (2015). Show, attend and tell: Neural image caption generation with visual attention. In *Proceedings of the 32nd International Conference on Machine Learning, ICML 2015, Lille, France, 6–11 July 2015* (pp. 2048–2057).

Yang, C., Liu, Z., Zhao, D., Sun, M., & Chang, E. Y. (2015). Network representation learning with rich text information. In *IJCAI* (pp. 2111–2117).

Yang, C., Sun, M., Zhao, W. X., Liu, Z., & Chang, E. Y. (2017). A neural network approach to jointly modeling social networks and mobile trajectories. *ACM Transactions on Information Systems*, *35*(4), 36:1–36:28.

Yu, Y., Wan, X., & Zhou, X. (2016). User embedding for scholarly microblog recommendation. In *Proceedings of the 54th Annual Meeting of the Association for Computational Linguistics, ACL 2016, August 7–12, 2016, Berlin, Germany, Volume 2: Short Papers*.

Zhang, Q., Wang, J., Huang, H., Huang, X., & Gong, Y. (2017). Hashtag recommendation for multimodal microblog using co-attention network. In *Proceedings of the Twenty-Sixth International Joint Conference on Artificial Intelligence, IJCAI 2017, Melbourne, Australia, August 19–25, 2017* (pp. 3420–3426).

Zhang, F., Yuan, N. J., Lian, D., Xie, X., & Ma, W. (2016). Collaborative knowledge base embedding for recommender systems. In *Proceedings of the 22nd ACM SIGKDD International Conference on Knowledge Discovery and Data Mining, San Francisco, CA, USA, August 13–17, 2016*, pages 353–362.

Zhao, W. X., Guo, Y., He, Y., Jiang, H., Wu, Y., & Li, X. (2014). We know what you want to buy: A demographic-based system for product recommendation on microblogs. In *The 20th ACM SIGKDD International Conference on Knowledge Discovery and Data Mining, KDD'14, New York, NY, USA, August 24–27, 2014* (pp. 1935–1944).

Zhao, W. X., Wang, J., He, Y., Nie, J., & Li, X. (2013). Originator or propagator?: Incorporating social role theory into topic models for twitter content analysis. In *22nd ACM International Conference on Information and Knowledge Management, CIKM'13, San Francisco, CA, USA, October 27–November 1, 2013* (pp. 1649–1654).

Zhao, W. X., Huang, J., & Wen, J.-R. (2016a). Learning distributed representations for recommender systems with a network embedding approach. *Information retrieval technology* (pp. 224–236). Cham: Springer.

Zhao, W. X., Li, S., He, Y., Chang, E. Y., Wen, J.-R., & Li, X. (2016b). Connecting social media to e-commerce: Cold-start product recommendation using microblogging information. *IEEE Transactions on Knowledge and Data Engineering*, *28*(5), 1147–1159.

Zhao, W. X., Li, S., He, Y., Wang, L., Wen, J., & Li, X. (2016c). Exploring demographic information in social media for product recommendation. *Knowledge and Information Systems*, *49*(1), 61–89.

Zhao, W. X., Wang, J., He, Y., Nie, J., Wen, J., & Li, X. (2015). Incorporating social role theory into topic models for social media content analysis. *IEEE Transactions on Knowledge and Data Engineering*, *27*(4), 1032–1044.

Zheng, L., Noroozi, V., & Yu, P. S. (2017). Joint deep modeling of users and items using reviews for recommendation. In *Proceedings of the Tenth ACM International Conference on Web Search and Data Mining* (pp. 425–434). ACM.

Zheng, Y., Tang, B., Ding, W., & Zhou, H. (2016). A neural autoregressive approach to collaborative filtering. In *Proceedings of the 33nd International Conference on Machine Learning, ICML 2016, New York City, NY, USA, June 19–24, 2016* (pp. 764–773).

Zhou, N., Zhao, W. X., Zhang, X., Wen, J.-R., & Wang, S. (2016). A general multi-context embedding model for mining human trajectory data. *IEEE Transactions on Knowledge and Data Engineering*, *28*(8), 1945–1958.

Chapter 10
Deep Learning in Natural Language Generation from Images

Xiaodong He and Li Deng

Abstract Natural language generation from images, referred to as image or visual captioning also, is an emerging deep learning application that is in the intersection between computer vision and natural language processing. Image captioning also forms the technical foundation for many practical applications. The advances in deep learning technologies have created significant progress in this area in recent years. In this chapter, we review the key developments in image captioning and their impact in both research and industry deployment. Two major schemes developed for image captioning, both based on deep learning, are presented in detail. A number of examples of natural language descriptions of images produced by two state-of-the-art captioning systems are provided to illustrate the high quality of the systems' outputs. Finally, recent research on generating stylistic natural language from images is reviewed.

10.1 Introduction

In this final technical chapter of the book, we will discuss a very important but often lightly treated topic in natural language processing (NLP)—natural language generation (NLG), which had been progressing quite slowly until the recent rise of deep learning. As briefly discussed in Chap. 3 in the context of dialog systems, NLG is the process of generating text from a meaning representation and can be regarded as the reverse of natural language understanding.

In addition to serving as an integral component of dialog systems, NLG also plays a key role in text summarization, machine translation, image and video captioning, and other NLP applications. Both the earlier general-purpose rule-based and machine learning-based NLG systems were reviewed in Chap. 3, mainly for the specific dialog

X. He (✉)
Microsoft Research, Redmond, WA, USA
e-mail: xiaohe@microsoft.com

L. Deng
Citadel, Chicago & Seattle, USA
e-mail: l.deng@ieee.org

L. Deng and Y. Liu (eds.), *Deep Learning in Natural Language Processing*, https://doi.org/10.1007/978-981-10-5209-5_10

system application. In a few earlier chapters, more recent developments of deep learning-based methods for NLG, including mainly those based on recurrent neural nets and on the encoder–decoder deep neural architecture, were also briefly surveyed. These deep learning models can be trained from unaligned natural language data and can produce longer, more fluent utterances than previous methods.

In this chapter, rather than providing a comprehensive review of general NLG technology, we limit our scope to NLG in a special application—generating natural language sentences from images, or image captioning. This very difficult task had not been possible until deep learning methods for encoding images and for subsequent generation of natural language became matured within only past 2 years or so. The success of deep learning in image captioning presents another powerful evidence for the impact of deep learning in NLP in addition to several other NLP applications described in detail in the preceding chapters.

Generating a natural language description from an image or image captioning is an emerging interdisciplinary problem at the intersection of computer vision and NLP, and it forms the technical foundation of many important applications, such as semantic visual search, visual intelligence in chatting robots, photo and video sharing in social media, and aid for visually impaired people to perceive surrounding visual content. Thanks to the recent advances in deep learning, tremendous progress of this specialized NLG task has been achieved in recent years. In the remainder of this chapter, we will first summarize this exciting emerging NLG area, and then analyze the key development and the major progress. We will also discuss the impact of this progress both on research and on industry deployment, as well as potential future breakthroughs.

10.2 Background

It has been long envisioned that machines one day can understand the visual world at a human level of intelligence. Thanks to the progress in deep learning (Hinton et al. 2012; Dahl et al. 2011; Deng and Yu 2014), now researchers can build very deep convolutional neural networks (CNN), and achieve an impressively low error rate for tasks like large-scale image classification (Krizhevsky et al. 2012; He et al. 2015). In these tasks, to train a model for predicting the category of a given image, one can first annotate each image in a training set with a category label from a predefined set of categories. Through such fully supervised training, the computer learns how to classify an image.

However, in tasks like image classification, the content of an image is usually simple, containing a predominate object to be classified. The situation could be much more challenging when we want computers to understand complex scenes. Image captioning is one of such tasks. The challenges come from two perspectives. First, to generate a semantically meaningful and syntactically fluent caption, the system needs to detect salient semantic concepts in the image, understand relationships among them, and compose a coherent description about the overall content of the image, which involves language and commonsense knowledge modeling beyond object recognition. In addition, due to the complexity of scenes in the image, it is difficult to represent all fine-grained, subtle differences among them with the simple attribute of category. The supervision for training image captioning models is a full description of the content of the image in natural language, which is sometimes ambiguous with a lack of fine-grained alignments between the subregions in the image and the words in the description.

Further, unlike image classification tasks, where one can easily tell if the classification output is correct or wrong after comparing it to the ground truth, there are multiple valid ways to describe the content of an image. It is not easy to tell if the generated caption is correct or not and to what degree. In practice, human studies are often employed to judge the quality of the caption given an image. However, since human evaluation is costly and time-consuming, many automatic metrics are proposed, which could serve as proxies mainly for speeding up the development cycle of the system.

Early approaches to image captioning can be divided approximately into two families. The first one is based on template matching (Farhadi et al. 2010; Kulkarni et al. 2015). These approaches start from detecting objects, actions, scenes, and attributes in images, and then fill them into a hand-designed and rigid sentence template. The captions generated by these approaches are not always fluent and expressive. The second family is grounded on retrieval-based approaches, which first select a set of the visually similar images from a large database, and then transfer the captions of retrieved images to fit the query image (Hodosh et al. 2013; Ordonez et al. 2011). There is little flexibility to modify words based on the content of the query image, since they directly rely on captions of training images and could not generate new captions.

Deep neural networks can potentially address both of these issues by generating fluent and expressive captions, which can also generalize beyond those in the train set. In particular, recent successes of using neural networks in image classification (Krizhevsky et al. 2012; He et al. 2015) and object detection (Girshick 2015) have motivated strong interest in using neural networks for visual captioning.

10.3 Deep Learning Frameworks to Generate Natural Language from an Image

10.3.1 The End-to-End Framework

Motivated by recent success of sequence-to-sequence learning in machine translation (Sutskever et al. 2014; Bahdanau et al. 2015), researchers studied an end-to-end encoder–decoder framework for image captioning (Vinyals et al. 2015; Karpathy and Fei-Fei 2015; Fang et al. 2015; Devlin et al. 2015; Chen and Zitnick 2015). Figure 10.1 illustrates a typical encoder–decoder-based captioning system (Vinyals

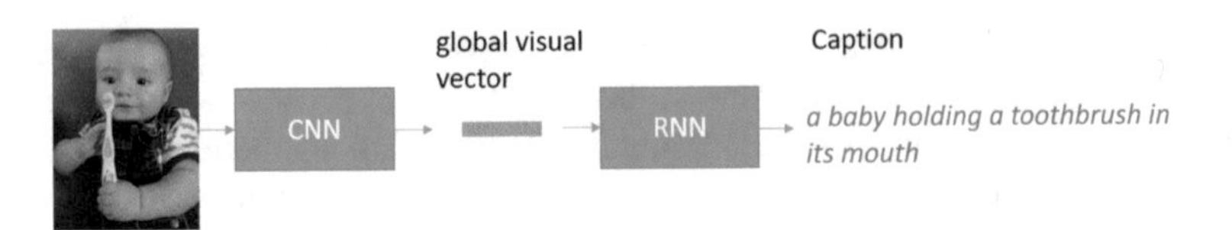

Fig. 10.1 NLG from an image using a CNN and RNN trained together in an end-to-end manner (figure from He and Deng 2017)

et al. 2015). In this framework, first the raw image is encoded by a global visual feature vector which represents the overall semantic information of the image, via deep CNN. As illustrated in Fig. 10.2, a CNN consists of several convolutional, max-pooling, response-normalization, and fully connected layers. Here, the CNN is trained for a 1000-class image classification task on the large-scale ImageNet dataset (Deng et al. 2009). The last layer of this AlexNet contains 1000 nodes, each corresponding to a category. Meanwhile, the second last fully connected dense layer is extracted as the global visual feature vector, representing the semantic content of the overall images. Given a raw image, the activation values at the second to the last fully connected layer are usually extracted as the global visual feature vector. This architecture has been very successful for large-scale image classification, and the learned features have shown to transfer to a broad variety of vision tasks.

Once the global visual vector is extracted, it is then fed into another recurrent neural network (RNN)-based decoder for caption generation, as illustrated in Fig. 10.3. At the initial step, the global visual vector, which represents the overall semantic meaning of the image, is fed into the RNN to compute the hidden layer at the first step. At the same time, the sentence-start symbol <s> is used as the input to the hidden layer at the first step. Then, the first word is generated from the hidden layer. Continuing this process, the word generated in the previous step becomes the input to the hidden layer at the next step to generate the next word. This generation process keeps going until the sentence-end symbol is generated. In practice, a long-short memory network (LSTM) (Hochreiter and Schmidhuber 1997) or gated recurrent unit (GRU) (Chung et al. 2015) variation of the RNN is often used, both of which have been shown to be more efficient and effective in training and capturing long-span language dependency (Bahdanau et al. 2015; Chung et al. 2015), and have found successful applications in action recognition tasks (Varior et al. 2016).

The representative set of studies using the above end-to-end framework include (Chen and Zitnick 2015; Devlin et al. 2015; Donahue et al. 2015; Gan et al. 2017a, b; Karpathy and Fei-Fei 2015; Mao et al. 2015; Vinyals et al. 2015) for image captioning and (Venugopalan et al. 2015a, b; Ballas et al. 2016; Pan et al. 2016; Yu et al. 2016) for video captioning. The differences of the various methods mainly lie in the types of CNN architectures and the RNN-based language models. For example, the vanilla RNN was used in Karpathy and Fei-Fei (2015), Mao et al. (2015), while the LSTM was used in (Vinyals et al. 2015). The visual feature vector was only fed into the RNN once at the first time step in Vinyals et al. (2015), while it was used at each time step of the RNN in Karpathy and Fei-Fei (2015). It is useful to point out that the

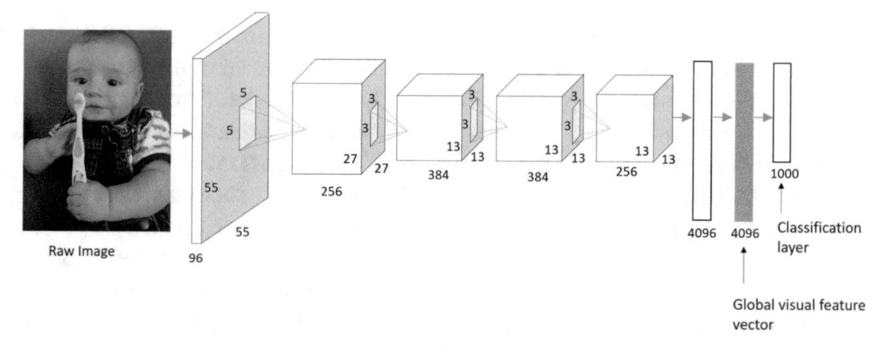

Fig. 10.2 A deep CNN (e.g., AlexNet) used as a front-end encoder of the image captioning system (figure from He and Deng 2017)

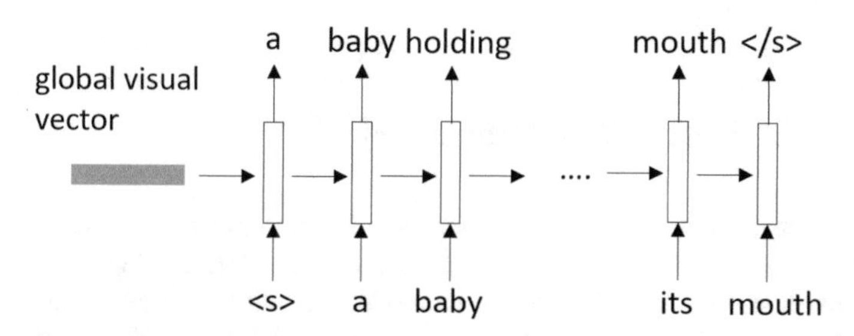

Fig. 10.3 An RNN used as a back-end decoder of the image captioning system (figure from He and Deng 2017)

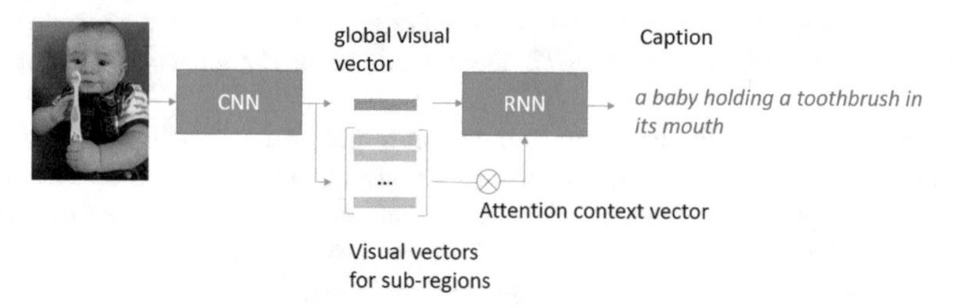

Fig. 10.4 The attention mechanism in the image captioning system's NLG process (figure from He and Deng 2017)

deep CNN, which is essential for the success of image-to-text applications described here, takes into account special translation-invariant properties of the image inputs.

Most recently, Xu et al. (2015) utilized an attention-based mechanism to learn where to focus in the image during caption generation. The attention architecture is illustrated in Fig. 10.4. Different from the simple encoder–decoder approach, the attention-based approach first uses the CNN to not only generate a global visual vector but also generate a set of visual vectors for subregions in the image. These subregion vectors can be extracted from lower convolutional layer in the CNN. Then in language generation, at each step of generating a new word, the RNN will refer to these subregion vectors, and determine the likelihood that each of the subregions is relevant to the current state to generate the word. Eventually, the attention mechanism will form a contextual vector, which is a sum of subregional visual vectors weighted by the likelihood of relevance, for the RNN to decode the next new word.

This work was followed by Yang et al. (2016), which introduced a review module to improve the attention mechanism and further by Liu et al. (2016), which proposed a method to improve the correctness of visual attention. More recently, based on object detection, a bottom-up attention model is proposed by Anderson et al. (2017), which demonstrates state-of-the-art performance on image captioning. In this framework, all the parameters, including the CNN, the RNN, and the attention model, can be trained jointly from the start to the end parts of the overall model; hence the name "end-to-end".

10.3.2 The compositional framework

Different from the end-to-end encoder–decoder framework just described, a separate class of image-to-text approaches uses an explicit semantic-concept-detection process for caption generation. The detection model and other modules are often trained separately. Figure 10.5 illustrates a semantic-concept-detection-based compositional approach proposed by Fang et al. (2015). This approach is akin to and motivated by the long-standing architecture in speech recognition, consisting of multiple composed modules of the acoustic model, the pronunciation model, and the language model (Baker et al. 2009; Hinton et al. 2012; Deng et al. 2013; Deng and O'Shaughnessy 2003).

In this framework, the first step in the caption generation pipeline detects a set of semantic concepts, as known as tags or attributes, that are likely to be part of the images' description. These tags may belong to any part of speech, including nouns, verbs, and adjectives. Unlike image classification, standard supervised learning techniques are not directly applicable for learning detectors since the supervision only contains the whole image and the human-annotated whole sentence of caption, while

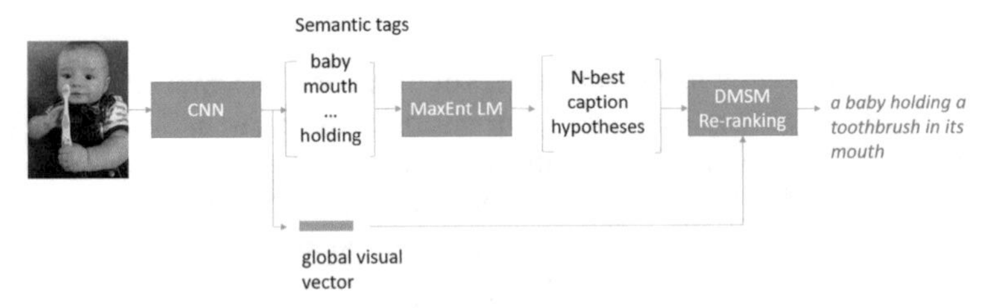

Fig. 10.5 A compositional approach based on semantic-concept-detection in image captioning (figure from He and Deng 2017)

the image bounding boxes corresponding to the words are unknown. To address this issue, Fang et al. (2015) proposed learning the detectors using the weakly supervised approach of multiple instance learning (MIL) (Zhang et al. 2005). While in Tran et al. (2016), this problem is treated as a multi-label classification task.

In Fang et al. (2015), the detected tags are then fed into an n-gram-based Max-Entropy language model to generate a list of caption hypotheses. Each hypothesis is a full sentence that covers certain tags and is regularized by the syntax modeled by the language model that defines the probability distribution over word sequences.

All these hypotheses were then re-ranked by a linear combination of features computed over an entire sentence and the whole image, including sentence length, language model scores, and semantic similarity between the overall image and an entire caption hypothesis. Among them, the image-caption semantic similarity is computed by a deep multimodal similarity model, a multimodal extension of the deep structured semantic model developed earlier for information retrieval (Huang et al. 2013). This "semantic" model consists of a pair of neural networks, one for mapping each input modality, image, and language, to be vectors in a common semantic space. Image-caption semantic similarity is then defined as the cosine similarity between their vectors.

Compared to the end-to-end framework, the compositional approach provides better flexibility in system development and deployment, and facilitates exploiting various data sources to optimizing the performance of different modules more effectively, rather than learn all the models on limited image-caption paired data. On the other hand, end-to-end model usually has a simpler architecture and can optimize different components of the overall system jointly for a better performance.

More recently, a class of models have been proposed to integrate explicit semantic-concept-detection in an encoder–decoder framework. For example, Ballas et al. (2016) applied retrieved sentences as additional semantic information to guide the LSTM when generating captions, while Fang et al. (2015), You et al. (2016), Tran et al. (2016) applied a semantic-concept-detection process before generating sentences. In Gan et al. (2017b), a semantic compositional network is constructed based on the probability of detected semantic concepts for composing captions. This line of methods also represents the current state-of-the-art in image captioning.

From the architectural and task-definition points of view, this type of compositional framework for image captioning and for speech recognition shares a number of common themes. Both of the tasks have the output of natural language sentences, with different inputs of image pixels in the former and of speech waves in the latter. The attribute detection module in image captioning plays a similar role to the phonetic recognition module in speech recognition (Deng and Yu 2007). The use of language model to transform the detected attributes in the image to a list of caption hypotheses in image captioning has the correspondence in the later stages of speech recognition that turn the acoustic features and phonetic units into a collection of lexically correct word hypotheses (via a pronunciation model) and then into a linguistically plausible word sequence (via a language model) (Bridle et al. 1998; Deng 1998). The final, re-ranking module in image captioning is unique in that the earlier module of attribute detection does not possess the global information of the full image, while to generate a meaningful natural sentence for the full image requires such information. In contrast, this requirement for matching global properties of input and output is not needed in speech recognition,

10.3.3 Other Frameworks

In addition to the two main frameworks for image captioning, other related frameworks also learn a joint embedding of visual features and associated captions, For example, Wei et al. (2015) have investigated to generate dense image captions for individual regions in images, and a variational autoencoder was developed in Pu et al. (2016) for image captioning. Further, motivated by the recent successes of reinforcement learning, image captioning researchers also proposed a set of reinforcement learning-based algorithms to directly optimize the captioning models for specific rewards. For example, Rennie et al. (2017) proposed a self-critical sequence training algorithm. It uses the REINFORCE algorithm to optimize an evaluation metric like CIDEr, which is usually not differentiable and therefore not easy to optimize by conventional gradient-based methods. In Ren et al. (2017), within the actor–critic framework, a policy network and a value network are learned to generate the caption by optimizing a visual semantic reward, which measures the similarity between the image and generated caption. Relevant to image caption generation, models based on the generative adversarial network (GAN) are proposed recently for text generation. Among them, SeqGAN (Yu et al. 2017) models the generator as a stochastic policy in reinforcement learning for discrete outputs like texts, and RankGAN (Lin et al. 2017) proposes a ranking-based loss for the discriminator, which gives better assessment of the quality of the generated text, and therefore leads to a better generator.

10.4 Evaluation Metrics and Benchmarks

The quality of the automatically generated captions is evaluated and reported in the literature in both automatic metrics and human studies. Commonly used automatic metrics include bilingual evaluation understudy BLEU (Papineni et al. 2002), METEOR (Denkowski and Lavie 2014), CIDEr (Vedantam et al. 2015), and SPICE (Anderson et al. 2016). BLEU (Papineni et al. 2002) is widely used in machine translation and measures the fraction of N-grams (up to 4-gram) that are in common between a hypothesis and a reference or set of references. METEOR (Denkowski and Lavie 2014) instead measures unigram precision and recall, but extends exact word matches to include similar words based on WordNet synonyms and stemmed tokens. CIDEr (Vedantam et al. 2015) also measures the n-gram match between the caption hypothesis and the references, while the n-grams are weighted by TF-IDF. SPICE (Anderson et al. 2016), instead, measures the F1 score of semantic propositional content contained in image captions given the references, and therefore, it gives the best correlation to human judgment. These automatic metrics can be computed efficiently, and therefore greatly speed up the development of image captioning algorithms. However, all of these automatic metrics are known to only roughly correlate with human judgment (Elliott and Keller 2014).

Researchers have created many datasets to facilitate the research of image captioning. The Flickr dataset (Young et al. 2014) and the PASCAL sentence dataset (Rashtchian et al. 2010) were created for facilitating the research of image captioning. More recently, Microsoft sponsored the creation of the COCO (Common Objects in Context) dataset (Lin et al. 2015), the largest image captioning dataset available to the public today. The availability of the large-scale datasets significantly prompted research in image captioning in the last several years. In 2015, about 15 groups participated in the COCO Captioning Challenge (Cui et al. 2015). The entries in the challenge are evaluated by human judgment. In the competition, all entries are assessed based on the results of M1—percentage of captions that are evaluated as better or equal to human caption, and M2—the percentage of captions that pass Turing test. Additional three metrics have been used as diagnostic and interpretation of the results: M3—Average correctness of the captions on a scale 1–5 (incorrect–correct), M4—average amount of detail of the captions on a scale 1–5 (lack of details—very detailed), and M5—percentage of captions that are similar to human description. More specifically, in evaluation, each task presents a human judge with an image and two captions: one is automatically generated, and the other is a human caption. For M1, the judge is asked to select which caption better describes the image, or to choose the same option when they are of equal quality. For M2, the judge is asked to tell which of the two captions are generated by human. If the judge chooses the automatically generated caption, or choose "cannot tell" option, it is considered to have passed Turing test.

The results, quantified by M1 to M5 metrics above, obtained from the top 15 image captioning systems in the 2015 COCO Captioning Challenge plus other recent top entries measured by automatic metrics have been summarized and analyzed in (He and Deng 2017). The success of these systems reflects the huge progress in this challenging task from perception to cognition achieved by deep learning methods.

10.5 Industrial Deployment of Image Captioning

Propelled by the fast progress in the research community, the industry started deploying image captioning services. In March 2016, Microsoft released the image captioning service as a cloud API to the public. To showcase the usage of the functionality, it also deployed a web application called CaptionBot (http://CaptionBot.ai), which captions arbitrary pictures users uploaded. More recently, Microsoft also deployed the caption service in the widely used product Office, specifically, Word and PowerPoint, for automatically generating alter-text for accessibility. Facebook also released an automatic image captioning tool that provides a list of objects and scenes identified in a photo. Meanwhile, Google open sourced their image captioning system for the community (https://github.com/tensorflow/models/tree/master/im2txt), as a step toward public deployment of the captioning service.

With all these industrial-scale deployment and open-source projects, a massive number of images and user feedbacks in real-world scenarios are being collected that serve as the ever-increasing training data to steadfastly improve the performance of the systems. This will in turn stimulate new research in deep learning methods for visual understanding and natural language generation.

10.6 Examples: Natural Language Descriptions of Images

In this section, we provide typical examples of generating natural language captions that describe the contents of digital images, using the various deep learning techniques described in the preceding sections.

Given a digital image, such as a photo shown in the upper part of Fig. 10.6, the machine-generated textual description of the contents of the image—"a woman in a kitchen preparing food"—together with the human-annotated description—"woman working on counter near kitchen sink preparing a meal"—are shown in the lower part of the figure. In this case, an independent human (a mechanical Turker) slightly prefers the machine-generated text. Among the many images from Microsoft COCO database, about 30% of images are of this type, i.e., whose captions by the system are preferred, or are viewed equally good as human-generated captions.

From Figs. 10.7, 10.8, 10.9 and 10.10, we provide several other examples where mechanical Turkers prefer machine-generated textual descriptions of images to human-annotated ones, or view them as equally good.

The image captioning system that provides the above examples has been implemented in CaptionBot via calling Microsoft Cognitive Services, which allows mobile phone users to upload any photo from the phone to obtain its corresponding natural language caption. Several examples are provided from Figs. 10.11, 10.12 and 10.13. In the last example, we include the result when the celebrity detection component is added to the captioning system.

10.7 Recent Research on Generating Stylistic Natural Language from Images

The natural language captions generated by deep learning systems from images, with numerous techniques and examples provided in the preceding sections, usually gave only a factual description of the image content (Vinyals et al. 2015; Mao et al. 2015; Karpathy and Fei-Fei 2015; Chen and Lawrence Zitnick 2015; Fang et al. 2015; Donahue et al. 2015; Xu et al. 2015; Yang et al. 2016; You et al. 2016; Bengio et al. 2015; Tran et al. 2016). The natural language style has often been overlooked in the

Machine-generated (but turker prefered)	a woman in a kitchen preparing food
Human-annotated (but turker not prefered)	woman working on counter near kitchen sink preparing a meal

Fig. 10.6 An example of image captioning in contrast with human annotation

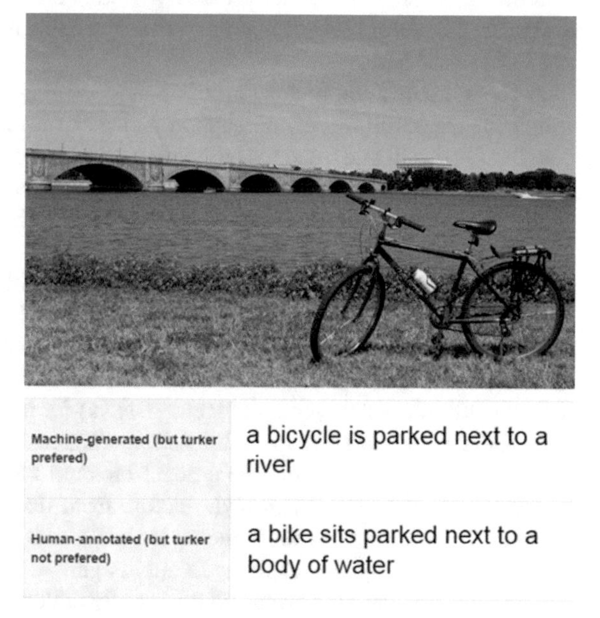

Machine-generated (but turker prefered)	a bicycle is parked next to a river
Human-annotated (but turker not prefered)	a bike sits parked next to a body of water

Fig. 10.7 Another example of image captioning in contrast with human annotation

Fig. 10.8 Another example of image captioning in contrast with human annotation

caption generation process. Specifically, the existing image captioning systems have been using a language generation model that mixes the style with other linguistic patterns of language generation, thereby lacking a mechanism to control the style explicitly. The recent research aims to overcome this deficiency (Gan et al. 2017a) and is reviewed here.

A romantic or humorous natural language description of an image can greatly enrich the expressibility of the caption and make it more attractive. An attractive image caption will add more visual interest to images and can even become a distinguishing trademark of the captioning system. This is particularly valuable for certain applications; e.g., increasing user engagement in chatting bots or enlightening users in photo captioning for social media.

Gan et al. (2017a) proposed the StyleNet, which is able to produce attractive visual captions with styles only using monolingual stylized language corpus (i.e., without paired images) and standard factual image/video–caption pairs. StyleNet is built upon the recently developed methods that combine convolutional neural networks (CNNs) with recurrent neural networks (RNNs) for image captioning. The work is also motivated by the spirit of multitask sequence-to-sequence training Luong et al. (2015). Particularly, it introduces a novel factored LSTM model that can be used to disentangle the factual and style factors from the sentences through multitask training. Then at running time, the style factors can be explicitly incorporated to generate different stylized captions for an image.

The StyleNet has been evaluated on a newly collected Flickr stylized image caption dataset, with the results demonstrating that the proposed StyleNet significantly outperforms previous state-of-the-art image captioning approaches, measured by a

Machine-generated (but turker prefered)	a kitchen with wooden cabinets and a sink
Human-annotated (but turker not prefered)	an ornate kitchen is designed with rustic wooden parts

Fig. 10.9 Another example of image captioning in contrast with human annotation

Machine-generated (but turker prefered)	a clock tower in the middle of the street
Human-annotated (but turker not prefered)	a statue with a clock on it near a parking lot

Fig. 10.10 A final example of image captioning in contrast with human annotation

Fig. 10.11 The image which automatically generates natural sentence of "I think it's a group of people standing in front of a mountain." using Microsoft Cognition Services

Fig. 10.12 The image which automatically generates natural sentence of "I think it's a view of a plane flying over a snow covered mountain." using Microsoft Cognition Services

"Sasha Obama, Malia Obama, Michelle Obama, Peng Liyuan et al. posing for a picture with Forbidden City in the background."

Fig. 10.13 The image which automatically generates a natural sentence using Microsoft Cognition Services with an added celebrity detection component

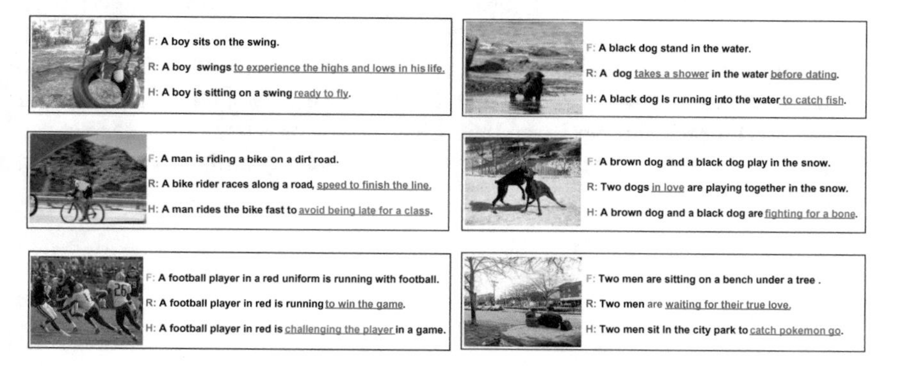

Fig. 10.14 Six examples of natural language captions generated by the StyleNet from images each with three different styles

set of automatic metrics and human evaluation. Some typical examples of stylistic caption generation are shown in Fig. 10.14, where it is observed that the caption with the standard factual style only describes the facts in the image in a dull language, while both the romantic and humorous style captions not only describe the content of the image but also express the content in a romantic or humorous way through generating phrases that bear a romantic (e.g., in love, true love, enjoying, dating, win the game, etc.) or humorous (e.g., find gold, ready to fly, catch Pokemon Go, bone, etc.) sense. Further, it has been found that the phrases that the StyleNet generates fit the visual content of the image coherently, making the caption visually relevant and attractive.

10.8 Summary

Natural language generation from images, or image captioning, is an emerging deep learning application that intersects computer vision and natural language processing. It also forms the technical foundation for many practical applications. Thanks to deep learning technologies, we have seen significant progress in this area in recent years. In this chapter, we have reviewed the key developments in image captioning that the community has made and their impact on both research and industry deployment. Two major frameworks developed for image captioning, both based on deep learning, are reviewed in detail. A number of examples of natural language descriptions of images produced by two state-of-the-art captioning systems are provided to illustrate the high quality of the systems' outputs.

Looking forward, while image captioning is a particular application of NLG in NLP, it is also a subarea in the image-natural language multimodal intelligence field. A number of new problems in this field have been proposed lately, including visual question answering (Fei-Fei and Perona 2016; Young et al. 2014; Agrawal et al. 2015), visual storytelling (Huang et al. 2016), visually grounded dialog (Das et al. 2017), and image synthesis from text description (Zhang et al. 2017). The progress in multimodal intelligence involving natural language is critical for building general artificial intelligence abilities in the future. The review provided in this chapter can hopefully encourage new students and researchers alike to contribute to this exciting area.

References

Agrawal, A., Lu, J., Antol, S., Mitchell, M., Zitnick, L., Batra, D., & Parikh, D. (2015). Vqa: Visual question answering. In *ICCV*.

Anderson, P., Fernando, B., Johnson, M., & Gould, S. (2016). Spice: Semantic propositional image caption evaluation. In *ECCV*.

Anderson, P., He, X., Buehler, C., Teney, D., Johnson, M., Gould, S., & Zhang, L. (2017). Bottom-up and top-down attention for image captioning and VQA. arXiv:1707.07998.

Bahdanau, D., Cho, K., & Bengio, Y. (2015). Neural machine translation by jointly learning to align and translate. In *Proceedings of ICLR*.

Baker, J., et al. (2009). Research developments and directions in speech recognition and understanding. *IEEE Signal Processing Magazine*, *26*(4),

Ballas, N., Yao, L., Pal, C., & Courville, A. (2016). Delving deeper into convolutional networks for learning video representations. In *ICLR*.

Bengio, S., Vinyals, O., Jaitly, N., & Shazeer, N. (2015). Scheduled sampling for sequence prediction with recurrent neural networks. In *NIPS* (pp. 1171–1179).

Bridle, J., et al. (1998). An investigation of segmental hidden dynamic models of speech coarticulation for automatic speech recognition. *Final Report for 1998 Workshop on Language Engineering, Johns Hopkins University CLSP*.

Chen, X., & Lawrence Zitnick, C. (2015). Mind's eye: A recurrent visual representation for image caption generation. In *CVPR* (pp. 2422–2431).

Chen, X., & Zitnick, C. L. (2015). Mind's eye: A recurrent visual representation for image caption generation. In *CVPR*.

Chung, J., Gulcehre, C., Cho, K., & Bengio, Y. (2015). Gated feedback recurrent neural networks. In *ICML*.

Cui, Y., Ronchi, M. R., Lin, T. -Y., Dollar, P., & Zitnick, L. (2015). Coco captioning challenge. In http://mscoco.org/dataset/captions-challenge2015.

Dahl, G., Yu, D., & Deng, L. (2011). Large-vocabulry continuous speech recognition with context-dependent DBN-HMMs. In *Proceedings of ICASSP*.

Das, A., et al. (2017). Visual dialog. In *CVPR*.

Deng, L. (1998). A dynamic, feature-based approach to the interface between phonology and phonetics for speech modeling and recognition. *Speech Communication, 24*(4),

Deng, L., & O'Shaughnessy, D. (2003). *SPEECH PROCESSING A Dynamic and Optimization-Oriented Approach*. New York: Marcel Dekker.

Deng, L., & Yu, D. (2007). Use of differential cepstra as acoustic features in hidden trajectory modeling for phonetic recognition. In *Proceedings of ICASSP*.

Deng, L., & Yu, D. (2014). *Deep Learning: Methods and Applications*. Breda: NOW Publishers.

Deng, J., Dong, W., Socher, R., Li, L. -J., Li, K., & Fei-Fei, L. (2009). Imagenet: A large-scale hierarchical image database. In *CVPR* (pp. 248–255).

Deng, L., Hinton, G., & Kingsbury, B. (2013). New types of deep neural network learning for speech recognition and related applications: An overview. In *Proceedings of ICASSP*.

Denkowski, M., & Lavie, A. (2014). Meteor universal: Language specific translation evaluation for any target language. In *ACL*.

Devlin, J., et al. (2015). Language models for image captioning: The quirks and what works. In *Proceedings of CVPR*.

Donahue, J., Anne Hendricks, L., Guadarrama, S., Rohrbach, M., Venugopalan, S., Saenko, K., & Darrell, T. (2015). Long-term recurrent convolutional networks for visual recognition and description. In *CVPR* (pp, 2625–2634).

Elliott, D., & Keller, F. (2014). Comparing automatic evaluation measures for image description. In *ACL*.

Fang, H., Gupta, S., Iandola, F., Srivastava, R. K., Deng, L., Dollár, P., Gao, J., He, X., Mitchell, M., Platt, J. C., et al. (2015). From captions to visual concepts and back. In *CVPR* (pp. 1473–1482).

Farhadi, A., Hejrati, M., Sadeghi, M. A., Young, P., Rashtchian, C., Hockenmaier, J., & Forsyth, D. (2010). Every picture tells a story: Generating sentences from images. In *ECCV*.

Fei-Fei, L., & Perona, P. (2016). Stacked attention networks for image question answering. In *Proceedings of CVPR*.

Gan, C., et al. (2017a). Stylenet: Generating attractive visual captions with styles. In *CVPR*.

Gan, Z., et al. (2017b). Semantic compositional networks for visual captioning. In *CVPR*.

Girshick, R. (2015). Fast r-cnn. In *ICCV*.

He, X., & Deng, L. (2017). Deep learning for image-to-text generation. In *IEEE Signal Processing Magazine*.

He, K., Zhang, X., Ren, S., & Sun, J. (2015). Deep residual learning for image recognition. In *CVPR*.

Hinton, G., Deng, L., Yu, D., Dahl, G., Mohamed, A. -r., Jaitly, N., Senior, A., Vanhoucke, V., Nguyen, P., Kingsbury, B., & Sainath, T. (2012). Deep neural networks for acoustic modeling in speech recognition. *IEEE Signal Processing Magazine, 29*.

Hochreiter, S., & Schmidhuber, J. (1997). Long short-term memory. *Neural Computation, 9*(8), 1735–1780.

Hodosh, M., Young, P., & Hockenmaier, J. (2013). Framing image description as a ranking task: Data, models and evaluation metrics. *Journal of Artificial Intelligence Research, 47*.

Huang, P., et al. (2013). Learning deep structured semantic models for web search using clickthrough data. *Proceedings of CIKM*.

Huang, T. -H., et al. (2016). Visual storytelling. In *NAACL*.

Karpathy, A., & Fei-Fei, L. (2015). Deep visual-semantic alignments for generating image descriptions. In *CVPR* (pp. 3128–3137).

Krizhevsky, A., Sutskever, I., & Hinton, G. (2012). Imagenet classification with deep convolutional neural networks. In *Proceedings of NIPS*.

Kulkarni, G., Premraj, V., Ordonez, V., Dhar, S., Li, S., Choi, Y., Berg, A. C., & Berg, T. L. (2015). Babytalk: Understanding and generating simple image descriptions. In *CVPR*.

Lin, K., Li, D., He, X., Zhang, Z., & Sun, M.- T. (2017). Adversarial ranking for language generation. In *NIPS*.

Lin, T. -Y., Maire, M., Belongie, S., Bourdev, L., Girshick, R., Hays, J., Perona, P., Ramanan, D., Zitnick, C. L., & Dollar, P. (2015). Microsoft coco: Common objects in context. In *ECCV*.

Liu, C., Mao, J., Sha, M., & Yuille, A. (2016). Attention correctness in neural image captioning. preprint arXiv:1605.09553.

Luong, M. -T., Le, Q. V., Sutskever, I., Vinyals, O., & Kaiser, L. (2015). Multi-task sequence to sequence learning. In *ICLR*.

Mao, J., Xu, W., Yang, Y., Wang, J., Huang, Z., & Yuille, A. (2015). Deep captioning with multimodal recurrent neural networks (m-RNN). In *ICLR*.

Ordonez, V., Kulkarni, G., Ordonez, V., Kulkarni, G., & Berg, T. L. (2011). Im2text: Describing images using 1 million captioned photographs. In *NIPS*.

Pan, Y., Mei, T., Yao, T., Li, H., & Rui, Y. (2016). Jointly modeling embedding and translation to bridge video and language. In *CVPR*.

Papineni, K., Roukos, S., Ward, T., & Zhu, W.-J. (2002). BLEU: A method for automatic evaluation of machine translation. In *ACL* (pp. 311–318).

Pu, Y., Gan, Z., Henao, R., Yuan, X., Li, C., Stevens, A., & Carin, L. (2016). Variational autoencoder for deep learning of images, labels and captions. In *NIPS*.

Rashtchian, C., Young, P., Hodosh, M., & Hockenmaier, J. (2010). Collecting image annotations using amazons mechanical turk. In *NAACL HLT Workshop Creating Speech and Language Data with Amazons Mechanical Turk*.

Ren, Z., Wang, X., Zhang, N., Lv, X., & Li, L. -J. (2017). Deep reinforcement learning-based image captioning with embedding reward. In *CVPR*.

Rennie, S. J., Marcheret, E., Mroueh, Y., Ross, J., & Goel, V. (2017). Self-critical sequence training for image captioning. In *CVPR*.

Sutskever, I., Vinyals, O., & Le, Q. V. (2014). Sequence to sequence learning with neural networks. In *NIPS* (pp. 3104–3112).

Tran, K., He, X., Zhang, L., Sun, J., Carapcea, C., Thrasher, C., Buehler, C., & Sienkiewicz, C. (2016). Rich image captioning in the wild. arXiv preprint arXiv:1603.09016.

Varior, R. R., Shuai, B., Lu, J., Xu, D., & Wang, G. (2016). A siamese long short-term memory architecture for human re-identification. In *ECCV*.

Vedantam, R., Lawrence Zitnick, C., & Parikh, D. (2015). Cider: Consensus-based image description evaluation. In *CVPR* (pp. 4566–4575).

Venugopalan, S., Rohrbach, M., Donahue, J., Mooney, R., Darrell, T., & Saenko, K. (2015a). Sequence to sequence-video to text. In *ICCV*.

Venugopalan, S., Xu, H., Donahue, J., Rohrbach, M., Mooney, R., & Saenko, K. (2015b). Translating videos to natural language using deep recurrent neural networks. In *NAACL*.

Vinyals, O., Toshev, A., Bengio, S., & Erhan, D. (2015). Show and tell: A neural image caption generator. In *CVPR* (pp. 3156–3164).

Wei, L., Huang, Q., Ceylan, D., Vouga, E., & Li, H. (2015). Densecap: Fully convolutional localization networks for dense captioning. *Computer Science*.

Xu, K., Ba, J., Kiros, R., Cho, K., Courville, A., Salakhudinov, R., Zemel, R., & Bengio, Y. (2015). Show, attend and tell: Neural image caption generation with visual attention. In *ICML* (pp. 2048–2057).

Yang, Z., Yuan, Y., Wu, Y., Salakhutdinov, R., & Cohen, W. W. (2016). Encode, review, and decode: Reviewer module for caption generation. In *NIPS*.

You, Q., Jin, H., Wang, Z., Fang, C., & Luo, J. (2016). Image captioning with semantic attention. In *CVPR*.

Young, P., Lai, A., Hodosh, M., & Hockenmaier, J. (2014). From image descriptions to visual denotations: New similarity metrics for semantic inference over event descriptions. In *Transactions of ACL*.
Yu, H., Wang, J., Huang, Z., Yang, Y., & Xu, W. (2016). Video paragraph captioning using hierarchical recurrent neural networks. In *CVPR*.
Yu, L., Zhang, W., Wang, J., & Yu, Y. (2017). Seqgan: Sequence generative adversarial nets with policy gradient. In *AAAI*.
Zhang, H., et al. (2017). Stackgan: Text to photo-realistic image synthesis with stacked generative adversarial networks. In *ICCV*.
Zhang, C., Platt, J. C., & Viola, P. A. (2005). Multiple instance boosting for object detection. In *NIPS*.

Chapter 11
Epilogue: Frontiers of NLP in the Deep Learning Era

Li Deng and Yang Liu

Abstract In the first part of this epilogue, we summarize the book holistically from two perspectives. The first, task-centric perspective ties together and categories a wide range of NLP techniques discussed in book in terms of general machine learning paradigms. In this way, the majority of sections and chapters of the book can be naturally clustered into four classes: classification, sequence-based prediction, higher-order structured prediction, and sequential decision-making. The second, representation-centric perspective distills insight from holistically analyzed book chapters from cognitive science viewpoints and in terms of two basic types of natural language representations: symbolic and distributed representations. In the second part of the epilogue, we update the most recent progress on deep learning in NLP (mainly during the later part of 2017, not surveyed in earlier chapters). Based on our reviews of these rapid recent advances, we then enrich our earlier writing on the research frontiers of NLP in Chap. 1 by addressing future directions of exploiting compositionality of natural language for generalization, unsupervised and reinforcement learning for NLP and their intricate connections, meta-learning for NLP, and weak-sense and strong-sense interpretability for NLP systems based on deep learning.

11.1 Introduction

Natural language processing (NLP) is a most important technology in our information age, constituting a crucial branch of artificial intelligence via understanding complex natural language in both spoken and text forms. The history of NLP is nothing short of fascinating, with three major waves closely paralleling those of the development of artificial intelligence. The current rising wave of NLP has been propelled by deep

L. Deng (✉)
Citadel, Chicago & Seattle, USA
e-mail: l.deng@ieee.org

Y. Liu
Tsinghua University, Beijing, China
e-mail: liuyang2011@tsinghua.edu.cn

L. Deng and Y. Liu (eds.), *Deep Learning in Natural Language Processing*, https://doi.org/10.1007/978-981-10-5209-5_11

learning over the past few years. As of the time of writing this epilogue in November of 2017, we see expansions of many deep learning and neural networks methods presented in this book in multiple directions, with no sign of slowing down.

Since we started this book project about one year ago, the NLP field has witnessed significant advances in both methods and applications, many empowered by deep learning. For example, unsupervised learning methods have very recently emerged in the literature; e.g. (Lample et al. 2017; Artetxe et al. 2017; Liu et al. 2017; Radford et al. 2017). In addition, excellent tutorial and survey materials have been published recently, offering new insight into numerous deep learning methods and comprehensive state of the art results for NLP; e.g. (Goldberg 2017; Young et al. 2017; Couto 2017; Shoham et al. 2017). These new developments and literature prompted us to make an excursion, in the later part of this final chapter of the book, to update and enhance what we wrote in Chap. 1 about the state-of-the-art and future directions of NLP. Let us start first with the main goal of summarizing the entire technical content of the book from novel and holistic perspectives next.

11.2 Two New Perspectives

This book starts with an introduction to the basics of NLP and deep learning, with a survey of the historical development of NLP characterized as three waves with representative research outlined: rationalism, empiricism (Brown et al. 1993; Church and Mercer 1993; Och 2003, etc.), and the current deep learning wave (Hinton et al. 2012; Bahdanau et al. 2015; Deng and Yu 2014, etc.). We stressed that deep learning technology for NLP is a paradigmatic shift from the NLP technologies developed from the previous two waves. The historical survey sets up the context to outline a selective few prominent successes of NLP tasks attributed with no controversy to deep learning (speech recognition and understanding, language modeling, machine translation, etc.), leading to much more detailed coverages of the applications of deep learning to ten core areas of NLP.

Each of Chaps. 2–10 (and part of Chap. 1) is devoted to one of the following NLP applications dominated by or impacted significantly by deep learning:

- Speech Recognition (part of Chap. 1)
- Spoken Language Understanding (Chap. 2)
- Spoken Dialogue (Chap. 3)
- Lexical Analysis and Parsing (Chap. 4)
- Knowledge Graph (Chap. 5)
- Machine Translation (Chap. 6)
- Question Answering (Chap. 7)
- Sentiment Analysis (Chap. 8)
- Social Computing (Chap. 9)
- Language Generation (Chap. 10)

To provide a summary that distills insight from the above chapters with a common thread of exploiting deep semantic representations, we review them below from two novel perspectives that cut across these separate chapters.

11.2.1 The Task-Centric Perspective

Here, we take the perspective based on machine learning paradigms (e.g., Deng and Li 2013) to analyze and cluster the NLP methods and applications in terms of the "tasks and paradigms" covered in the entire book into four major categories.

The **first category** is *classification*, the most popular task in supervised machine learning. Text classification has a long history in NLP, with highly successful applications including email spam detection and sentiment analysis. Sentiment analysis (Chap. 8) is covered in detail in the book, where deep learning methods equipped inherently with the capability of semantic composition of large chunks of texts (e.g., sentences, paragraphs, and documents) have been shown to produce excellent results. Two of the three main problems in spoken language understanding (Chap. 2)—domain detection, intent determination, (and slot filling)—both fall into the category of text classification. Further, the current deep learning methods for question answering and machine comprehension (Chap. 7) can also be regarded as classification. This is a more sophisticated type of classification problem in that context information needs to be provided to constrain the complexity of answer classes in current approaches. As pointed out in Chap. 7, future research will need to relax such constraints in order to achieve understanding and reasoning from text and then to solve the question-answering problem in more principled ways.

The **second category** of NLP tasks is *(sequence-based) structured prediction.* This is also called sequential pattern recognition/classification (He et al. 2008), in contrast to the first category of classification, where the output is a single entity with no sequential structure. Prominent examples of structured prediction in machine learning have been drawn mostly from NLP applications. We have covered many of them in this book, including slot filling in conversational language understanding (Chap. 2), speech recognition (Chap. 1), word segmentation and part of speech tagging in lexical and text analysis (Chap. 4), machine translation (Chap. 6), natural language generation from images (Chap. 10), and advanced versions of question answering (Chap. 7). Note a popular NLP application, document or text summarization, is also well suited to sequence-to-sequence learning and prediction in this category but we do not have this application covered in the book.

The **third category** of NLP tasks from the perspective of machine learning is *higher-order structured prediction (e.g., tree-based and graph-based).* As discussed in Chap. 1, high-order structure is a distinctive characteristic of natural language. Our book dedicates a full chapter to present deep learning models for the text parsing problem formulated as high-order structured prediction (Chap. 4). It shows that deep learning models can be used effectively to augment or replace statistical models in the traditional graph-based and transition-based frameworks. Further, they also demonstrate strong representation power of neural networks which goes beyond the function of mere modeling. A separate chapter (Chap. 5) is also devoted to graph-based structured prediction and learning, where deep learning techniques are used to embed entities and relations for knowledge graph representation. Deep learning is also used to represent relation instances in relation extraction for knowledge graph construction and to represent heterogeneous evidences for entity linking. Exploitation of deep learning to knowledge graphs holds a promising future since such higher-order graph structures are expected to provide a solid foundation for principled ways of question answering, text understanding, and common sense reasoning. All of these are challenging NLP applications that require deep semantic processing, missing in most of the current NLP systems. Two of three main elements of social computing (Chap. 9), modeling user social connection structures and recommendation, also involve graph-based learning and prediction via network embedding accomplished by deep learning. Such network embedding facilitates an automatic and unsupervised feature engineering procedure in numerous social network analysis tasks including network reconstruction, link prediction, node classification, and node clustering/visualization.

While the above three categories of NLP tasks motivated by machine learning methods can be broadly grouped into supervised deep learning or pattern recognition, the **fourth category**, sequential decision-making, goes beyond supervised learning. The modern dialogue systems surveyed in Chap. 3 of this book make use of sequential decision-making process, as part of deep reinforcement learning in a key component of the dialogue systems—dialogue manager. The output of the dialogue manager component is natural language to be received by the user in performing multi-turn conversations with the dialogue system. This type of NLP task—sequential decision-making—is very different from supervised learning in the other three categories summarized above. The difference is that there is no teaching signal at each turn of the dialogue informing whether the natural language output, as the "action" in the decision-making process of "managing" the dialogue. Rather, the overall goal of the dialogue is measured by whether the dialogue is completed satisfactorily to the user and whether the number of turns is desirable for the user. This type of "teacher" signals is far more remote than that in supervised learning and is more challenging from the technology standpoint.

11.2.2 The Representation-Centric Perspective

An alternative perspective, the representation-centric one, can be used to summarize, analyze, and to distill insights from a diverse set of NLP methods and applications described across all chapters in this book.

Throughout this book, two basic types of natural language representations have been used. The **first type** is *symbolic, localist, or one-hot* representation, adopted pervasively during the rationalist and empiricist waves in the NLP history discussed in Chap. 1. The most common example of symbolic presentation is bag-of-words and N-grams for text, where words and text are treated as arbitrary symbols and their (term) frequencies are extracted and exploited. For improving bag-of-words and N-grams, weights based on inverse document frequencies can be added, forming a vector-space model. Further improvements in symbolic representations of text include topic models, where each topic is modeled as a distribution over words and each document modeled as a distribution over topics. In most of the chapters in this book, various types of symbolic representations discussed above have been used as baseline systems to compare with deep learning-based systems exploiting sub-symbolic semantic representations. For example, in sentiment analysis from text (Chap. 8), one popular baseline system based on symbolic presentations of text makes use of sentiment dictionary. The dictionary consists of two sets of word list: the positive set and the negative set. By symbolically counting positive versus negative words in a document, sentiment values associated with all words can be determined.

Symbolic representations are often manually constructed, for example, by coding into a computer the meanings of symbolic words via manually specifying relationships between the words. Knowledge graphs (Chap. 5) are a common way of compiling such symbolic relationships over entities. Typical knowledge graphs of this sort have been described and used in Chap. 5 in detail, serving as the basic data sources with which knowledge-based representation learning using neural network methods would subsequently proceed.

Improvements over the entity-based knowledge graphs (e.g., WordNet, Freebase, etc.) are semantics-based networks such as FrameNet, ConceptNet, and YAGO. The slot filling task in spoken language understanding (part of Chap. 2) and its use in dialogue systems (part of Chap. 3) have been based on FrameNet in the empiricist approach to language understanding developed during the second wave of NLP.

The **second type** of semantic representation of natural language text is *sub-symbolic or distributed* representation, where each word, phrase, sentence, paragraph, or a full document is represented as a dense *embedding* vector with each element corresponding to and influencing not just one linguistic entity but many of them. In all NLP applications presented in each of the chapters of this book, the use of such distributed representations have been described to implement state of the art systems, often contrasting the counterpart baseline system built with symbolic representations with high-dimensional sparse vectors for linguistic entities. Note that while all deep learning systems are based on distributed representations, shallow machine learning methods can rely on either symbolic or distributed representations. Section 9.2 of Chap. 9 has provided an informative review on both symbolic and distributed representations of user-generated textual content for use in social computing. Associations of these two types of representations are made in Chap. 9 with various NLP approaches including traditional (symbolic), shallow learning, and deep learning ones.

The most important common thread cutting across all chapters in this book is the pervasive use of distributed representations of text with various sizes (e.g., word, phrase, sentence, paragraph, and document) as basic as well as automatically learned intermediate features for solving NLP problems. In particular, the compositional property of natural language, from low-level units (e.g., words) to high-level ones (e.g., documents), is exploited to build deep learning architectures in the form of hierarchical neural networks for representation learning in a naturally justifiable manner. The embedding vectors at different linguistic granularity levels constructed using deep models are learned typically by unsupervised methods, where no label information is provided by human. Rather, the "label" information is implicitly captured from context of the text, giving rise to *distributional* properties of the derived distributed representation. One pioneering success of such unsupervised deep learning approaches in NLP is language modeling using recurrent neural networks, as reviewed in Chap. 6 and other places in the book. This type of unsupervised learning is often called **(contextual) predictive learning** and has recently spread its popularity from word sequence prediction in NLP to video sequence prediction (Villegas et al. 2017; Lotter et al. 2017).

The embedding vectors with fully distributed representations learned by unsupervised contextual prediction can be "fine-tuned" and learned in an end-to-end manner if the ultimate NLP tasks are clearly specified and sufficient amounts of label data are available for the training. Spoken language understanding in dialogue systems (Chaps. 2 and 3), machine translation (Chap. 6), question answering (Chap. 7), sentiment analysis (Chap. 8), recommendation in social computing (Chap. 9), and image captioning (Chap. 10) presented in this book all contain successful examples of this type of end-to-end learning bootstrapped from unsupervised representation learning.

11.3 Major Recent Advances in Deep Learning for NLP and Research Frontiers

In Chap. 1 of this book written several months ago, we analyzed a few well-known challenges of deep learning that are general in machine learning as well as specific to NLP. From that analysis, we then discussed research directions for future advances in NLP including the frameworks for neural-symbolic integration, exploration of better memory models and better use of knowledge, as well as better deep learning paradigms including unsupervised and generative learning, multimodal and multitask learning, and meta-learning. Due to the rapid progress in both deep learning and its tight connections to NLP, here we provide an update and elaboration on our earlier analysis.

11.3.1 Compositionality for Generalization

A common drawback of current deep learning under supervised settings is that it requires a large amount of training data with labels. In the NLP context, this drawback results from the difficulty of deep learning methods in handling long-tail phenomena since natural language data generally follows a power-law distribution. That is, any large size of natural language training data will always leave cases the training data cannot cover. This is an intrinsic problem for the localist or symbolic representation in any learning system. However, this difficulty provides an excellent research direction for deep learning approaches as they are based on distributed representations free from the data coverage problem, at least in principle. The research frontier lies in how to design new deep learning architectures and algorithms that can effectively exploit compositional properties of the distributed representations capable of disentangling the underlying factors of variation in natural language data. Recent work on the feasibility of such approaches for video and image data (Denton and Birodkar 2017; Gan et al. 2017) gives promises to solving the generalization problem without formidable amounts of data for natural language data. As a first step, the very recent study reported by Larsson and Nilsson (2017) developed disentangled representations that are shown to be effective in manipulating the sentiment of natural language while preserving the semantics. The proposed algorithm generalizes better than all sentiment analysis techniques surveyed in Chap. 8 of this book.

11.3.2 Unsupervised Learning for NLP

Several months ago, we wrote in Chap. 1 about the preliminary promising work on unsupervised learning with novel methods of exploiting sequential output structure, relationships between inputs and outputs, and advanced optimization methods to eliminate the need for costly parallel corpora (which pair data with labels for each training token) in training prediction systems (Russell and Stefano 2017; Liu et al. 2017). Since then, similar types of unsupervised learning have been more recently scaled up to large-scale machine translation tasks (Artetxe et al. 2017; Lample et al. 2017; Hutson 2017). (Chapter 6 did not include this new progress in machine translation, which was published in November 2017 after the chapter was written.)

The two unsupervised learning methods published for machine translation in Artetxe et al. (2017) and Lample et al. (2017) both use back translation and denoising in the respective training systems. The training is performed without pairing inputs and outputs, with the same setting as the earlier work on non-NLP tasks described in Chen et al. (2016) and Liu et al. (2017) which made use of output structure and the relationship between input (image) and output (text). The back translation step proposed in both Lample et al. (2017) and Artetxe et al. (2017) is a more elegant way of exploiting the relationship between input (source text) and output (target text), taking advantage of the similarity of information rates in input and output (i.e., both being natural language text). More specifically, in back translation, a sentence in input source language is approximately translated into the output target language, which is then translated back into the source. If the back-translated sentence is not identical to the source, the deep neural network then learns to adjust its weights so that next time they will become closer. The denoising step in both studies serves a similar function but it is limited to one language only by adding noise to a sentence and then recovering the original clean version using denoising auto-encoders. The main idea is to build a common latent space between the source and target languages and to learn to translate by reconstructing in both source and target domains. Effective exploitation of the relationship between the source (input) and target (output) domains enables huge cost saving in creating paired source and target sentences for training the machine translation systems.

Another interesting recent study related to unsupervised learning in NLP, sentiment analysis in particular, comes from Radford et al. (2017). The original goal of the study was to explore the properties of byte-level LSTM language models for predicting the next character from given texts (Amazon reviews). Accidentally and somewhat surprisingly, one of the neurons in the multiplicative LSTM trained in an unsupervised way was found to be able to accurately classify the reviews as positive or negative. When the same model is tested on another sentiment data, Stanford Sentiment Treebank, the model also did extremely well.

11.3.3 Reinforcement Learning for NLP

The initial success of unsupervised learning for machine translation reported recently in Artetxe et al. (2017) and Lample et al. (2017), as summarized above, is reminiscent of the success of the self-play strategy in the setting of reinforcement learning in AlphaGo Zero without human data, reported also recently in Silver et al. (2017). With self-play, AlphaGo becomes its own teacher, where a deep neural network is trained to predict AlphaGo Zero's own move selections and also the winner of AlphaGo's games. This prediction is possible because there is a distant teacher informing who wins and who loses in the self-play, which guides the reinforcement learning algorithm. For unsupervised machine translation, back translation serves the same role as self-play in AlphaGo Zero, except there is no analogous teacher to win–loss information. However, if one replaces the win–loss signal used in reinforcement learning by a measure of how good the back-translated sentence is to the original source sentence, such a measure can be used as an objective function to guide unsupervised learning for the weight parameters in deep neural networks.

The above comparison points to the potential of reinforcement learning, which has developed a set of powerful algorithms, for existing and new NLP applications. Reinforcement learning is particularly promising if the NLP problems can be elegantly formulated to enable the use of the concept of "self-play" or the input–output relationship to define distant teaching signals. Successes in this research frontier would add powerful methods from reinforcement learning to overcome a key aspect of the current bottleneck in NLP and deep learning: They are grounded principally on pattern recognition and supervised learning paradigms and thus require large amounts of labeled data and lack reasoning abilities.

A typical reinforcement learning scenario in NLP is dialogue systems. As surveyed in Chap. 3 of this book, dialogue management was one of the first major successes in reinforcement learning in NLP, where standard tools of Markov decision processes and their partial observed versions to handle uncertainty were used. In the recent past, deep neural networks controlled and trained by reinforcement learning have been applied to all three types of dialogue systems or chatbots (intelligent assistants) (Deng 2016; Dhingra et al. 2017). While the "rewards" for reinforcement learning have been reasonably clearly defined in terms of a heuristic combination of task completion (or otherwise), the number of turns in the dialogue, the level of engagement between the chatbot and the user, etc., the requirement for large amounts of conversation data remains a challenging problem. Given that good "world" models or simulators for human–chatbot conversations are very hard to develop, the common requirement for large training data in reinforcement learning would not be easily overcome until appropriate formalisms incorporating the concept of "self-play" are established. The progress in this research frontier is gaining greater urgency as chatbot conversations are expected to become more realistic in practical applications.

Other more recent developments of applying reinforcement learning to NLP problems include the SeqGAN method for creative text generation, via effectively training sequence generative adversarial networks by policy gradient (Yu et al. 2017). A related recent study on using another popular method in reinforcement learning, actor-critic algorithm, is reported in Bahdanau et al. (2017). The analysis of the method and experimental results showed promise in many natural language generation tasks including machine translation, caption generation, and dialogue modeling. Further, reinforcement learning also finds its effectiveness in solving NLP problems of text-based gaming and predicting popular treads in text forums (e.g., Reddit discussion threads). Specifically, the experiments reported in recent literature (He et al. 2016; He 2017) show that separate modeling of state and action spaces, both taking the form of natural language, is capable of extracting semantic information from text rather than simply memorizing strings of text. Another application of reinforcement learning to NLP published recently is in text summarization (an important NLP task but since deep learning only started very recently in tackling text summarization, we have not covered it in this book.). In (Paulus et al. 2017), it was shown that in the neural encoder–decoder model, when standard word prediction using supervised learning is combined with the global sequence prediction training with reinforcement learning, the resulting summarized texts become more readable.

Finally, we observe with high interest the recent success of applying reinforcement learning in generating structure queries from natural language (Zhong et al. 2017). This NLP task, which was called "slot filling" in Chap. 2 of this book, is the core of language understanding within a restricted domain. It was handled in the past using structured supervised learning as surveyed in Chap. 2. The research frontier of spoken language understanding and dialogue systems would be advanced if reinforcement learning can demonstrate its consistent superiority in many practically useful domains as described in Chaps. 2 and 3.

11.3.4 Meta-Learning for NLP

Meta-learning has very different scopes and definitions for different researchers (as can be witnessed at the NIPS Symposium on Meta-learning held in December 2017.) Here, we adopt the general view in Vilalta and Drissi (2002). That is, meta-learning aims to build self-adaptive and continual learners that improve their bias dynamically through experience by accumulating knowledge about learning. Meta-learning is a hallmark of intelligent beings, which can be rightfully characterized as having the ability to continually improve one's own learning capabilities through experience as well as knowledge acquisition.

In Chap. 1 of this book written several months ago, we briefly outlined some initial progress of meta-learning in several non-NLP applications such as hyper-parameter optimization, neural network architecture optimization, and fast reinforcement learning. We also pointed out that meta-learning is a powerful emerging artificial intelligence and deep learning paradigm, which is a fertile research area expected to impact real-world NLP applications.

In the recent past, huge advances in meta-learning applications have been made, notably in navigation and locomotion (Finn et al. 2017a), robotic skills (Finn et al. 2017b), improved active learning (Anonymous-Authors 2018b), and one-shot image recognition (Munkhdalai and Yu 2017). Applications of meta-learning to NLP tasks are starting to appear, which we briefly review here.

In Anonymous-Authors (2018a), meta-learning is applied to continually adapt word embeddings, which subsequently are used for solving down-stream NLP tasks. Given the knowledge learned from a number of previous domains and a small corpus in the new domain, the proposed method can effectively generate word embeddings in the new domain in an incremental manner by leveraging an effective algorithm and a meta-learner. The meta-learner provides word context similarity information at the domain level. Experimental results show the effectiveness of the proposed meta-learning method in forming embeddings in the new domain from a small corpus and the old domain's knowledge for three NLP tasks: text classification (for product type), binary semantic classification, and aspect extraction.

The same goal of leveraging embeddings across several domains for improving down-stream task performance in a new domain can be achieved by a different meta-learning method proposed in another recent study (Bollegala et al. 2017). In this study, an unsupervised, locally linear method is developed to learn the embeddings for a new domain, which are called *meta-embeddings*, from a given set of pretrained *source embeddings* in previous domains. Experimental results on four NLP tasks—semantic similarity, word analogy, relation classification, and short-text classification—show that the new meta-embeddings significantly outperform prior methods in several benchmark datasets.

Yet another interesting recent work on applying meta-learning to the NLP task of questioning answering (from images) is reported by Anton and van den Hengel (2017). The deep learning model is initially trained on a small set of questions and answers and is provided with an additional support set of examples at test time. Given this setting, the model must learn to learn, that is, to exploit the additional data on-the-fly or incrementally and continually without the need for retraining the model. The deep learning model proposed in this work is shown to take advantage of the meta-learning scenario. It demonstrates strong performance in improved recall of rare answers. It also provides better sample efficiency and a unique capability of learning to produce novel answers. The research challenge is to extend the current use of the support set of questions/answers as reported in this study to the future use of more comprehensive datasets obtained from large knowledge bases and web searches.

Finally, we note the very recent study on a highly interesting problem of continuous adaptation of deep learning systems under rapidly nonstationary and adversarial environments, cast in the (gradient-based) meta-learning setting (Al-Shedivat et al. 2017). The novel method is designed to treat a nonstationary task as a sequence of stationary tasks, thus turning the problem into a multitask learning one. Then, multiple deep learning systems (i.e., multi-agents) are trained to exploit the dependencies between consecutive tasks to the extent that the fast nonstationarity exhibited at test time can be effectively handled. The general meta-learning paradigm is adopted, which learns a high-level procedure used to generate a good policy. This is done each time the environment changes. That is, the agents meta-learn to anticipate the changes in the environment and update their policies accordingly.

An important characteristic of the multi-agent environments is that they are nonstationary from the perspective of any individual agent since all actors are learning and changing concurrently (Lowe et al. 2017; Foerster et al. 2017). The results show that when an agent adopts a policy designed assuming other adversarial agents treat it as a competitor, then this policy becomes superior to those that do not make this assumption. The main reason for such superiority is that in this competitive multi-agent setting, the agents have a model of the realistic environment that allows them to exploit the dependencies between consecutive quasi-stationary tasks (modeled as a Markov chain) such that they can handle similar nonstationarities at execution time. More specifically, meta-learning provides optimal updates of an agent's policy with respect to transitions between pairs of tasks, enabling few-shot execution-time adaptation that would otherwise degrade as the environment diverges from training time.

While meta-learning methods for continuous adaptation in rapidly nonstationary and competitive environments have been designed for and applied to robotics and games as reported in Al-Shedivat et al. (2017), the implications for potential future NLP and related applications are profound. This is especially so for a selected few NLP application areas (e.g., finance), where the application environments are highly competitive. Such fast competitions necessarily induce highly nonstationary environments, making the signals extracted for intended NLP applications from recent past to lose their effectiveness quickly. As an exciting research frontier for NLP, modeling such environments for advanced NLP systems using the meta-learning framework is expected to help extend the effectiveness of extracted signals derived from NLP analysis and other means.

11.3.5 Interpretability: Weak-Sense and Strong-Sense

The successes of deep learning models, especially those in NLP applications, often come at a cost of interpretability due to the continuous representations and hierarchical nonlinearity of neural networks. The "black box" quality of most deep neural networks makes them notoriously difficult to control and debug. This difficulty often leads not only to the high cost of developing neural models for NLP but also to the rejection of deploying such non-interpretable models in practice. An obvious case in point is dialogue systems as discussed in Chap. 3. To this date, the majority of dialogue systems under deployment in industry are not based on deep learning despite its technical superiority. Rather, rule-based systems are still common in practice due principally to the ability to interpret, to debug, and to control.

For other NLP applications such as question answering and reading comprehension, similar challenges are prevalent. As an example, almost all existing datasets designed for question answering and reading comprehension research are equipped with a set of undesirable characteristics, including notably the requirement that the answers to questions have to be restricted to an entity or a span from the existing reading text. This turns the difficult text understanding problem that would require often complex reasoning into a supervised pattern recognition problem with the black box quality that requires no reasoning from and no interpretation of the read texts. In making real advances in this research front, more advanced datasets and deep learning methods must be developed to assess and facilitate research toward real, human-like reading comprehension with interpretability as proposed in Nguyen et al. (2017).

Starting from around 2010, most successes of deep learning have been demonstrated in pattern classification and recognition tasks. Extending these successes over the past two years or so, the more complex reasoning process in many current deep learning-based question answering and reading comprehension methods has relied on multiple stages of memory networks with attention mechanisms and with clean supervision information for classification. These artificial memory elements are far away from the human memory mechanism, and they derive their power mainly from the labeled data (single or multiple answers as labels) which guides the learning of network weights using a largely supervised learning paradigm. This is completely different from how human does reasoning. If we were to ask the current neural reasoning models trained on question–answer pairs to do another task such as recommendation, dialogue, or language translation that are away from the intended classification task (i.e., answering questions expressed in a prefixed vocabulary), they would completely fail.

While to succeed in this endeavor requires long-term research efforts, during 2017 we have seen encouraging preliminary progress toward this goal by first making the trained models interpretable (without injecting the goal of interpretability in the training process). Interpretability in the weak sense here is loosely defined as being able to draw insights from the already trained neural models that can provide indirect explanation of how the models perform the desired NLP tasks (such as machine translation). In Ding et al. (2017), in order to interpret neural machine translation by visualization, relevance scores are computed to quantify how much a particular neuron in a hidden layer contributes to neurons in another hidden layer using the proposed layer-wise relevance propagation method. The relevance scores are a direct measure of how much one neuron affects a down-stream neuron, indirectly showing inner workings of the trained neural model. As another example, while little is known about what end-to-end neural translation models learn about source and target languages during the training process, the recent study reported by Belinkov et al. (2017) carefully analyzed the representations learned by neural translation models at various levels of granularity. The quality of the representations for learning morphology is evaluated through part of speech and morphological tagging tasks. This data-driven, quantitative evaluation sheds light on important aspects in the neural translation system in terms of its ability to capture word structure. In yet another recent study (Trost and Klakow 2017), to overcome the difficulty of interpreting word embedding vectors due to the continuous and high-dimensional nature, clustering on word embeddings is carried out to create a hierarchical tree-like structure. The hierarchy is shown to give geometrically meaningful representations of the original relations between the words, thus providing a more human-interpretable way to explore the neighborhood structure in the otherwise non-interpretable embedding vectors.

The weak-sense interpretability studied above is relatively easy to achieve. The deep learning models with strong-sense interpretability, i.e., those that are constructed and trained with interpretability as part of the training objective, are much harder to build but are more useful. Neural-symbolic integration discussed earlier in Sect. 6 in Chap. 1 at a greater length pertains to the general principle for achieving strong-sense interpretability. This principle, inspired by cognitive science (Smolensky et al. 2016; Palangi et al. 2018; Huang et al. 2018) strives for a natural "harmony" between the powerful continuous neural representations and the intuitive symbolic representations more amenable to human understanding and logical reasoning using natural language.

The strong-sense interpretability in deep NLP systems would enable powerful practical applications, e.g., to accomplish multiple NLP tasks of question answering, recommendation, dialogue, and translation, etc. mentioned earlier but require either no labeled data or the labels for at most a small number of tasks. This would be possible because the systems would have true understanding and reasoning abilities, unlike the current NLP systems that rely largely on supervised pattern recognition.

One specific benefit of such deep NLP systems with the strong-sense interpretability is that human users would trust the responses from these systems since they can provide logical reasoning (in the symbolic or natural language form) behind the responses. For instance, an NLP system for reading comprehension may answer correctly a question about who murders a victim after reading a thriller book. But if the logical reasoning steps (as the thought process inside the brain of a detective) are also provided along with the answer, then the answer would be more trusted. A related, simpler example is to learn to solve algebraic problems while showing the steps toward the solution. The recent study (Ling et al. 2017) made a successful attempt in doing so. The work addresses the specific problem of generating rationales for math problems, where the task is to not only obtain the correct answer of the problem but also generate a description of the method used to solve the problem. Experiments show that the proposed (strongly interpretable) method outperforms earlier neural models in both the fluency of the rationales that are generated and the ability to solve the problem. Another very recent study reported in Lei (2017) is also aimed at the strong-sense interpretability in deep NLP systems. Methods are developed to learn to extract pieces of input text as justifications tailored to be short and coherent, which are at the same time sufficient for making the same prediction. Experiments on the

NLP task of multi-aspect sentiment analysis demonstrate the desired goal of making the neural predictions justifiable and thus interpretable to human users.

Although very preliminary work on deep learning for NLP with strong-sense interpretability has started only within the past one or two years, the line of direction discussed in section represents an exciting frontier for NLP research in the current deep learning era.

11.4 Summary

NLP and deep learning are both progressing fast. Over the past three years, especially over the past several months since the earlier parts of this book were completed, deep learning has been increasingly becoming a central paradigm and methodology in solving a wide range of NLP problems. Therefore, this epilogue chapter, completed by the end of 2017, serves not only the standard role of summarizing the full book but also of somewhat unusual roles of updating the most recent progress on deep learning in NLP and of updating our views on the research frontiers of NLP in the deep learning era.

The first part of this chapter summarizes the book holistically from two perspectives: the task-centric one and the representation-centric one. These perspectives are inspired by machine learning paradigms and by cognitive science, respectively. In the second part of this chapter, we update the most recent advances in deep learning as applied to NLP, mainly those during the latter part of 2017 not surveyed in earlier chapters. And supported by these rapid recent advances, we subsequently expand our earlier writing on the research frontiers of NLP in Chap. 1 by addressing future directions in five areas: (1) Compositionality of natural language for generalization; (2) Unsupervised learning for NLP; (3) Reinforcement learning for NLP; (4) Meta-learning for NLP; and (5) Neuro-symbolic integration and interpretability for NLP systems based on deep learning.

Deep learning offers a powerful tool to harness large amounts of computation and data for end-to-end learning and information distillation. Armed with ever more sophisticated distributed representations (e.g., McCann et al. 2017), ever more delicate modular design of functional blocks (e.g., hierarchical attentions), and highly efficient gradient-based learning methods, deep learning has become a dominant paradigm and new state of the art methodology for an increasing number of NLP problems. In addition to the many of them we have surveyed in Chaps. 1–10 plus the updated new NLP problems discussed in earlier parts of this chapter solved fully or partially by deep learning *individually*, we also see the power of a single deep learning to *jointly* solve many NLP tasks by growing a neural network (Hashimoto et al. 2017). Moreover, the very difficult NLP task under extremely noisy conditions—sentiment analysis on Twitter text—has recently been conquered by deep learning to a large extent (Cliche 2017).

In Sect. 5 of Chap. 1, we discussed and analyzed a number of limitations in current deep learning technology, especially those relevant to NLP methods and applications.

As evident in all previous chapters and in earlier parts of this chapter, there have been rapidly increasing improvements of the capabilities of deep learning methods while the identified limitations have been overcome one by one, either partially or completely. As new advances in deep learning move from the supervised paradigm to those of unsupervised, reinforcement, and meta-learning, and as the deep models become increasingly more complex, new fundamental insights into why and how deep learning works extremely well in many tasks and when it may not work so well in other tasks are needed. This is a grand challenge and research frontier in deep learning research, especially for NLP.

Given the amazingly rich deep learning methodology and vibrant research activities devoted to almost all areas of NLP, as compiled in this book, we have confidence that the trend we currently see will continue. We expect more and better deep learning model architectures to appear, and we also expect new NLP applications to be enabled by deep reinforcement learning, unsupervised learning, and meta-learning which would go beyond what we summarized earlier in this chapter.

As a final note to this chapter and the full book, we outline here some of the recent popular discussions on a principled extension of the scope of (vanilla) deep learning (the main topic of this book) to a more general one, called differentiable programming, especially those relevant to NLP. The essence of the generalization is to make the deep neural networks (as the computation graphs for parameterized functional blocks) from being *fixed* to being *dynamic*. That is, after the generalization, the network architecture consisting of many differentiable modules can now be created on the fly in a data-dependent manner. In this differentiable programming paradigm, the deep neural network architectures, including memory, attention, stacks, queue, and pointer modules as we have seen in many chapters in this book, are composed procedurally with logic expressions, conditionals, assignments, and loops. This type of flexibility has been a goal provided by many of the current deep learning frameworks (e.g., PyTorch, Tensorflow, Chainer, MXNet, CNTK, etc., and for the latter see Chap. 14 in Yu and Deng 2015). Once fully developed with highly efficient compilers being built, we will have a brand new breed of software, which, instead of being characterized by the traditional control structures in regular programming such as loops and conditionals, will be established by assembling graphs of parameterized functional blocks, each of which is a neural network in itself. Crucially, all parameters (e.g., neural network weights and the parameters defining network nonlinearities and memory modules) in the assembled graphs can be trained automatically from data using highly efficient, gradient-based optimization methods. This is because no matter how complex the assembled graphs are, differentiability ensures that they can be learned end-to-end via back-propagation.

Differentiable programming has ushered in an exciting field of technology built on top of our existing software stack that is now parameterized, differentiable, and learnable with high efficiency. It represents not only a paradigm to bridge the gap between general algorithms and ways of implementing deep learning, but also a path toward artificial general intelligence where symbolic processing and neural-centric deep learning are harmoniously integrated. This new way of thinking deep learning has special relevance to NLP. First, while being developed at relatively later stages in human cognition, symbolic processing has high efficiency in logical reasoning and is easily interpretable, both desirable in many NLP applications. With tensor-product-like encoding schemes that aim to unify neural and linguistic-structured representations, the high efficiency in learning in complex, flexible, and dynamically constructed neural networks offered by differentiable programming would complete the best of both symbolic and neural worlds. Second, the dynamic nature of NLP models is becoming increasingly more prevalent in NLP methods, as we have demonstrated in each of the chapters in this book. This is due to the very nature of the studied object of NLP—language and text, which has inherently variable dimensions, e.g., the (input) lengths and structures in documents, sentences, or words. The popularity is also due to the capabilities of existing deep learning frameworks in supporting the dynamically varied neural network architectures tailored to the variable-dimensioned text inputs. Finally, natural language has recently been shown to be a very useful latent space over which optimization can be carried out to solve various kinds of difficult machine learning problems (Andreas et al. 2017). The discreteness of language would not allow end-to-end learning to take advantage of differentiability as a requirement for differentiable programming. However, a relaxation technique based on approximations via a *proposal* model has overcome this difficulty, enabling broader opportunities for exploiting naturally occurring language data to improve machine learning and NLP tasks.

In summary, equipped with the generalized deep learning or differentiable programming framework, more powerful, flexible, and advanced deep learning architectures are expected in the near future to solve the remaining difficult NLP tasks that we posed as research frontiers in this and previous chapters. The new success beyond what we have presented in this book will push us closer to artificial general intelligence of which NLP is an integral part.

References

Al-Shedivat, M., Bansal, T., Burda, Y., Sutskever, I., Mordatch, I., & Abbeel, P. (2017). Continuous adaptation via meta-learning in nonstationary and competitive environments. In arXiv:1710.03641v1.

Andreas, J., Klein, D. & Levine, S. (2017). Learning with latent language. arXiv:1711.00482.

Anonymous-Authors (2018a). Lifelong word embedding via meta-learning. *submitted to ICLR*.

Anonymous-Authors (2018b). Meta-learning transferable active learning policies by deep reinforcement learning. *submitted to ICLR*.

Anton, D. T. & van den Hengel (2017). Visual question answering as a meta learning task. In arXiv:1711.08105v1.

Artetxe, M., Labaka, G., Agirre, E., & Cho, K. (2017). Unsupervised neural machine translation. In arXiv:1710.11041v1.

Bahdanau, D., Cho, K., & Bengio, Y. (2015). Neural machine translation by jointly learning to align and translate. In *Proceedings of ICLR*.

Bahdanau, D., et al. (2017). *An actor-critic algorithm for sequence prediction*. ICLR: Proc.

Belinkov, Y., Durrani, N., Dalvi, F., Sajjad, H., & Glass, J. (2017). *What do neural machine translation models learn about morphology?*. ACL: Proc.

Bollegala, D., Hayashi, K., & ichi Kawarabayashi, K. (2017). Think globally, embed locally locally linear meta-embedding of words. arXiv:1709.06671v1.
Brown, P. F., Della Pietra, S. A., Della Pietra, V. J., & Mercer, R. L. (1993). The mathematics of statistical machine translation: Parameter estimation. *Computational Linguistics*.
Chen, J., Huang, P., He, X., Gao, J., & Deng, L. (2016). Unsupervised learning of predictors from unpaired input-output samples. In arXiv:1606.04646.
Church, K. & Mercer, R. (1993). Introduction to the special issue on computational linguistics using large corpora. *Computational Linguistics*, *9*(1),
Cliche, M. (2017). Bb twtr at semeval-2017 task 4: Twitter sentiment analysis with cnns and lstms. *Proc. the 11th International Workshop on Semantic Evaluations*.
Couto, J. (2017). Deep learning for NLP, advancements and trends in 2017. *blog post at https://tryolabs.com/blog/2017/12/12/deep-learning-for-nlp-advancements-and-trends-in-2017/.*
Deng, L. (2016). How deep reinforcement learning can help chatbots. *Venturebeat*.
Deng, L., & Li, X. (2013). Machine learning paradigms for speech recognition: An overview. *IEEE Transactions on Audio, Speech, and Language Processing*, *21*(5), 1060–1089.
Deng, L. & Yu, D. (2014). *Deep Learning: Methods and Applications*. NOW Publishers.
Denton, E. & Birodkar, V. (2017). Unsupervised learning of disentangled representations from video. In *NIPS*.
Dhingra, B., Li, L., Li, X., Gao, J., Chen, Y.-N., Ahmed, F., et al. (2017). *Towards end-to-end reinforcement learning of dialogue agents for information access*. ACL: Proc.
Ding, Y., Liu, Y., Luan, H., & Sun, M. (2017). *Visualizing and understanding neural machine translation*. ACL: Proc.
Finn, C., Abbeel, P., & Levine, S. (2017a). Model-agnostic meta-learning for fast adaptation of deep networks.
Finn, C., Yu, T., Zhang, T., Abbeel, P., & Levine, S. (2017b). One-shot visual imitation learning via meta-learning. arXiv:1709.04905v1.
Foerster, J. N., Chen, R., Al-Shedivat, M., Whiteson, S., Abbeel, P., & Mordatch, I. (2017). Learning with opponent-learning awareness.
Gan, Z. et al. (2017). Semantic compositional networks for visual captioning. In *CVPR*.
Goldberg, Y. (2017). *Neural Network Methods for Natural Language Processing*. Morgan & Claypool Publishers.
Hashimoto, K., Xiong, C., Tsuruoka, Y., & Socher, R. (2017). A joint many-task model: Growing a neural network for multiple NLP tasks. In *Proceedings of EMNLP*.
He, J. (2017). Deep reinforcement learning in natural language scenarios, Ph.D. Thesis, University of Washington, Seattle.
He, J., Chen, J., He, X., Gao, J., Li, L., Deng, L., & Ostendorf, M. (2016). Deep reinforcement learning with a natural language action space. In *ACL*.
He, X., Deng, L., & Chou, W. (2008). Discriminative learning in sequential pattern recognition. 25(5).
Hinton, G., Deng, L., Yu, D., Dahl, G., Mohamed, A.-r., Jaitly, N., Senior, A., Vanhoucke, V., Nguyen, P., Kingsbury, B., & Sainath, T. (2012). Deep neural networks for acoustic modeling in speech recognition. *IEEE Signal Processing Magazine*.
Huang, Q., Smolensky, P., He, X., Deng, L., & Wu, D. (2018). *Tensor product generation networks for deep NLP modeling*. NAACL: Proc.
Hutson, M. (2017). Artificial intelligence goes bilingual — without a dictionary. In *Science*.
Lample, G., Denoyer, L., & Ranzato, M. A. (2017). Unsupervised machine translation using monolingual corpora only. In arXiv:1711.00043v1.
Larsson, M. & Nilsson, A. (2017). Disentangled representations for manipulation of sentiment in text. *Proc. NIPS Workshop on Learning Disentangled Representations: from Perception to Control*.
Lei, T. (2017). Interpretable neural models for natural language processing. *Ph.D. Thesis, Massachusetts Institute of Technology*.

Ling, W., Yogatama, D., Dyer, C., & Blunsom, P. (2017). *Program induction by rationale generation: Learning to solve and explain algebraic word problems*. EMNLP: Proc.
Liu, Y., Chen, J., & Deng, L. (2017). Unsupervised sequence classification using sequential output statistics. In *NIPS*.
Lotter, W., Kreiman, G., & Cox, D. (2017). *Deep predictive coding networks for video prediction and unsupervised learning*. ICLR: Proc.
Lowe, R., Wu, Y., Tamar, A., Harb, J., Abbeel, P., & Mordatch, I. (2017). Multi-agent actor-critic for mixed cooperative-competitive environments.
McCann, B., Bradbury, J., Xiong, C., & Socher, R. (2017). *Learned in translation: Contextualized word vectors*. NIPS: Proc.
Munkhdalai, T., & Yu, H. (2017). *Meta networks*. ICML: Proc.
Nguyen, T. et al. (2017). MS MARCO: A human generated MAchine Reading COmprehension dataset. arXiv:1611,09268.
Och, F. (2003). Maximum error rate training in statistical machine translation. *Proceedings of ACL*.
Palangi, H., Smolensky, P., He, X., & Deng, L. (2018). *Deep learning of grammatically-interpretable representations through question-answering*. AAAI: Proc.
Paulus, R., Xiong, C., & Socher, R. (2017). A deep reinforced model for abstractive summarization. arXiv:1705.04304.
Radford, A., Józefowicz, R., & Sutskever, I. (2017). Learning to generate reviews and discovering sentiment. arXiv:1704.01444.
Russell, S. & Stefano, E. (2017). Label-free supervision of neural networks with physics and domain knowledge. In *Proceedings of AAAI*.
Shoham, Y., Perrault, R., Brynjolfsson, E., & Clark, J. (2017). *Artificial Intelligence Index — 2017 Annual Report*. Stanford University.
Silver, D. et al. (2017). Mastering the game of go without human knowledge. In *Nature*.
Smolensky, P. et al. (2016). Reasoning with tensor product representations. arXiv:1601,02745.
Trost, T., & Klakow, D. (2017). *Parameter free hierarchical graph-based clustering for analyzing continuous word embeddings*. ACL: Proc.
Vilalta, R. & Drissi, Y. (2002). A perspective view and survey of meta-learning. *Artificial Intelligence Review, 25*(2), 77–95.
Villegas, R., Yang, J., Hong, S., Lin, X., & Lee, H. (2017). *Decomposing motion and content for natural video sequence prediction*. ICLR: Proc.
Young, T., Hazarika, D., Poria, S., & Cambria, E. (2017). Recent trends in deep learning based natural language processing. arXiv:1708.02709.
Yu, D. & Deng, L. (2015). *Automatic Speech Recognition: A Deep Learning Approach*. Springer.
Yu, L., Zhang, W., Wang, J., & Yu, Y. (2017). SeqGAN: Sequence generative adversarial nets with policy gradient. In *AAAI*.
Zhong, V., Xiong, C., & Socher, R. (2017). Seq2SQL: Generating structured queries from natural language using reinforcement learning. In arXiv:1709.00103.

Glossary

Attention mechanism Inspired by human visual attention, the attention mechanism is able to help the neural network learn what to "focus" on when making predictions.
Averaged perceptron The averaged perceptron (AP) is an extension of the standard perceptron algorithm. It uses the averaged weight and bias which are estimated by each training instance.
Back-propagation The back-propagation algorithm efficiently calculates the gradients in a neural network by applying the chain rule of differentiation starting from the network output and propagating the gradients backward.
Belief tracker A statistical model that estimates the user's goal at every step of the dialog.
Bidirectional recurrent neural network A bidirectional recurrent neural network (BiRNN) uses a finite sequence to predict or label each element of the sequence based on the element's past and future contexts. This is done by concatenating the outputs of two RNNs, one processing the sequence from left to right, the other one from right to left.
Compound value typed Compound value typed (CVT) is a special data type used in Freebase to represent complex, structured data.
Combinatory categorial grammar The combinatory categorial grammar (CCG) is a syntax formalism which assigns lexical categories to phrases and derives new categories via application, composition, and type-raising.
Cocke-Younger-Kasami The Cocke–Younger–Kasami algorithm (alternatively called CKY) is a parsing algorithm for context-free grammars, named after its inventors, John Cocke, Daniel Younger, and Tadao Kasami. It employs bottom-up parsing and dynamic programming.
Dialog manager A dialog manager is a component of a dialog system, responsible for the state and flow of the conversation.
Dialog state tracker A component of spoken dialog systems that creates a "tracker" that can predict the dialog state for new dialogs to understand a user request and complete a related task with a clear goal within a limited number of dialog turns.
Dropout Dropout is a regularization technique for neural networks that prevents overfitting by randomly setting a fraction of neurons to 0 at each training iteration.

L. Deng and Y. Liu (eds.), *Deep Learning in Natural Language Processing*, https://doi.org/10.1007/978-981-10-5209-5

End-to-end dialog systems Training approach for dialog systems which does not require feature engineering (only architecture engineering) can be transferred to different domains and does not require supervised data for each module.
Goal-oriented dialog system A goal-oriented dialog system needs to understand a user request and complete a related task with a clear goal within a limited number of dialog turns.
Information extraction Information extraction (IE) is a task of automatically extracting structured information from unstructured and/or semi-structured machine-readable documents.
Latent semantic indexing Latent semantic indexing (LSI) is a dimensionality reduction technique that projects queries and documents into a space with latent semantic dimensions.
Limited-memory BFGS Limited-memory BFGS is a limited-memory quasi-Newton optimization algorithm that approximates the Broyden–Fletcher–Goldfarb–Shanno (BFGS) algorithm.
Long short-term memory Long short-term memory networks aim to prevent the vanishing gradient problem in recurrent neural networks by using a memory gating mechanism.
Machine comprehension Machine comprehension (MC) is an extension to the traditional question answering, it is to answer users' questions only from a given document.
Margin-infused relaxed algorithm The margin-infused relaxed algorithm (MIRA) is an online algorithm for multi-class classification problems, where the current training example is classified correctly with a margin against incorrect classifications at least as large as their loss.
Maximum likelihood estimation Maximum likelihood estimation (MLE) is a method for parameter estimation of statistical models and it finds the parameters which can maximize the likelihood of the observation data.
Maximum spanning tree A maximum spanning tree (MST) is a spanning tree of a weighted graph having maximum weight. It can be computed by negating the weights for each edge and applying Kruskal's algorithm.
Minimum error rate training Minimum error rate training (MERT) is a training algorithm that searches for the optimal weights of SMT sub-model features to minimize a given error measure, or maximize a given translation metric such as BLEU and TER.
Minimum risk training Minimum risk training (MRT) is a training algorithm that finds parameters of the model to minimize the empirical risk of the training data.
Multiple layer Perceptron A multilayer perceptron (MLP) is a class of feed-forward artificial neural network which can distinguish data that is not linearly separable. An MLP usually consists of at least two nonlinear layers.
Named entity recognition Named entity recognition (NER) is a task of locating and classifying name entities in natural language documents into predefined categories such as the names of people, organizations, locations, etc.

Natural language generation Natural language generation (NLG) is the natural language processing task of generating natural language from a machine representation system such as a knowledge base or a logical form.
Neural machine translation Neural machine translation (NMT) is an MT paradigm that models the translation process with neural networks in an end-to-end manner.
Out-of-vocabulary Out-of-vocabulary (OOV) denotes the set of words that do not appear in the existing predefined vocabulary.
Part of speech A part of speech (POS) is a category of words which generally display similar behavior. In terms of syntax, they play similar roles within the grammatical structure of sentences. In terms of morphology, they undergo inflection for similar properties.
Point-wise mutual information Point-wise mutual information (PMI) is a measure of association used in information theory and statistics.
Principal component analysis Principal component analysis (PCA) is a mathematical procedure that transforms a number of (possibly) correlated variables into a (smaller) number of uncorrelated variables called principal components.
Semantic parsing Semantic parsing (SP) is a task of translating natural languages into formal meaning representations.
Softmax The softmax function is used to convert a vector of raw scores into class probabilities at the output layer of a neural network.
Spoken dialog systems A spoken dialog system (SDS) is a computer system that is capable of conversing with a human with voice. It has two essential components that do not exist in a written text dialog system: a speech recognizer and a text-to-speech module (written text dialog systems usually use other input systems provided by an OS).
Spoken language understanding Spoken language understanding (SLU) is a subtopic of natural language processing in artificial intelligence that has largely been coined for targeted understanding of human speech directed at machines.
Statistical machine translation Statistical machine translation (SMT) is an MT paradigm that generates translation with a statistical model whose parameters are learnt from parallel corpus.
Semantic role labeling Semantic role labeling (SRL) (also known as shallow semantic parsing) is a task consisting of the detection of the semantic arguments associated with predicates of a sentence and their classification into their specific semantic roles.
User-generated content User-generated content (UGC) is any type of content that has been created by users of a system or service and made available to the public on that system.
User goal The task of recognizing and interpreting the information seeking behavior of a user.
User simulator A statistical model acting as a user in the dialog system is an efficient and effective way to train and evaluate the performance of a (spoken) dialog system.

Printed in the United States
By Bookmasters